A Practical Design Guide for

Coastal Rubble Mound Structures

A Practical Design Guide for **Coastal Rubble Mound Structures**

Louay A Mohammad
COWI A/S, Denmark

World Scientific

NEW JERSEY • LONDON • SINGAPORE • BEIJING • SHANGHAI • HONG KONG • TAIPEI • CHENNAI • TOKYO

Published by

World Scientific Publishing Co. Pte. Ltd.
5 Toh Tuck Link, Singapore 596224
USA office: 27 Warren Street, Suite 401-402, Hackensack, NJ 07601
UK office: 57 Shelton Street, Covent Garden, London WC2H 9HE

Library of Congress Cataloging-in-Publication Data
Names: Mohammad, Louay A., author.
Title: A practical design guide for coastal rubble mound structures /
Louay A. Mohammad, COWI A/S, Denmark.
Description: New Jersey : World Scientific, [2023] | Includes bibliographical references and index.
Identifiers: LCCN 2022047667 | ISBN 9789811261718 (hardcover) |
ISBN 9789811261725 (ebook for institutions) | ISBN 9789811261732 (ebook for individuals)
Subjects: LCSH: Breakwaters--Design and construction. | Sea-walls--Design and construction.
Classification: LCC TC333 .M64 2023 | DDC 627/.24--dc23/eng/20221205
LC record available at https://lccn.loc.gov/2022047667

British Library Cataloguing-in-Publication Data
A catalogue record for this book is available from the British Library.

For any available supplementary material, please visit
https://www.worldscientific.com/worldscibooks/10.1142/13007#t=suppl

Desk Editors: Balasubramanian Shanmugam/Steven Patt

Typeset by Stallion Press
Email: enquiries@stallionpress.com

About the Author

Dr. Louay A. Mohammad started his professional career after receiving his B.Sc. degree in civil engineering from University of Basrah (Basrah, Iraq) in 1977. This was followed by a Master's degree, M.Sc., in hydraulics, hydrology and coastal dynamics from University of Strathclyde (Glasgow, Scotland, UK) in 1982. He continued his research in the same field to be awarded a Doctor of Philosophy, Ph.D. in 1987 from the University of Strathclyde. The author then became a lecturer in hydraulics and coastal engineering at the University of Basrah, Iraq, supervising also many postgraduate research projects.

The author's academic life came to an end after joining COWI A/S in Copenhagen, Denmark (a world leading multidisciplinary engineering design consultancy) in 1999 as a senior coastal/marine design engineer. He was involved in the design of a wide variety of coastal structures from various worldwide locations. Further to that, he was engaged in a good working relation with DHI (Danish Hydraulic Institute, situated in Copenhagen) where many COWI designs underwent physical model testing.

After few years of office design work, Dr. Mohammad decided to grab the opportunity to be directly involved in the most exciting coastal projects at the time, or ever will be, when NAKHEEL Properties of United Arab Emirates requested technical assistance from COWI. He was assigned in 2004 to be a Technical Expert Advisor to NAKHEEL, the developer of Palm Jumeirah, Palm Jebel Ali, The World and Palm Deira Projects in Dubai, UAE. However, in 2007 the author elected to join NAKHEEL to be the General Manager–Coastal and Marine Works for Palm Deira, the world's largest offshore reclamation project. A revision to the project masterplan was instigated with significant optimisations on different aspects achieved. A design review of the various coastal structures in this iconic project was part of his duties, together with careful supervision of the construction, which comprised coastal structures, dredging, reclamation and soil improvement works.

Due to the global economic crisis in 2009, NAKHEEL were forced to drastically scale down their operations and practically stop all new construction activities. The author has therefore moved on to join Archirodon Construction (Oversea) Co. S.A., an international construction company with many large marine/coastal projects in its portfolio. He became the Engineering Manager for the North Africa operations, based in Cairo, Egypt, for the period 2010–2015. The company then restructured its operations, and he was relocated to Dubai, UAE, for the period 2015–2020 to be the engineering manager of some key projects in the region.

Dr. Mohammad has been wearing many hats in his career, a researcher, an academic, a design consultant, a stake holder representative and an engineering manager/designer for a construction company. Therefore, he knows very well what is required from the design engineer in order to produce a good design and at the same time satisfying and meeting the expectations of all the different parties involved in the work. With such a long, diverse and rich background, the author was able to extract useful, practical and important information to be presented in this book. Some of his earlier work has already been published or presented in conferences and seminars.

Notation

A	Total surface area of armour layer panel parallel to slope	(m^2)
A_a	Area covered by one armour unit	(m^2)
CAU	Concrete armour unit	(-)
CG	Centre of gravity	(-)
CH	Chainage	(-)
CHL	Chainage line	(-)
CL	Core-loc unit	(-)
C_r	Wave reflection coefficient	(-)
C-S	Cross-section	(-)
D	Characteristic length of CAU (unit height)	(m)
DWL	Design water level	(m)
D_b	Grain diameter of base material	(m)
D_f	Grain diameter of granular filter	(m)
D_H	Horizontal distance between the CG of two adjacent units	(m)
D_n	Nominal unit diameter, or equivalent cube size, $D_n = \left(\frac{W}{\gamma}\right)^{1/3}$	(m)
D_{n15}	Nominal diameter, or equivalent cube size, $D_{n15} = (M_{15}/\rho_r)^{1/3}$	(m)

D_{n50}	Nominal diameter, or equivalent cube size, $D_{n50} = (M_{50}/\rho_r)^{1/3}$	(m)
D_{n85}	Nominal diameter, or equivalent cube size, $D_{n85} = (M_{85}/\rho_r)^{1/3}$	(m)
D_{na50}	D_{n50} of the armour rock	(m)
D_{50}	Sieve diameter, diameter of stone that exceeds the 50% value of the sieve curve	(m)
D_{85}	85% value of sieve curve	(m)
D_{15}	15% value of sieve curve	(m)
D_H	Distance between the CGs of two consecutive CAUs in the same row	(m)
D_V	Distance between the CGs of two consecutive rows along the CAU armour slope	(m)
DHWL	Design high water level	(m)
DLWL	Design low water level	(m)
F_D	Drag force	(N)
F_F	Resultant force	(N)
F_I	Inertia force	(N)
F_L	Drag force	(N)
GMSL	Global mean sea level	(-)
g	Acceleration due to gravity (9.81)	(m/s^2)
H	Wave height	(m)
HAT	Highest astronomical tide	(m)
HWL	High water level	(m)
$H_{1/3}$	Significant wave height based on time domain analysis, average of highest 1/3-rd of all wave heights	(m)
H_s	Significant wave height, $H_s = H_{1/3}$	(m)
$H_s/\Delta D_{n50}$	Stability number	(-)
h	Water depth	(m)
i	Hydraulic gradient	(-)

K_D	Stability coefficient in Hudson formula, $W = \frac{\rho_r \cdot g \cdot H_S^3}{K_D \cdot cot\alpha \cdot \Delta^3}$	(-)
K_S	CAU shape coefficient	(-)
k_t	Layer thickness coefficient	(-)
L	Wave length	(m)
LAT	Lowest astronomical tide	(m)
LWL	Low water level	(m)
M	Mass of armour unit	(kg)
M_5	Mass of particle for which 5% of the granular material is lighter	(kg)
M_{15}	Mass of particle for which 15% of the granular material is lighter	(kg)
M_{50}	Mass of particle for which 50% of the granular material is lighter	(kg)
M_{85}	Mass of particle for which 85% of the granular material is lighter	(kg)
M_{em}	Effective mean mass (of a standard grading), i.e. the arithmetic average of all pieces excluding those that fall below ELL for the grading.	(kg)
M_{emll}	Lower limit of M_{em}	(kg)
M_{emul}	Upper limit of M_{em}	(kg)
M_y	Mass of stone exceeded by *y*% on mass distribution curve	(kg)
MHWN	Mean high water neap	(m)
MHWS	Mean high water spring	(m)
MLWN	Mean low water neap	(m)
MLWS	Mean low water spring	(m)
MSL	Mean sea level	(m)
N	Number of armour units per unit area ($1/m^2$)	(-)
N_a	Number of armour units	(-)
n	Number of layers	(-)
n_v	Volumetric porosity of rock material	(-)

OD	Pipe outer diameter	(-)
P	Notional permeability factor	(-)
p	Volumetric porosity ratio	(-)
R_c	Crest freeboard, height of crest relative to still water level	(m)
R.C.	Reinforced concrete	(-)
RMS	Rubble mound structure	(-)
SOL	Setting out line	(-)
SOP	Setting out point	(-)
SLR	Sea level rise	(-)
TP	Tetrapod unit	(-)
t_a	Armour layer thickness	(m)
V	Volume of concrete armour unit	(m^3)
V_c	Volume of concrete per unit area	(m^3/m^2)
W	Unit weight	(N)
WL	Water level	(-)
α	Structure slope angle	(°)
β	Angle of wave attack with respect to the structure	(°)
γ	Unit weight of concrete (γ_c), or rock (γ_r)	(N/m^3)
Δ	Relative buoyant density of material, $\Delta = (\rho_r/\rho_w) - 1$	(-)
ρ_{bulk}	Mass bulk density (including voids)	(kg/m^3)
ρ_c	Mass density of concrete	(kg/m^3)
ρ_r	Mass density of rock	(kg/m^3)
ρ_w	Mass density of water	(kg/m^3)
ϕ	Packing density coefficient	(-)

Chapter 1

Introduction

1.1. Scope and Purpose

Coastal engineering and coastal rubble mound structures have been the subject of numerous books with theoretical approaches and textbooks with little to offer on the practical side of the subject. Design manuals and guidelines, e.g. CIRIA, CUR, CETMEF (2007) and CEM (2003/08/11), managed to cover some practical aspects, however, young design engineers are still in need of practical examples and guidelines on how best to convert the design calculations (i.e. size of armour, crest height, … etc.) into a constructible, optimised, robust structure.

This book aims, with the help of many drawings/sketches/photos, to provide examples from actual design cases with many practical details and comments. There will be no theoretical analysis or discussion to design formulae or methods, only where it is considered necessary.

Widening the horizons of young engineers beyond the traditional theoretical textbooks is an issue of high importance. The book will open many windows on the practical side of the design in an attempt to compensate for lack of experience and to offer a glimpse on some innovative design approaches.

1.2. Marine Civil Engineering

Marine/coastal civil engineering encompasses the planning, design, and construction of various types and combinations of structures. Young engineers normally rely heavily on various relevant formulae to calculate the structure dimensions and to size its components. If these formulae are empirically derived, then they will be valid in certain ranges, and if they are theoretical, then these are often used with many simplifications. Therefore, a good understanding of the physical processes with clear insight and comprehension of the design case is vital to producing a practical, safe, constructible, and economical design.

1.3. Reclamation Projects

Reclamation projects are increasingly taking a significant part in coastal projects all around the world. Therefore, part of this book is dedicated to explaining different, and important, aspects of the reclamation projects, starting from the planning, to construction, to edge structures. Special attention is paid to the planning and layout of the reclamation projects with practical examples.

The location of these projects in shallow or deep waters will greatly influence the selection of the edge perimeter structure type, its design and construction. It is important to take the right approach and to select the appropriate structures as the cost impact for any improper decision is usually great.

1.4. Design Integrity

Designers in design consultancies have always two hats ready to wear depending on the client, one when it is the owner of the project, and the other one if it is the construction contractor. Therefore, they are "sometimes" inclined to produce an over-designed work if the client is the owner of the project to show that they are looking after his interests. The produced design in this case is usually known as the FEED design, that is "Front End Engineering Design", which is equivalent to what is known as Preliminary/Basic Design. This design will be issued by the owner to invite tenders. The issued package normally includes design basis, design criteria and the already performed studies, other requirements and possible additional site studies and investigations will be requested to be carried out by the contractor as part of the detailed design.

On the other hand, if the client of the design consultancy is the construction company, then the designers will do their best, usually with some demands and recommendations from the contractor, to cut corners and save cost or minimise construction activities. The design process for the contractor may have two directions. The first direction it to develop a new and more cost-effective concept to be submitted as a design alternative to the Client's FEED design. The second direction is to build on the FEED design after adjusting some elements in the design to improve the design or make it suitable to the contractor's proposed construction methodology.

In trying to impress one side or satisfy the other, the designer will, probably unwittingly, likely produce a design that either can safely be optimised or needs to be improved to reach a well-balanced and safe design to construct. Example designs and illustrations are presented in the book to elaborate on this issue.

Chapter 2

Coastal RMS Projects

2.1. Introduction

Coastal RMS (Rubble Mound Structures) Projects could generally be defined as the projects which are being exposed to wave action. The location of these projects could be categorised in relation to the shoreline, being on the shoreline, in the nearshore zone with a connection to the shoreline, or near/far from shoreline and detached. Projects in the coastal zone will be influenced, apart from the waves, by a number of other effects. Some of these effects are the cross-shore and longshore sediment transport, currents, temporal/long-term changing sea levels (i.e. tidal and due to global warming). Note that, regardless of the location, most of the coastal projects involve reclamation, or sand filling, in varying degrees, and for different purposes.

2.2. Natural Shoreline

Coastal projects along the coastline could basically be for coastal protection, as shown in Figure 2.1, where a rock revetment is used. In Figure 2.2, a combination of hard and soft protection is used, providing protection as well as a recreation beach area. The natural shoreline in Figure 2.3 has been adjusted to create two types of recreation areas, a promenade and a beach.

Figure 2.1. Rock revetment (hard protection).

Figure 2.2. Detached groynes (hard+soft protection).

Figure 2.3. Abu Dhabi, UAE, corniche developments.

2.3. Land Extension Reclamation

Reclaimed land extensions are frequently required to enlarge prime waterfront properties, construct runways to airports near congested city centres and for port extensions where the port area cannot be extended inland anymore. This type of reclamation is to be connected to the mainland, either through a continuous connection to the shoreline, or by a causeway or a bridge.

Different examples are presented in this section for recreational, residential, and industrial developments. The first, in Figure 2.4, are recreational and residential developments, some have continuous connection to the shoreline and others are connected by a causeway. The main distinct feature which differentiates those developments from the industrial ones is having beaches as part of the edge perimeter. This type of land extension could be small or very large, as shown in Chapter 3.

A good example for an industrial development is Jebel Ali Port in Dubai, U.A.E. Figure 2.5 shows where the stage-1 of the port was constructed inland by dredging two basins, in the 1970s. Forty years later an offshore reclamation is being constructed as stage-2 to extend the port and add more berthing quays and storage areas. The sheltering breakwater to the port basin will also serve as the connecting road to the new extension.

Figure 2.6 shows an example of a limited industrial land extension, with continuous connection to land, required to accommodate additional industrial facilities. Whereas in Figure 2.7 the industrial land extension is an island very close to the shoreline and connected to land by a causeway.

Figure 2.4. Recreational and residential developments along the coast of Dubai (Google Earth).

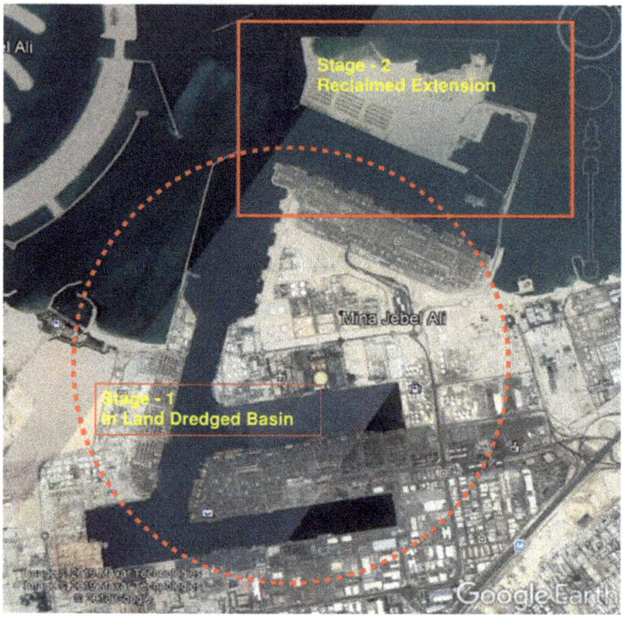

Figure 2.5. Jebel Ali Port, Dubai, U.A.E. (Google Earth).

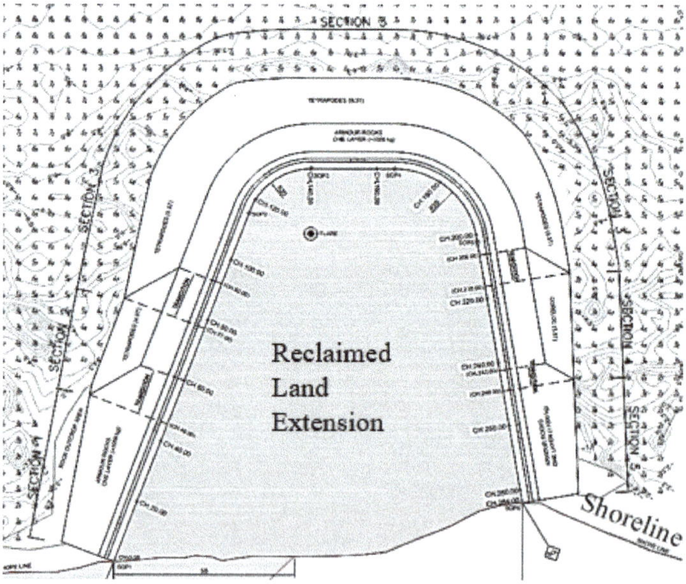

Figure 2.6. Example of a limited land extension — continuous connection.

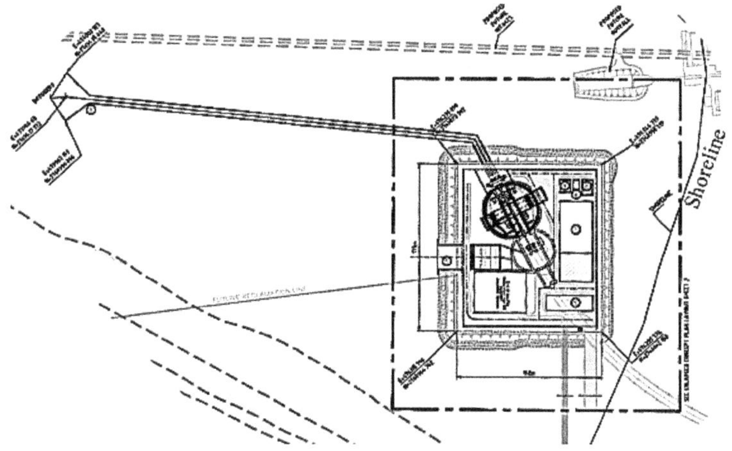

Figure 2.7. A pumping station island with outfall pipelines connected to the shore by a causeway.

Land extension schemes must fulfil the developer/industrial project requirements with input from coastal engineers starting from the concept stage to the detailed design. Extensive coastal studies are to be carried out covering exposure to waves, water circulation and littoral drift. More examples are presented and discussed in Chapter 3.

2.4. Nearshore/Offshore Artificial Islands

Artificial islands are increasing in numbers, primarily reclaimed, for industrial needs but also for recreational and residential use. These islands are designed with varying sizes and with different edge perimeter protection arrangements as these are, to some degree, dictated by the water depth at the project site. Few examples are given in this section to illustrate the variety in these offshore islands with respect to edge perimeter protection.

A maritime traffic control centre, shown in Figure 2.8, is to be placed on an offshore island and to be accessed by sea and by air. The island is to be constructed on a nearshore sandbar and is to be reclaimed by using the sand dredged from the access channel and the inner basin. Rubble mound breakwaters and revetments are to be used for the island perimeter protection.

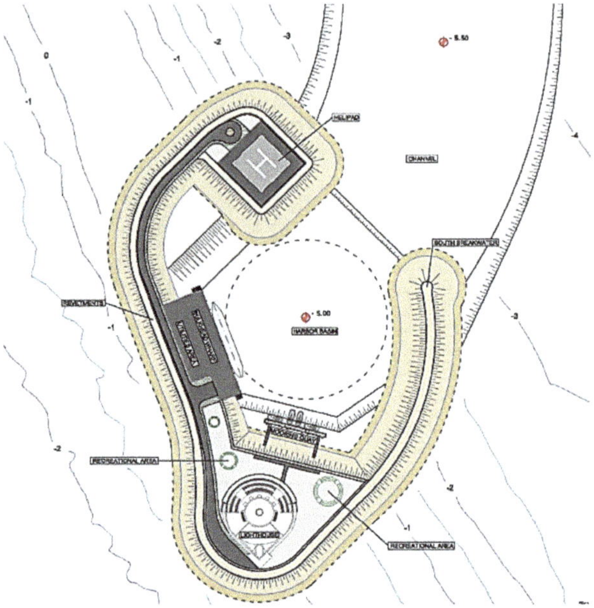

Figure 2.8. Offshore Island — rubble mound protection (Seabed level at 0.0 to –2.0 mCD).

A different example is shown in Figure 2.9 for a nearshore artificial island. It is to be constructed from the dredging of an access channel to a nearby new port. The island is in a relatively shallow area and not exposed to high waves. Therefore, the selected design was to use soft edges, i.e. beaches, confined and protected by a series of groynes, with the longest beach front facing the nearby shoreline.

The next example in Figure 2.10 is for an artificial island where the seabed level is at –15.0 mCD. This island is a support and service base for an offshore oil field. The island is to be constructed on a reclaimed platform with its perimeter protected by rubble mound revetments.

All the above given examples are for industrial/service projects whereas the one in Figure 2.11 is a recreational and residential project. It is the World project in Dubai, with 10 km and 8.0 km for the major and minor axes, respectively, and where the seabed level is at –8.0 to –11.0 mCD. The project is also shown in Figure 3.12, at a close distance from the shoreline. The project consists of two major elements, a perimeter rubble mound breakwater and reclaimed sand islands inside. With

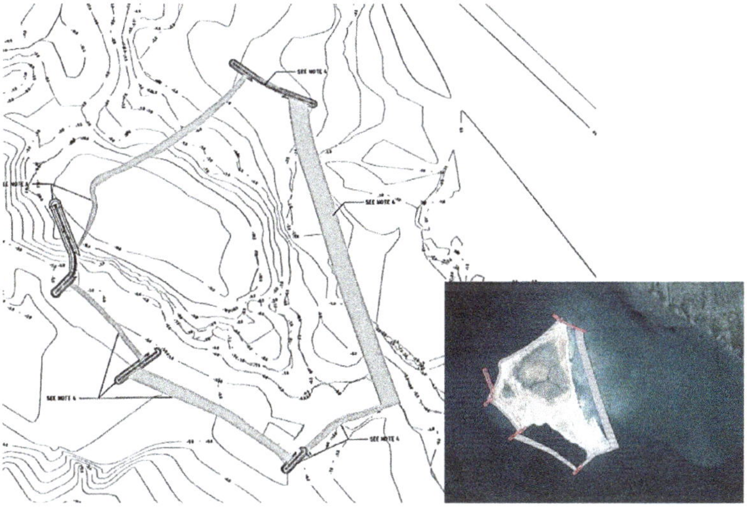

Figure 2.9. Nearshore Island — groynes and beaches (Seabed level at –5.0 to –8.0 mCD).

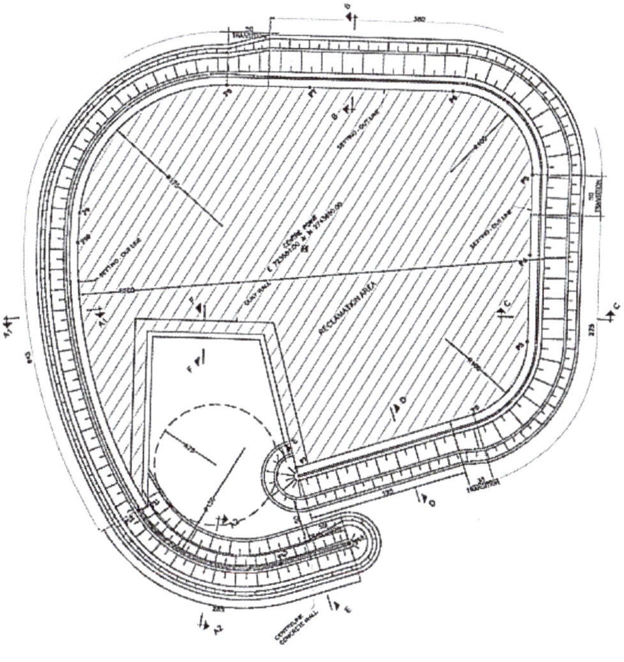

Figure 2.10. Offshore Island — rubble mound protection (Seabed level at –15.0 mCD).

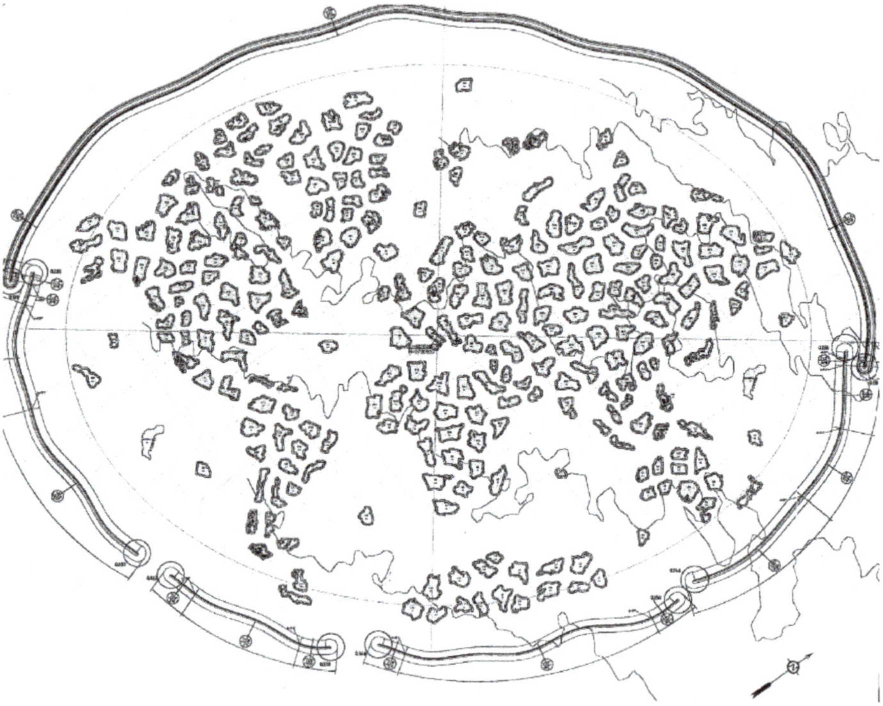

Figure 2.11. Nearshore islands — Outer rubble mound breakwater (Seabed level at –8.0 to –11.0 mCD).

Figure 2.12. Offshore island — caisson protection — preliminary layout and configuration (Seabed level at — 26.0 mCD).

no waves reaching the reclaimed sand islands, it requires no edge protection. In fact most of the islands were reclaimed as one large island then channels were dredged through this large island to create smaller individual islands.

The last example in Figure 2.12 is for the Energy Island in the North Sea which is to be constructed as a hub and a collection point/transformer station for the surrounding offshore wind farms. The design up to the time of writing this text has not been completed, but the location has been agreed upon, which is at a seabed level of –26.0 mCD. At such a water depth it is most cost-beneficial to use large caissons, which also have some construction advantages in the North Sea harsh environment.

2.5. Sheltering Structures

Coastal projects of any of the three basic categories (Sections 2.2–2.4) may require a sheltering breakwater, a revetment or a combination of the two. These are frequently required to shelter a semi-enclosed basin from open sea waves, or to shelter the entrance to inland waterways or industrial canals, see Figure 2.13, as in ports and marinas. The overall layout, basin size and configuration depend on the intended use and capacity of the port/marina.

Figure 2.14 presents a basin for a port, with a breakwater as sheltering structure in some stretches and at the entrance, and a revetment at stretches where some of the port facilities were constructed on a reclaimed land.

Offshore detached breakwaters, in their largest form, are used to provide shelter for deep water terminals, see Figure 2.15. The extent and configuration of the breakwater required to achieve the design wave agitation at the loading platform is to be investigated with the help of wave numerical modelling,

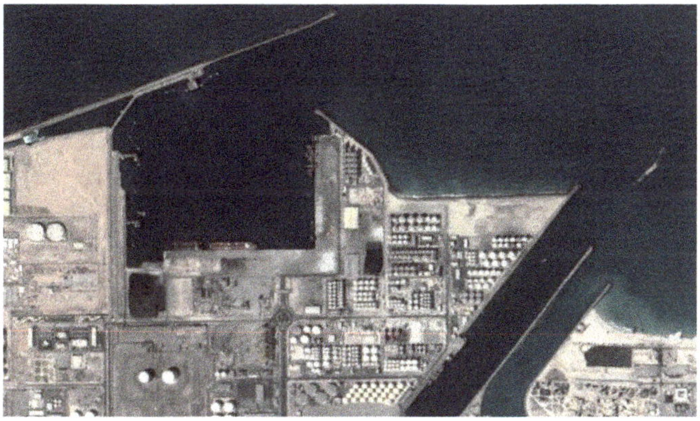

Figure 2.13. Sheltering breakwaters to a port basing and canal entrance.

Figure 2.14. A combined breakwater-revetment sheltering structure.

Figure 2.15. An offshore detached breakwater for sheltering LNG terminal.

short, dredged canal to create an inland marina while utilizing the surrounding area as residential and commercial developments. Whereas the second example involves dredging a relatively longer canal to connect the dead-end of a natural creek to the sea creating a long waterfront stretch. Figure 3.6 also shows three selected major developments on Dubai Creek, two of them on the natural creek and one on the dredged extension.

The construction of this type of Land–Water interaction is normally carried out through land excavation, completing the edge structures and then allowing the water to fill the excavated areas. Two examples are shown in Figures 3.7 and 3.8.

In some projects it is however preferred to use a Cutter Suction or backhoe dredger, as shown in Figure 3.9, excavating its way inside the land mass to create the required water body.

Figure 3.7. Excavation of the Arabian Canal in Dubai.

Figure 3.8. Construction of the Dow Wharfage in Dubai Creek.

Figure 3.9. Cutter Suction and Backhoe Dredgers (Typical).

3.2.3. *Land into water*

This type of developments has recently gained worldwide momentum by utilising dredged sand to reclaim coastal land. A comprehensive review for reclamation projects' planning, construction methodology, fill requirements, etc. are well detailed by Hoff *et al.* (2012).

Two examples from the Middle East are shown in what follows, Figure 3.10 shows the Pearl development in Qatar and Figure 3.12 shows the Dubai coastline with many mega coastal projects. These two examples represent two distinct categories, the Pearl is located in very shallow waters, whereas the Dubai developments are in medium to deep waters.

The construction of projects in shallow waters is relatively easier than in deep waters as it could, partially or completely, be carried out in the dry as shown in Figure 3.10 for the Pearl project. Shallow waters come with an added advantage of giving more freedom to the landscaping and architecture in utilizing a variety of edge perimeter structures without much concern for the construction cost or difficulty due to water depth.

For the coastal projects in Dubai, as shown in Figure 3.12, in general, the reclamation methodology is as described in the following. For depths in the order of 12 m or more, large trailer hopper dredgers could directly dump the sand in position as shown in Figure 3.11(a). For lesser depths, between 5.0 to 12.0 m, the hopper dredgers could deliver the sand either by rainbowing or pumping as in Figure 3.11(b). For shallower depths less than 5.0 m, pumping is normally used from the hopper dredgers or from cutter suction dredgers if these are used to dredge the sand from adjacent seabed areas as in Figure 3.11(c).

(a)

(b)

Figure 3.10. The Pearl development, Qatar: (a) Ariel photo after completion (Google Earth); (b) Construction photos.

The sand delivery process as described above has been used in all of Dubai waterfront projects with seabed depth contours are as shown in Figures 3.13 and 3.14 for the coastline and for Palm Deira, respectively.

The Dubai coastline early masterplan is as shown in Figure 3.1, and the updated one in Figure 3.12(a) is after revising Palm Deira and the Water Front projects. Figure 3.12(b) is an aerial image showing the actual

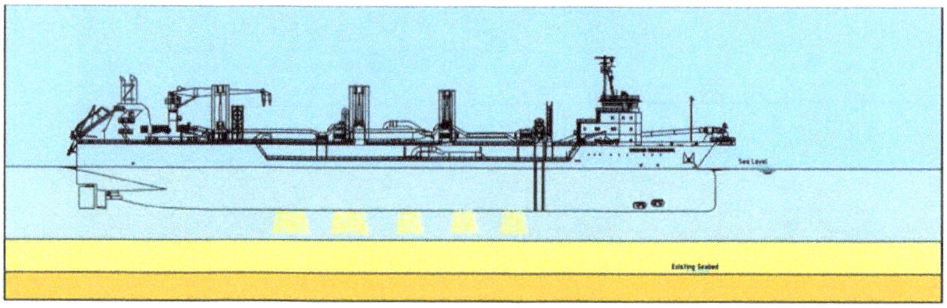

(a)

(b)

(c)

Figure 3.11. Reclamation in deep waters: (a) Sand dumping; (b) Sand rainbowing; (c) Sand pumping.

(a)

(b)

Figure 3.12. Dubai coastline developments: (a) Updated planned projects (with regards to Palm Deira and Waterfront); (b) Satellite image of Dubai coastline, Nov. 2019 (Google Earth).

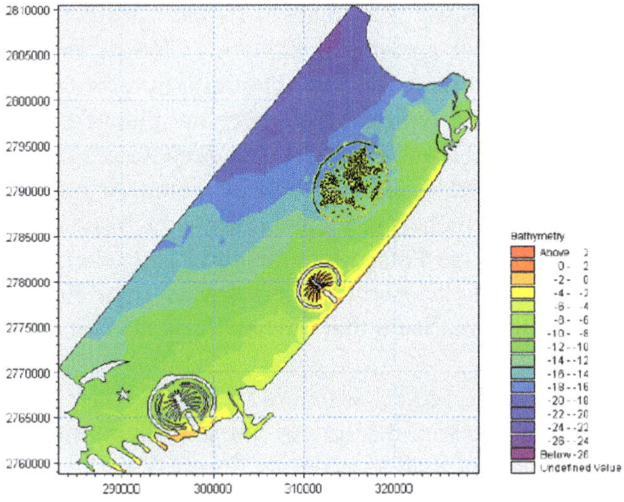

Figure 3.13. Seabed depth contours (m CD) for Dubai waters.

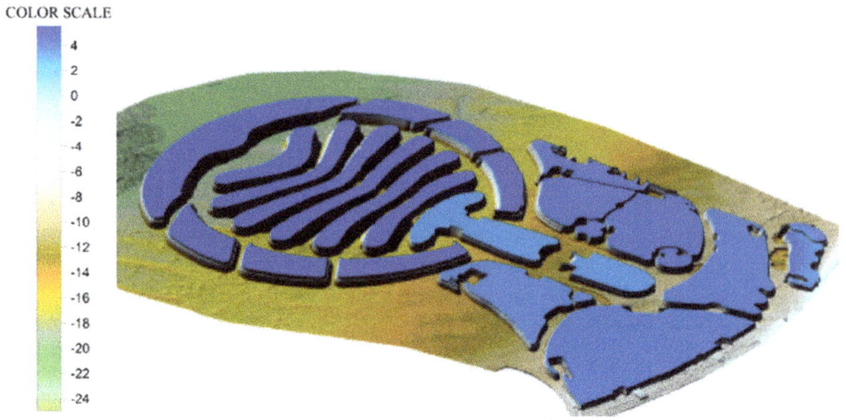

Figure 3.14. Seabed depth contours (m CD) for Palm Deira.

constructed projects as the Waterfront and Palm Deira projects were not completed as per the plans due to global recession in the property market in 2008. The optimisation process for the planned Palm Deira masterplan is presented in Section 3.3.

3.3. Palm Deira Case Study

Palm Deira is the largest man-made offshore project (Mohammad, 2008(a)). It spreads over an area of 8490 x 15260 m as per the initial masterplan layout, see Figure 3.15(a). The initially proposed masterplan for the project presents a typical case where the planners and architects have overlooked the fact that the project is in deep waters. The layout plan is drawn as if the project is in very shallow waters or on dry land. The starting point of the planners is to maintain the same basic elements of the first two palm projects, i.e. Palm Jumeirah and Palm Jebel Ali. These elements are: the Crescent (the perimeter breakwater), the Fronds, the Spin and the Trunk, without realising that the new project is on an entirely different scale.

For such an offshore project which is located in water depths ranging between 7 to 23 m, various disciplines are to collaborate to produce a workable efficient layout plan as given by Figure 3.3.

From a coastal engineering point of view, some key points need to be investigated and revised in the initial masterplan layout in order to achieve a workable optimised layout, (Mohammad, 2008(b)), and these shall be discussed in the following and as per the marking on Figure 3.15(a):

1. *Uneconomical breakwater at deep waters*: The construction of a breakwater almost at the tip of the Palm, which is at a water depth of 23m, is very costly and not practical. Hence, revetments shall be used as the preferred type of protection.
2. *Intricate vertical edge structure*: The construction of vertical edge structures at 16–23m water depths is extremely costly, see Section 4.3.1.
3. *Narrow fronds*: The concept of narrow fronds used for Palm Jumeirah and Palm Jebel Ali worked because of the relatively shallower depths, however, in Palm Deira with much deeper waters wider fronds are to be considered. The wider frond will considerably optimize the use of sand in producing larger surface areas, as illustrated in Figure 3.16, and also maintain navigable channels in between the fronds.
4. *Trunk/Spine — circulation barrier*: The typical design of the two smaller palms (i.e. Jumeirah and Jebel Ali) cannot just be copied on such a larger scale project. The long trunk/spin prevented the east–west tidal movement of water between the fronds. This resulted in a long and unacceptable pollutant residence time.
5. *Small undefined basins*: Small basins were included in the plan presumably to be marinas; however, these basins are too small to accommodate a marina.
6. *Disproportionally large water bodies with small land masses*: Land masses and water bodies are to be carefully distributed to achieve an optimised use of the reclamation sand. The reclamation efficiency could be expressed by a "Reclamation Ratio" indicator defined as follows:

$$\text{Reclamation Ratio, RR} = 1{:}\ X$$

where "X" is the required sand volume for each square meter of surface area. This concept is illustrated in Figure 3.16 for the frond design.

With reference to Figure 3.12 (showing the projects' plans) and Figure 3.13 (showing the water depth at the projects), a comparison diagram is prepared between the Palm trio [Palm Jumeirah (PJU), Palm Jebel Ali (PJA) and Palm Deira (PD)] shown in Figure 3.17.

7. *Wharves — wrong dimensions and configuration*: The layout plan included many finger piers as dhow wharfage; however, these were sketched with wrong dimensioning for proper operational requirements.
8. *Incorrect approach channel to Hamriya Port*: Hamriya Port requires a wider and a better oriented approach channel and these have not been provided.

After adopting all the necessary adjustments and corrections to the above-mentioned points, an optimised layout plan has been developed, as shown in Figure 3.15(b). By using almost the same volume of reclamation sand, an increase of about 50% in surface area has been achieved. Note that, and as a result of this optimisation process, the crest of the Palm was pulled back by nearly 3000 m, from a depth contour of –24.0 mCD to –21.0 mCD, which meant a huge gain to the developer as the sand is to be

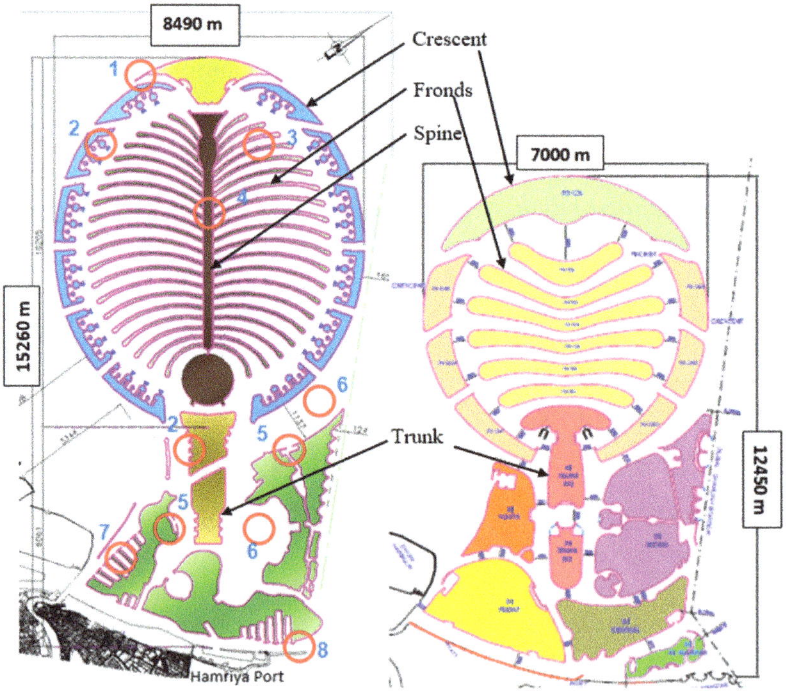

(a) *Initial layout plan* (b) *Optimised layout plan*

Figure 3.15. Palm Deira layout plan (Mohammad, 2008(a)).

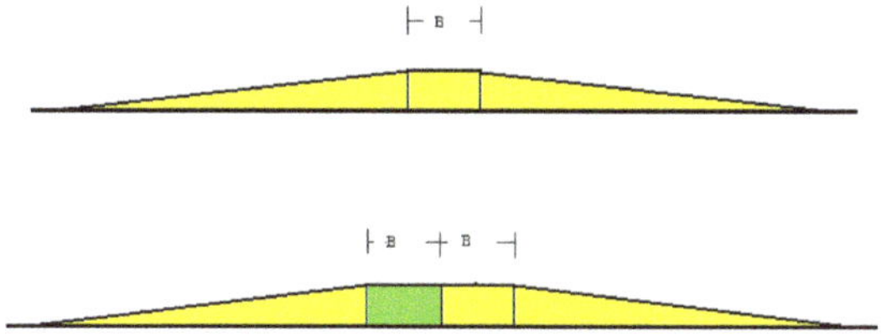

Figure 3.16. Wide frond concept.

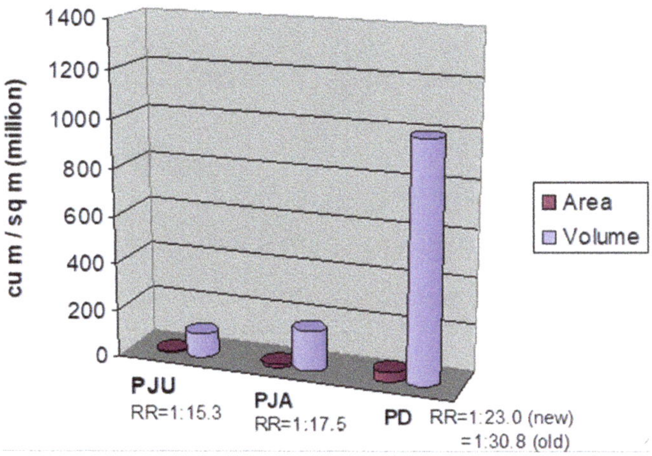

Figure 3.17. The Palm trio reclamation ratios.

used more efficiently in shallower areas with an increased surface area being the prim asset for the developer.

Unfortunately, the Palm Deira project stopped after the global financial crisis in 2008, consequently it has been revised 10 years later in 2019 in utilizing the already semi-completed parts of the project. As shown in Figure 3.18, the project is called now "Deira Islands" which is composed of five islands. Revetments were completed to protect the exposed parts of the islands while closing the large water body in the middle by two breakwaters at the top. It is however to note that this is probably not the final shape of the project as the land connections between the islands are still to be replaced by bridges, as per the original plan of the Palm Deira project. This is of course required to improve the water circulation in large, enclosed body of water.

Figure 3.18. Deira Islands — May 2020 (Google Earth).

3.4. Reclamation Environmental Impact

At the start of a waterfront development project, especially if it involves reclamation, a full EIA (Environment Impact Assessment) is to be carried out to evaluate any possible impact on the local environment. However, this is to be carried out after surveying and recording the site baseline (existing) conditions.

The envisaged environmental adverse impacts are:

- Impacts from dredging and reclamation on water quality and marine ecology in terms of physical loss of habitat, introduction of suspended solids into the water column, and oils spills and discharge of wastewater.
- Impacts on air quality from NO_2, SO_2 and CO_2 emissions from marine vessels.
- Noise impacts from trucks and other construction machinery.

While some impacts of the dredging and reclamation phase are unavoidable, such as the physical loss of habitat, most impacts will diminish in intensity once construction work is completed. Where possible, mitigation measures are recommended to minimize, contain, prevent, or offset any of the main adverse impacts, whether being short- or long-term impacts. In large projects a continuous water quality monitoring programme is recommended to be implemented during the construction period and even beyond.

One of the immediate short-term impacts during the reclamation process is the increase in water turbidity due to sediment dispersion in surrounding waters, as shown in Figure 3.19. An example illustrating the extent of sediment dispersion is shown in (a), however this could be greatly minimised and contained by encircling the project area by a silt curtain (with the curtain extending deep below the water surface), as shown in (b). The need for silt curtains should carefully be evaluated depending on the local water currents and the expected suspended sediment deposition areas.

Another source for an immediate short-term impact is the possible generation of foam when the newly placed sand comes in contact with water. This is usually observed when the reclamation material is calcareous (carbonate) sand and not pure quartz sand. Carbonate deposits on the seabed are normally formed by accumulation of skeletal remains of small marine organisms and shells which will be mixed with the sand during the dredging operation. The phenomenon of foam generation is usually very alarming to environmental authorities, especially in severe cases and if visible by the public. The foam is an ecologically harmless by-product of the reclamation activities. Its contents are mostly water and air with very little solid sedimentary material. The foam will break up in few days, a

(a) sediment dispersion (b) use of silt curtain

Figure 3.19. Sediment dispersion due to reclamation activities.

(a) foam on a sand transport barge

(b) foam at a reclamation project

(c) use of a floating boom

Figure 3.20. Foam generated by calcareous sand.

process which could be accelerated by spraying the foam with water. The sprays need to be adjusted to limit the water velocity so that it does not create more foam by stirring up the fine sediment already present in the water.

The foam could be encountered during sand transportation, as shown in Figure 3.20(a), or after placing the sand, as shown in (b). The spread of the foam could be contained to the project area by using a floating boom (oil boom) as shown in (c). This type of boom has a short curtain as the boom is intended to stop oil or foam floating on the surface.

Chapter 4

Selection and Design of Edge Perimeter Structures

4.1. Introduction

The type and configuration of coastal developments' edge structures are to be selected by the landscape planners, architects and coastal engineers to achieve, economically, the desired function and appearance. Therefore, these structures must not be looked at as "shore protection structures" only as they have to fulfil other requirements set by all the disciplines of the design team.

The following elements are normally taken into consideration to select the edge structure:

- Wave exposure
- Water depth
- Construction methodology
- Functional requirements
- Aesthetic requirements
- Cost

The subject of edge perimeter structures and the development of the design from concept to a detailed stage are well presented in Ch. 6 of CIRIA, CUR, CETMEF (2007). However, certain important issues which are relevant to the theme of this book shall be discussed in the following sections.

4.2. Coastal Structures

Coastal structures, in general, could be grouped into four categories:

4.2.1. *Breakwaters*

There are many different types of breakwaters, as follows, and illustrated in Figure 4.1 (after CIRIA, CUR, CETMEF, 2007):

- Mound breakwaters (basically a large heap of granular material with quarry stone or concrete blocks as armour).
- Monolithic breakwaters (for example: caissons, block wall, masonry structure, etc.).
- Composite breakwaters (combining a monolithic element with a mound berm).
- Unconventional breakwaters (for example: floating, pneumatic, pile, etc.).

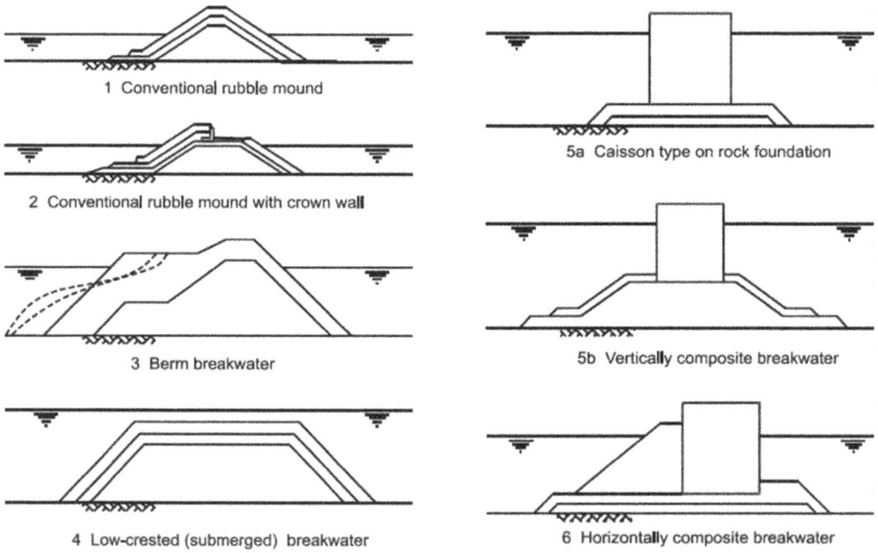

Figure 4.1. Typical cross-sections of mound/monolithic/composite breakwaters.

More comprehensive details and design guidelines are given by d'Angremond and van Roode (2004), Takahashi (2002) and Oumeraci *et al.* (2001).

4.2.2. *Revetments*

Protecting reclamation perimeter sand slopes, earth embankments, soft cliffs, etc. from wave and/or current action is generally achieved by constructing a revetment.

Revetments could either be classified according to their function or to used construction material (McConnell and Allsop, 1998; CEM, Part VI, 2011), as briefly listed in the following according to their functional class:

- Narrow clay embankments — man-made coastal defence.
- Wide sand/shingle ridges — natural or man-made.
- Cliff/reclamation protection — formed from imported or in-situ material.

Revetments could be constructed by using a variety of materials as an armour layer, these are summarised in the following, and illustrated in Figure 4.2:

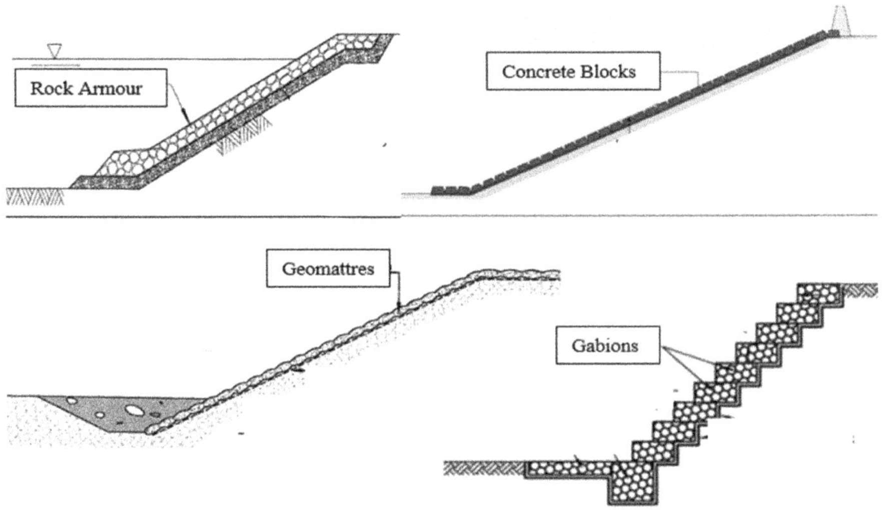

Figure 4.2. Typical cross-sections of selected revetments with different armour types.

- Rock (riprap/rock armour/pitching)
- Concrete armour units
- Articulated concrete mattresses
- Geomattress
- Gabions
- Grouting (using asphalt/cement)
- In-situ cast concrete/asphalt

4.2.3. *Bulkheads*

Bulkheads are structures used to retain fills along a water perimeter, i.e. retaining walls with soil on one side and water on the other. A bulkhead's primary function is therefore to retain fill at locations, normally, with only limited wave action. A bulkhead however needs to be designed with sufficient protection against erosion from the mild to moderate wave action and currents generated by tides or vessels.

A selection of various types of bulkheads is shown in Figure 4.3 (after PIANC, WG 38, 2001).

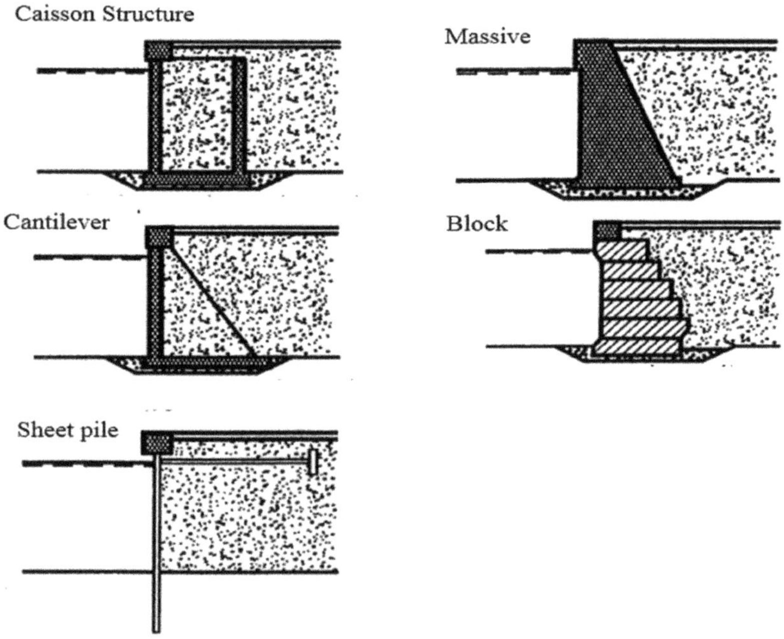

Figure 4.3. Typical cross-sections of selected types of bulkheads.

4.2.4. *Beaches*

Beaches are mostly natural or could be manmade. The natural beaches do require, in some locations, to be protected against excessive erosion. Whereas the manmade ones are created as recreational facilities (see Port Works Design Manual-Part 5, 2003 and Mangor, 2007). Rubble mound structures are normally used as sheltering structures to provide sufficient protection or for enclosing the beach to be within a defined toe line. The basic types of structures are listed as follows and illustrated in Figure 4.4.

- **Groynes:** shore connected breakwaters often perpendicular to the shore, protecting the beach from wave action. However, sometimes these breakwaters could be hooked or curved, or may have a T-head.
- **Detached breakwaters:** these are constructed in series, spaced along the shoreline, and close to the shore to protect a stretch of shoreline from low to moderate wave action.

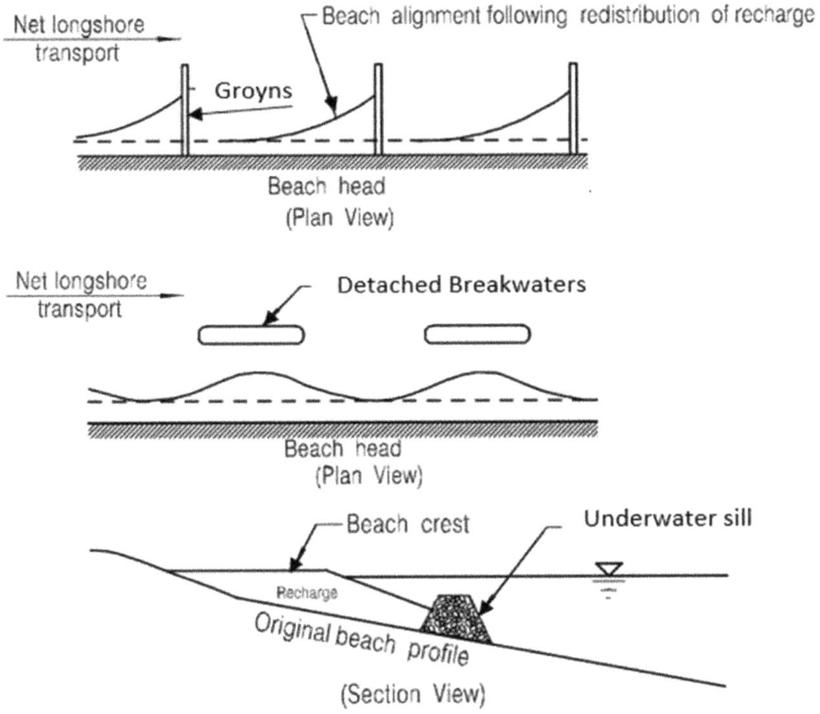

Figure 4.4. Typical beach protection schemes.

- **Offshore sills (perched beaches):** submerged or semi-submerged, shore parallel offshore sills are used to reduce the rate of offshore sand movement and to cause some incoming waves to break.

4.3. Section Design

4.3.1. *General*

The design of a coastal structure section for a particular stretch, which is part of a larger project, should be tuned to the project as a whole and as part of an integrated design process. A number of points, listed as follows, needs to be carefully considered:

(a) Wave exposure i.e. exposed/sheltered to waves from open sea.
(b) Water depth.
(c) Requirement for any particular type of structure, i.e. revetment/vertical wall/beach.
(d) Construction methodology of the reclamation/land fill.
(e) Construction methodology of the coastal structures, i.e. land based/marine equipment.
(f) Interface to adjacent structures.
(g) Functional requirements.
(h) Aesthetical requirements, if any.
(i) Construction material.
(j) Cost.
(k) Design life.

Combining and translating all the above-mentioned points into a successful design will not rely on engineering bases and guidelines only, but a significant part of the design development depends on the designer creativity and experience.

The selection of the type of coastal structure, i.e. bulkhead/revetment/beach, shall primarily be determined by the intended use and function of the structure, probably with other requirements to be considered, see Section 3.2. Few sketches given in what follows will help explain the effects and consequences of some of the above-mentioned points.

Figure 4.5 illustrates basic revetment configuration types. The bund type is usually used in low water depths, whereas the cover type could be

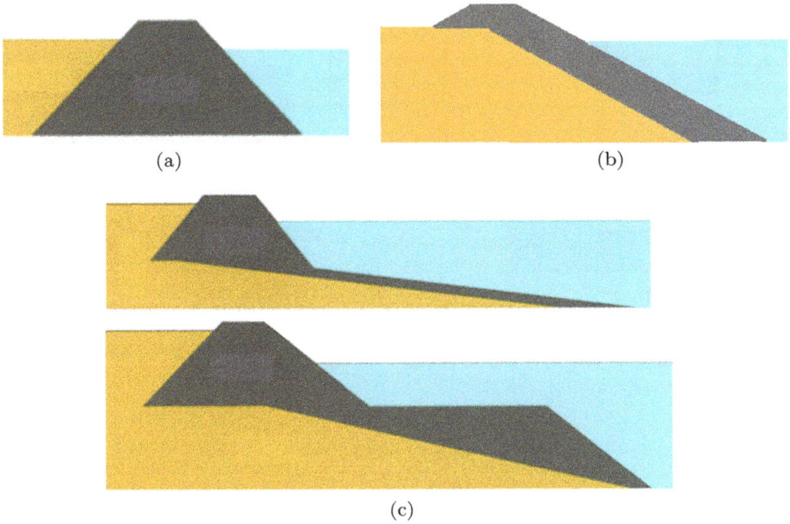

Figure 4.5. Basic revetment configuration types: (a) Bund type; (b) Cover type; (c) Bund-cover type.

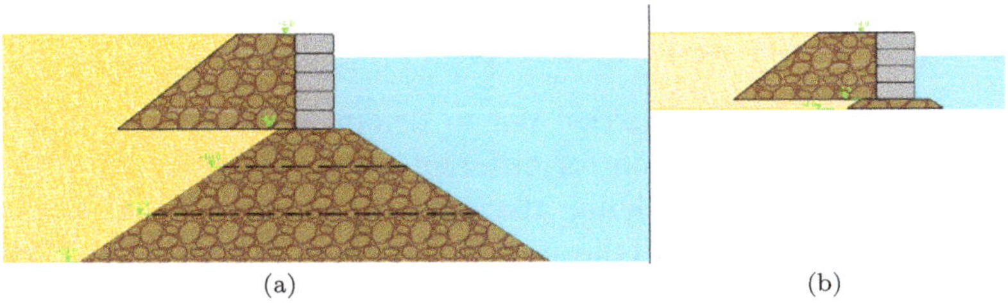

Figure 4.6. Possible block wall sections for shallow and deep-water depths.

used in all depths. The mixed type, i.e. bund-cover type, is usually used in medium to deep water depths.

Figure 4.6 shows the requirement for a large rock bund if a block vertical wall is to be constructed in deep waters, see Point 2 of Section 3.3.

Figure 4.7 indicates possible adjustments which need to be considered to a revetment section from (a) calm waters to (e) rough waters with high waves.

Figure 4.8 shows a possible design for a manmade perched beach normally used if a beach is required in relatively deep waters.

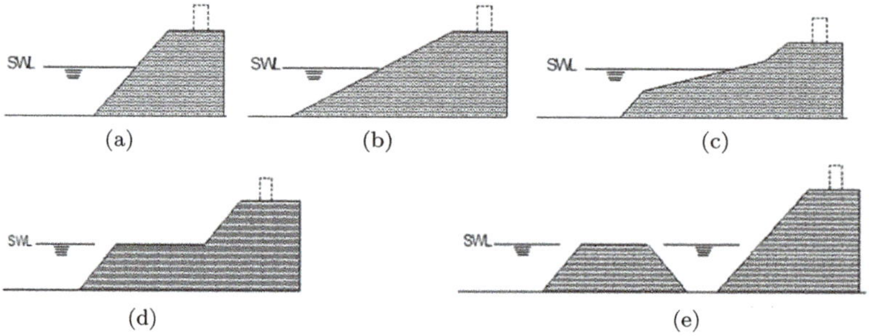

Figure 4.7. Wave height (a-low to e-high) possible effect on revetment section.

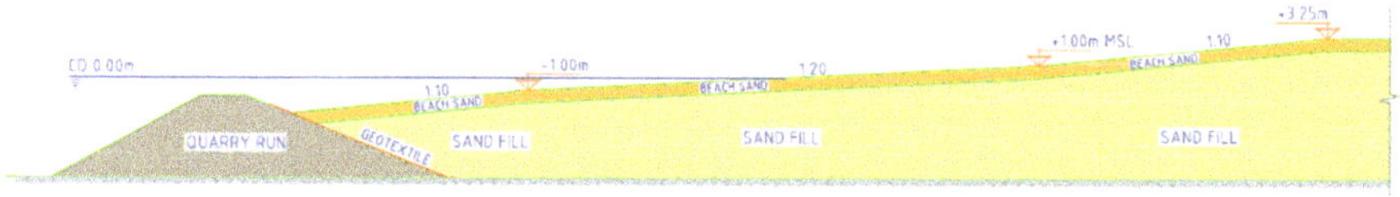

Figure 4.8. Typical perched beach cross-section.

4.3.2. *Design wave and water level*

Wave modelling shall be carried out to determine the impinging wave heights, periods and direction for different water levels and for different return periods. After completing the modelling runs from all principal wind directions, the extent of the appropriate design waves could be delineated along the project edge line. Structural stability and functional performance compliance with the design criteria requirements will then be verified accordingly.

For more details on the selection of the design conditions, reference could be made to Wallingford (2000) and Goda (2000).

Two examples of wave modelling runs are shown in Figure 4.9, for a project given earlier in Figure 2.6 as an illustration of a small land extension. This set of results is from one direction only, similar sets are to be available from all principal wind directions.

The next step is to divide the project edge line into segments, as shown in Figure 4.10, where each segment will have a section design that corresponds to the highest wave height in that segment. Note that the extent of each segment must be checked and confirmed to satisfy the results for all the model runs. For accurate values of the impinging waves,

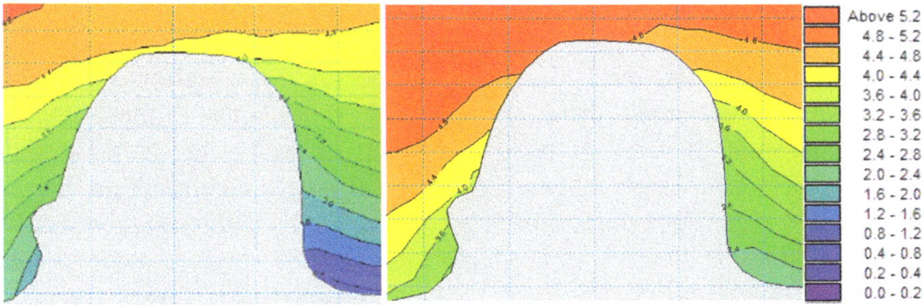

(a) (b)

Figure 4.9. Open sea waves approaching the project at low and high waters (Wave heights, Hs, as indicated in the colour scale): (a) Low water; (b) High water.

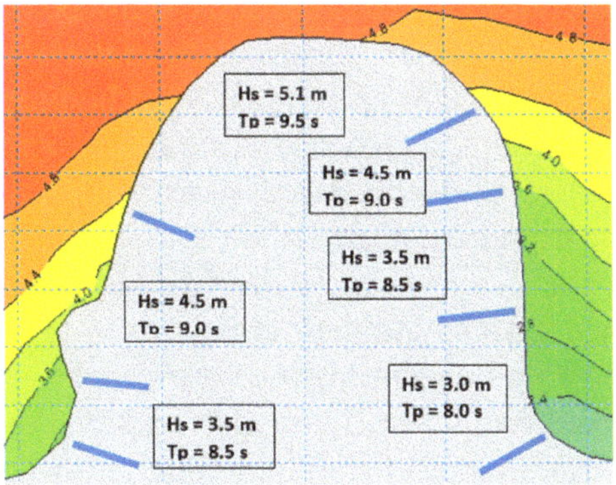

Figure 4.10. Design waves along the project edge line.

some extraction points for wave data could be provided by the wave modelling at specific selected locations. These locations are normally chosen at the approximate toe line and along the edge.

There are two approaches to the selection of the design conditions. The conservative approach, that is, by adopting the design high water level (i.e. tidal level with surges and long-term rise) with the extreme modelled wave heights, as shown in Figure 4.9. This means each event has the design return period. Alternatively, and more realistically, a joint probability analysis is to be carried out to determine the design conditions for the required return period. That is the design conditions (water level and wave height) has a joint probability of occurrence equal to the project's design event return period.

4.4. Design Stages

After completing the work described in the previous sections, the full design shall be executed in stages. The number, or names, of these stages and the amount of work or details required in each stage could differ between different design schemes, or it will follow the contract guidance if stipulated there. As a minimum, it is expected to have three stages specifying the amount of work to be performed in each stage to achieve the completed design work. The commonly used three basic stages are presented in the following sections.

4.4.1. *Concept design*

A sketch design of relevant cross-sections shall be carried out for all stretches. At this stage, different construction materials could be investigated and different section configurations could be proposed. These concept designs shall go through rigorous discussions with the construction team to accommodate their preferred construction methodology and equipment. The procurement team shall also be consulted to provide information about the availability of the various proposed construction materials in the local market and their cost.

Figure 4.11 shows a concept layout for a breakwater with some key cross-sections marked on it. The figure also includes two alternatives for cross-section D (the stretch just behind the round head). Alternative 1 uses concrete armour units (Accropodes) as the main armour for the seaside slope, while Alternative 2 uses rock armour, but the two alternatives are basically the same on the harbour side in using the rock armour.

The pros and cons of each alternative shall be listed, as shown in Table 4.1, for comparative evaluation. In this case, Alternative 1 (with CAU) was selected due to mainly the preferred land-based construction methodology.

Further to the points presented in the comparison table, other issues need to be assessed at this stage as it might affect the selection of the cross-section, or it might impose other requirements which are to be satisfied. Such additional issues are actually project specific; however, some of the possible issues are presented as follows:

- *Seabed geotechnical properties* — a proper and detailed geotechnical investigation should be carried out before starting the evaluation of the

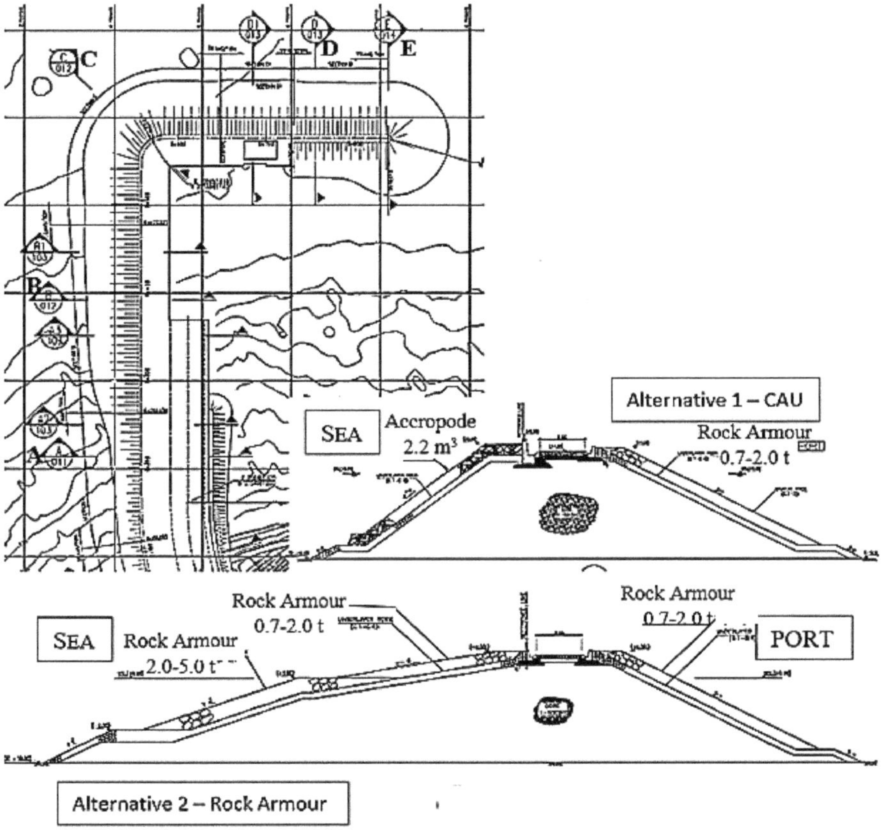

Figure 4.11. A concept layout and two concept alternatives for cross-section D.

possible concept designs. In extreme cases, as in a river delta, the top layer of the seabed is usually made of very soft deposits which probably cannot support the weight of the coastal structure.

Two examples are shown in Figure 4.12: in (a) the breakwater is constructed on an imported and vibro-compacted sand layer which is placed as a replacement for the original soft soil, whereas in (b) the soil replacement is limited with no vibro-compaction required. Another example is shown in Figure 14.35 where the breakwater is also constructed on soft soil deposits, however, the designer preferred to make the structure very wide. By distributing the weight over a wider area

Table 4.1. Comparison of alternative cross-sections 1 & 2.

	Item	Pros	Cons	Comments
Alt. 1	Construction Methodology	Complete land based construction		Minimum influence of weather and sea conditions.
	Construction Material	Less rock quantity		As rock material need to be transported by sea to the project site, less quantity means less cost.
	Armour		Concrete Armour Units (CAU) are more expensive than rocks.	Additional cost includes units royalties, moulds, casting, curing and placement.
Alt. 2	Construction Methodology		Combination of land based and waterborne construction.	More influenced by weather and sea conditions with risk of delay.
	Construction Material		Large quantities of rock material.	Higher cost (material +transport).
	Armour	Less expensive		

instead of replacing the seabed soil, with flat side slopes and wide berms, the slip failure possibility is very much reduced.

- *Aesthetics* — this is usually an important aspect to be considered in commercial land development projects where certain requirements made by the developers are to be met. For example, the type of armour units, allowable overtopping discharge and crest elevation, use of a crest wall, etc.
- *Hydraulic properties of the cross-section* — certain hydraulic properties of the structure may become a design input due to limitations in the project design criteria. Wave reflection and wave transmission are two of the likely properties which might need to be carefully assessed in breakwater design as it might increase the wave agitation inside a port basin beyond the allowable limit.
- *Return period of the design event* — the adopted design event return period for coastal structures is normally 1 in 50–100 years. However,

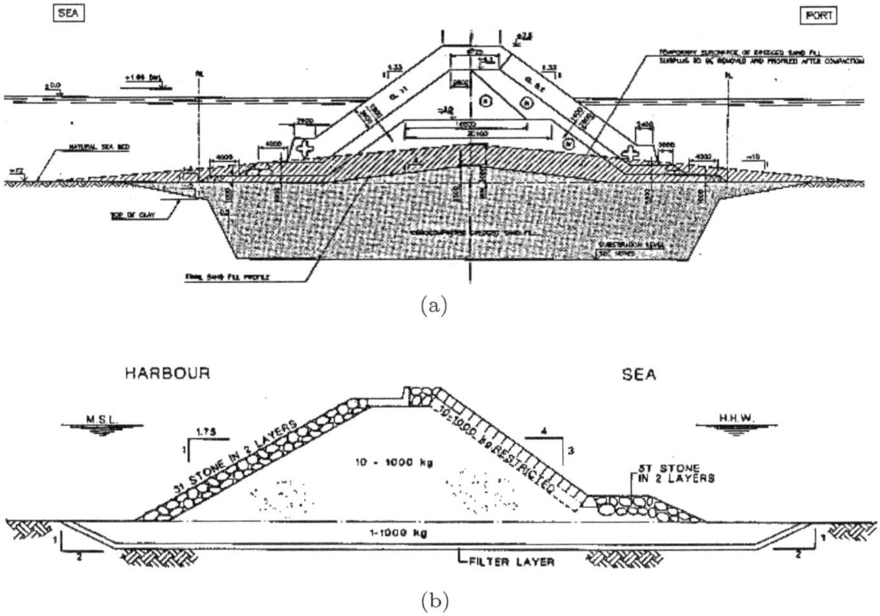

Figure 4.12. Seabed soil replacement under the breakwater: (a) Extensive soil replacement; (b) Limited soil replacement.

for some structures with strategic importance, the design return period could be 1:1000, or even more, and this could have some bearing on the cross-section design.

4.4.2. *Preliminary design*

The Concept Design stage shall be finalised after selecting the preferred cross-section design for all the main stretches for the whole project. All principal cross-sections shall be completed and a more detailed layout plan for the project is to be sketched accordingly.

With reference to the example given in Section 4.3.2, a typical cross-section design for the outer stretch of the revetment is given in Figure 4.13.

It is important to note that the project Setting Out Line (SOL) needs to be located and marked on the cross-section, as it shall clearly appear on the layout plan, as shown in Figure 4.14. This line should have a chainage

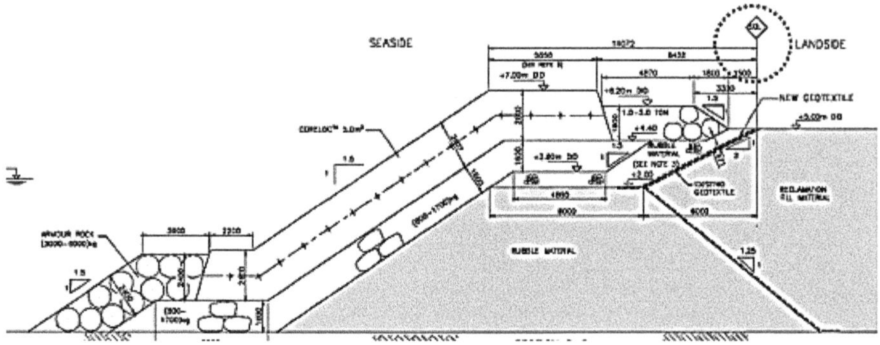

Figure 4.13. Typical cross-section design for the outer stretch of the revetment.

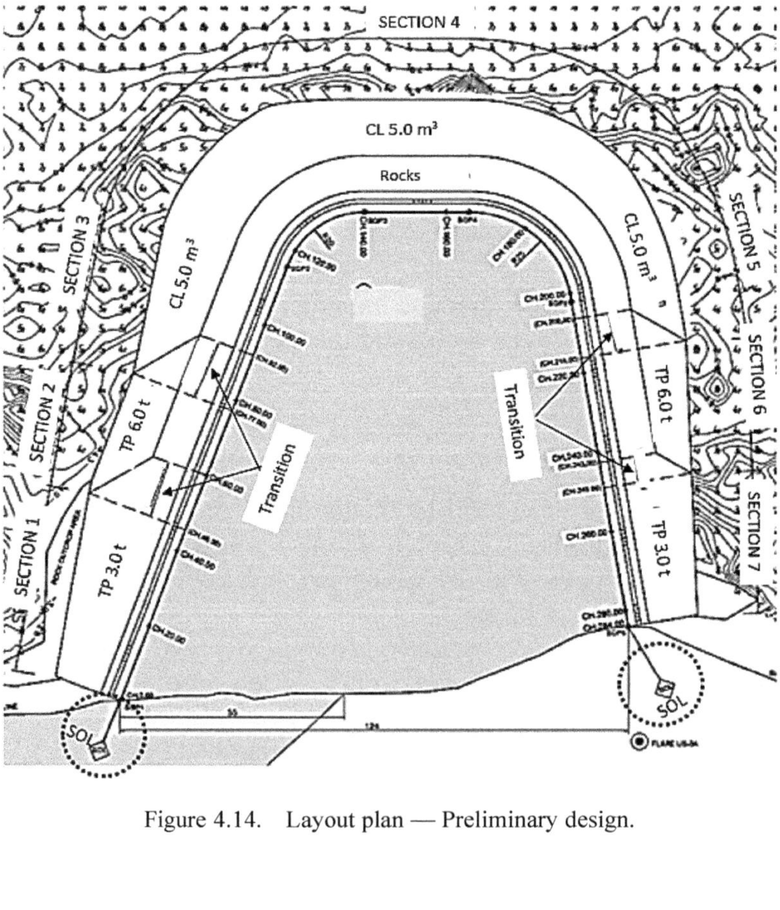

Figure 4.14. Layout plan — Preliminary design.

marked along the full length of the line; hence it is sometimes termed as the "Chainage Line, CHL". Accordingly, each cross-section will be correctly placed on the layout plan for each corresponding stretch.

The SOL shall be fixed by a set of Setting Out Points (SOP), with geodetic coordinate system in order to fix the start, end and every change in direction in the path of the SOL.

The extent of each cross-section shall be identified on the layout plan with transitions between the different cross-sections. At this stage no details are required for the tie-ups with the existing coastal structures, transitions, level changes between the different cross-sections or the exact location of the toe line.

If it is required to verify the design by physical modelling, then it is at this stage that the design of the most exposed cross-section will go through a 2-D flume modelling. For 3-D basin modelling, if required, probably few more details need to be added to the design to make it ready for the testing, especially if certain features are to be checked.

4.4.3. *Detailed design*

At this stage all design documents are to be produced in their final form, this includes a set of drawings, design reports, technical specifications and all relevant studies which could include metocean study (for wave climate and design waves), wave agitation inside the basin, morphology, water circulation, EIA (environmental impact assessment), navigation, vessel mooring, etc.

The drawings shall include final details of all cross-sections, transitions, tie-ups and a detailed layout plan with relevant levels and level transitions. The exact location of the toe line shall be given as it changes with seabed bathymetry.

An example for the expected level of details is shown in Figure 4.15, however, more details regarding the "detailed layout" shall be given in Chapter 5.

By comparing Figures 4.14 and 4.15 the following could be observed:

- Number of principle sections is optimised and reduced from 7 to 6. Accordingly the number of transitions is also reduced.
- All levels of flat surfaces are to be identified on the detailed plan, normally with using the level circle symbol.
- All sloped surfaces are to be identified by long-short lines which define the direction of inclination.
- The center point and radius of any arch in the SOL should be defined.

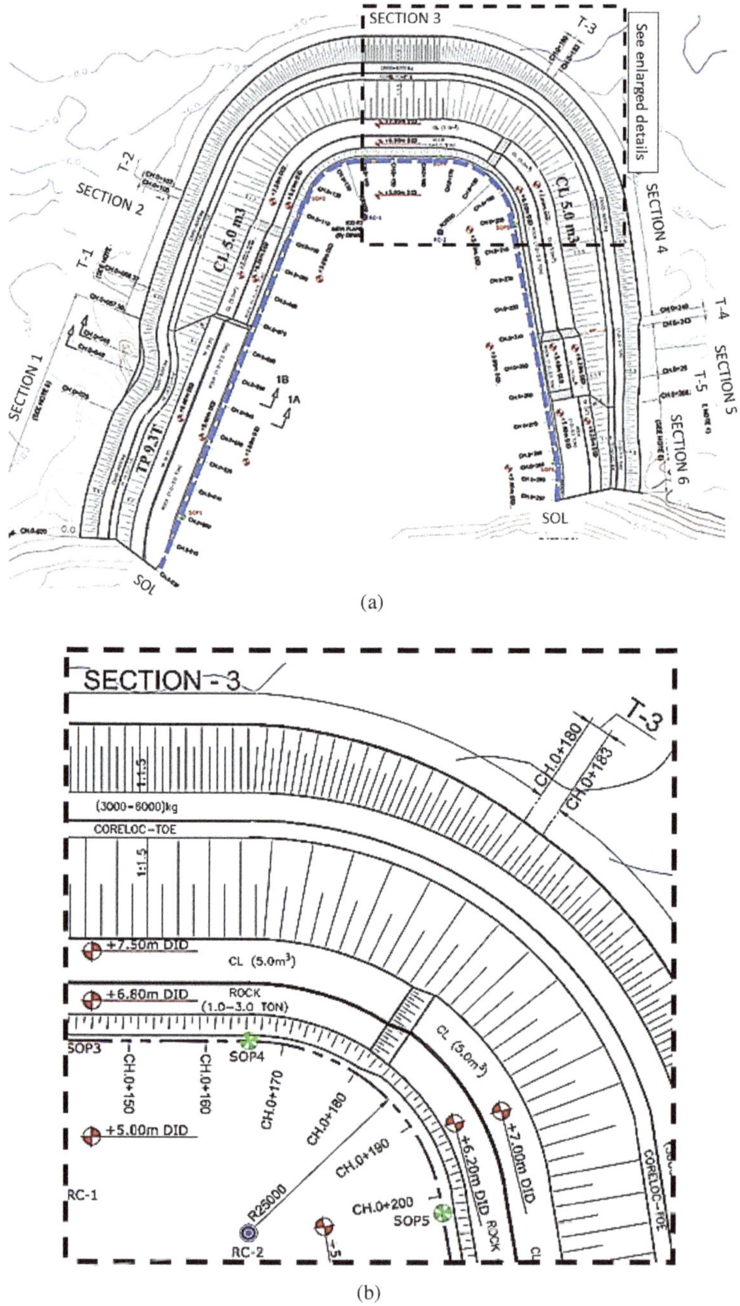

(a)

(b)

Figure 4.15. Layout plan — Detailed design.

Chapter 5

Detailed Layouts

5.1. Introduction

Few examples shall be presented in this chapter to illustrate what is required from the designer to effectively define and delineate the layout by a Setting Out Line, SOL. The degree of complexity ranges from a simple single SOL line to multiple lines. It is also important to note that the SOL must also indicate the chainage from a well-defined starting point as the SOL will serve as a Setting Out Line for the project and also as a Chainage Line. The chainage is required for planning, progress monitoring, supervision, inspection and approval of construction.

All SOLs must be well defined by a number of Setting Out Points, SOPs. The coordinates, in a global or project specific system, shall be given on the drawings.

5.2. Single SOL

Figure 5.1 shows an example for a simple breakwater consisting of a trunk, an elbow and a roundhead, a typical section of the trunk is also given in the figure to indicate the location of the SOL in the section. This simple layout requires a single SOL/chainage line with four SOP points at the chainage line start and end, and at both ends of the curve for the elbow. Note that the radius of the curve shall be given in the drawing.

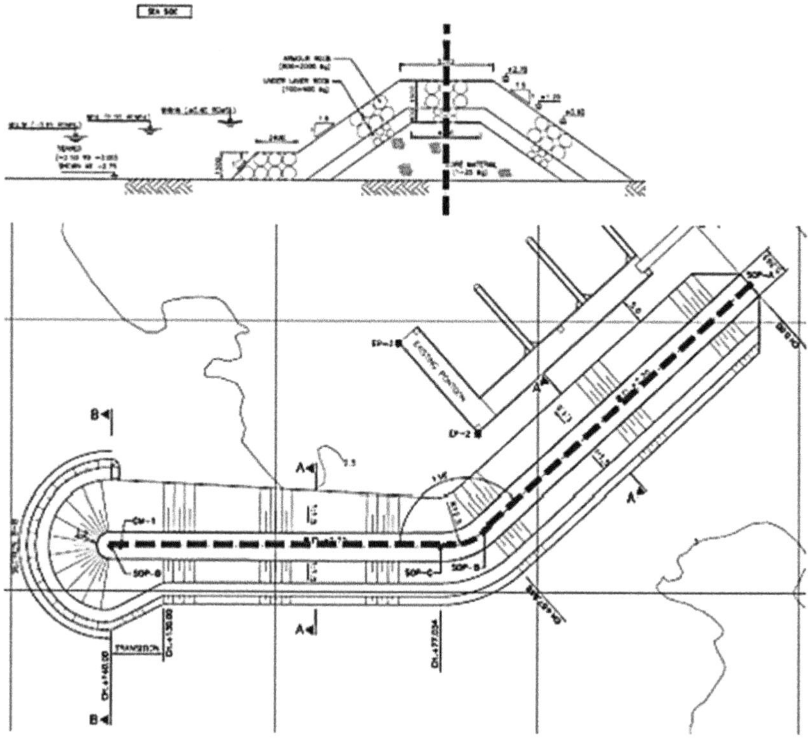

Figure 5.1. Simple layout with one SOL.

5.3. Multiple SOLs

A slightly more complex layout is shown in Figure 5.2 with two SOLs, together with one illustrative cross-section. The blue SOL will be used for the construction of the outer part of the breakwater (effectively a revetment) and its roundhead at the outermost part. The brown SOL is to be used for the structures of the inner side of the breakwater including the inner revetment at the root of the breakwater and the following block quay walls. The construction of the outer and the inner sides are mostly independent of each other, as long as the outer protection precedes the inner side. However, it is highly important to have the two SOLs related to each other. It is however to note that each of these two SOLs mush have its own chainage, and the chainage number is to be preceded by the designated letter of the relevant SOL (for example: B-CH. 250).

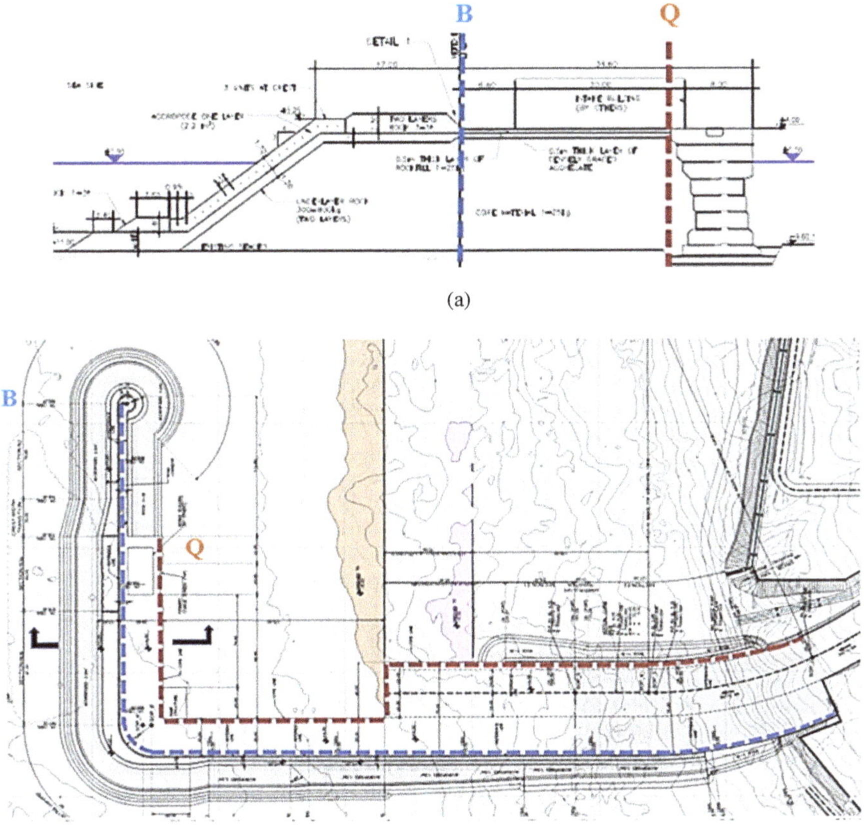

Figure 5.2. Layout with two SOLs: (a) Cross-section in the breakwater and the platform; (b) Plan.

A more composite layout with multiple SOLs is presented in Figure 5.3, with one illustrative cross-section. Four SOLs, of course, each with its chainage, are required to fully capture the various parts of the project. All these SOLs, as mentioned above, are to be well linked to each and the position of each SOL is to be clearly indicated in any relevant cross-section. It is also to be emphasised that the alignment of all these lines is to be well defined by SOP points. The blue SOL, (B), for the main breakwater, the brown SOL, (S), for the spur breakwater, the magenta SOL, (A), for the access road and its revetment, the green SOL, (L), for

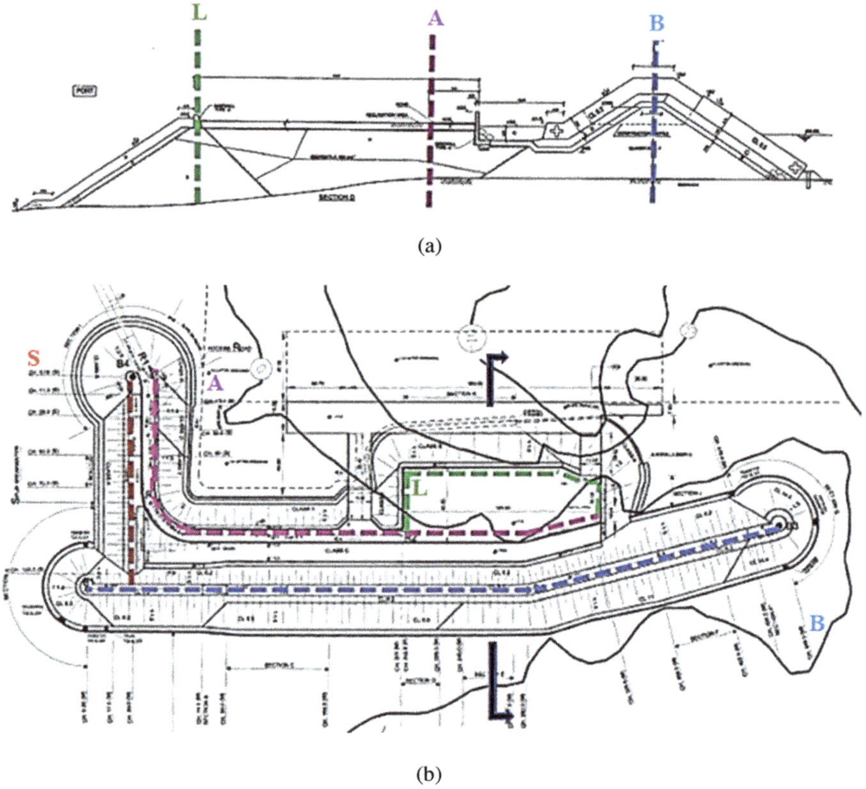

Figure 5.3. Layout with multiple SOL lines: (a) Cross-section in the breakwater, platform and revetment; (b) Plan.

the loading platform and all its relevant structures. As mentioned in the previous example the chainage for these SOLs is to include the designated name/letter for that SOL.

As each of the presented layouts, simple or compound, consist of many different cross-sections, every adjacent two are to be connected to each other in a "transition" zone. This will be discussed in Chapter 6, to clarify and illustrate the various issues which are to be taken into consideration in designing the transitions.

Chapter 6

Transitions, Tie-ups and Junctions

6.1. Introduction

A transition is the connection zone between two different cross-sections of the newly constructed structure. This connection will require some changes in the two cross-sections to facilitate the merging between them without affecting the integrity of the structure. These changes in the cross-sections may be required in any of the x–y–z directions to avoid abrupt changes in any direction and to not create any weak spots in the structure. A poorly designed transition can negatively influence the stability at the transition stretch, therefore transition zones usually represent the vulnerable part of the structure where damage often starts.

A tie-up is the termination/start of the new structure as it meets the shoreline or an existing structure. Changes are normally made in the cross-section of the new structure to have it terminated/merged with the shoreline/existing structure.

A junction is when there is a sudden transition between two different types of structures, for example when a revetment section meets a vertical quay wall.

6.2. Transitions

Transitions in rubble mound structures are normally required if there is a change in the following:

- Type of armour
- Size of armour
- Side slope
- Toe configuration
- Crest level and/or configuration
- Alignment, straight and curves stretches

One of the main reasons for the changes in the cross-section is the change in water depth and/or in the impinging wave height on the structure. Whenever there is sufficient reduction in wave loading or exposure which justifies a reduction in armour size or use of different armour type or there is a reduction in crest height/width, then a new cross-section must be introduced.

At transitions, the armour profile/face is preferably kept at the same level, or as gradual as possible, across the transition zone. This is because any sudden change in profile surface level, i.e. a step, will induce extra hydraulic loading on the protruding armour. However, it is to note that the profile level issue is more important for a rock armour than it is for concrete armour units (CAU), as having a slight drop in profile level across the interface line between two CAUs of different sizes could be tolerated. Reducing the CAU size along a coastal structure is expected, however, it needs to be carried out at stages. The reduction at each transition should not be more than two standard sizes of the unit, this is to ensure sufficient interlocking between all the units.

Transitions between different armour units (type or size) are advised to be inclined at 45° across the side slope with the lighter armour resting on top of the heavier units. At the transition, there is usually a change in underlayer rock size, which is to be compatible with the armour size, and this will also contribute to the reduction in height from the heavy to light armour side. Various options need to be considered to eliminate, or at least reduce, the height difference, these could include having a different transition line for the underlayer, using three layers (instead of two) of underlayer at the start of the transition, then reducing it to two at some distance within the transition zone, or the core profile could be adjusted to compensate for the height difference (and this is the most hated option by the construction team as it is time consuming).

A few illustrative examples are shown in Figure 6.1, in (a) the armour rocks are in three layers with a small difference in total layer thickness between the upper and lower parts of the slope as the rock size is changing

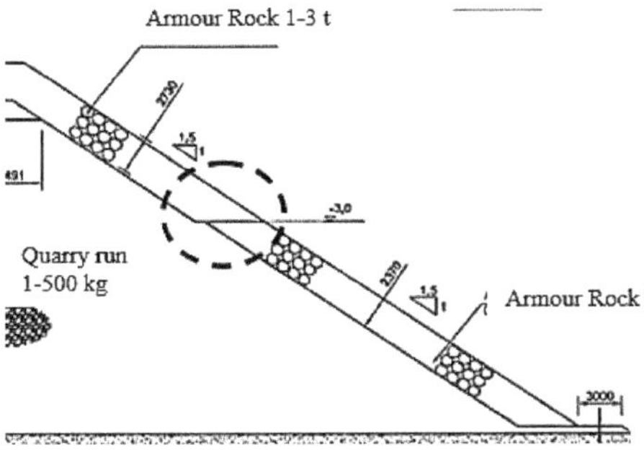

(a) *A small difference with quarry run as the base layer*

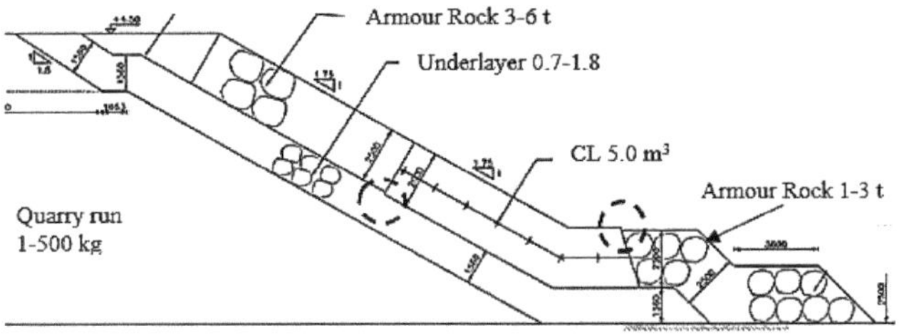

(b) *A small difference with rock underlayer*

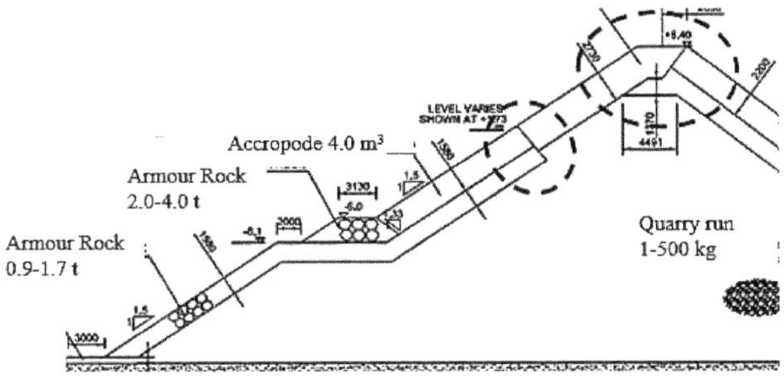

(c-1) *Large difference at slope and crest - basic*

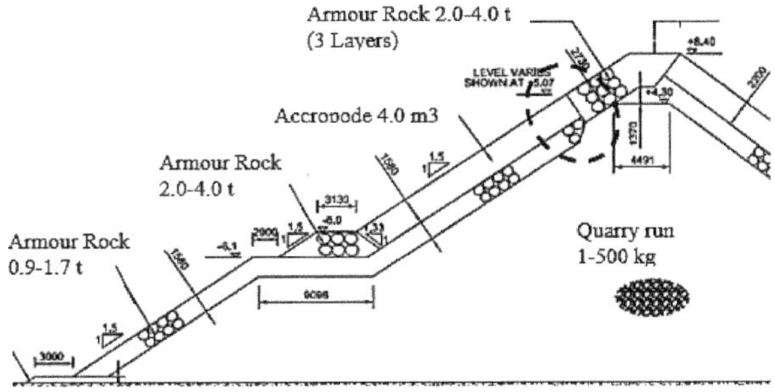

(c-2) *Large differences at slope and crest - recommended*

Figure 6.1. Examples of layer thickness difference at transitions in cross-sections at -3.0 mCD level. It is correctly kept on the inside as the outer rock slope surface is preferred to be kept as smooth as possible (which is, in general, not the case if the armour is CAU). The designer need not to elaborate on such a minor difference especially with quarry run as the base layer, such a small step will be smoothed out during construction, however, it is preferred to show the change in the drawing as a slope rather than a sharp step. This is because some supervising engineers would like to follow the design drawings to the letter. In (b) there is also a slight difference on the inside of the armour layer, however the base is not quarry run, it is a filter layer with a uniform thickness of $2D_{n50}$ along the layer. Therefore, the small difference is better to be made on the outside slope, noting that there is no harm in having a slight difference between the Core loc and rock armour. Such a small difference in thickness is also shown to take place at the toe. Two transitions are shown in (c), one on the slope and the other on the crest of the section. In (c-1) the designer showed the difference in the armour layer (and underlayer if applicable) thickness in the slope as a step and left it for the construction team to smooth it out, however this is not recommended with such a large difference. Therefore, it is better to show the construction team how to smooth the difference out by reducing the underlayer from two to one layer, as such a difference cannot be smoothed by smoothing the quarry run. As for the crest, where the lee side armour meets the seaside armour, it is normal that it will require more than one construction stage to have it finalised, as is in this case. It is normal to have the seaside armour to continue over the crest for low crest levels, and for the lee side armour to continue over the crest for higher crest levels.

A simple illustration for transitions on a layout is shown in Figure 6.2, the first between two Core-loc unit sizes, the larger one used for the outer stretch (including the roundhead and the elbow) and a smaller unit used for the nearshore trunk of the breakwater. The second transition is between the smaller Core-loc units and rock armour, which connects the breakwater trunk to the shoreline. Both of these transitions are at 45° across the side slope.

A transition along rock armour sections is shown in Figure 6.3, where Section B is exposed to moderate waves and Section C is exposed to more severe wave conditions that require heavier armour and a flatter slope than those of Section B. Therefore, the transition shall be three-fold, the first between two different armour sizes, and this will be at approximately 45° angle with longitudinal axis of the revetment. The second part of the transition is the change from 1:1.5 to 1:1.75 side slopes. The third change

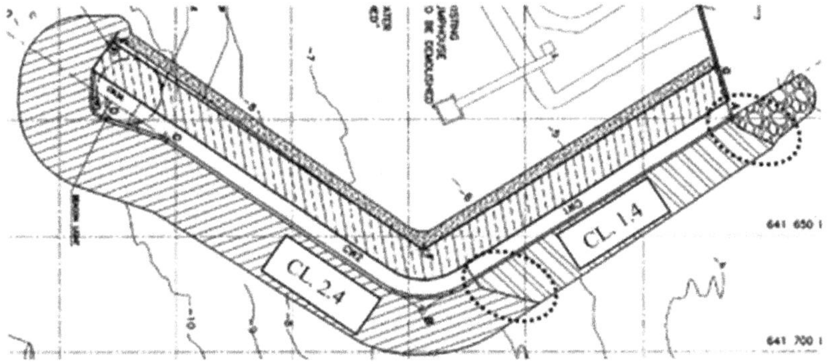

Figure 6.2. Transitions between different sizes of CAUs.

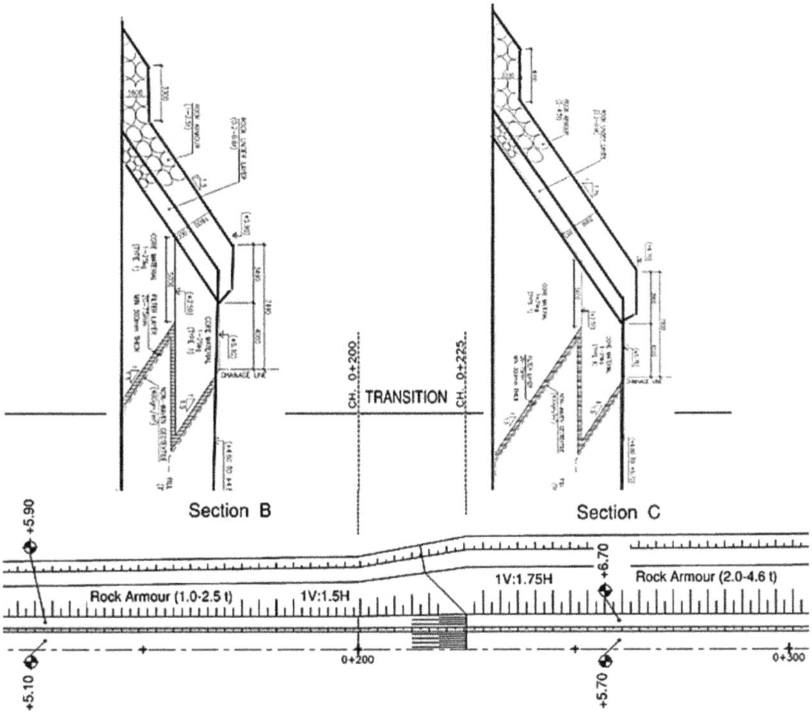

Figure 6.3. Transition between two rock armour sections.

between the sections is the difference of 0.60 m in crest level, which shall be gradually connected. It is important to note that the transition between the sections shall be made as gradual as possible, preferably over a distance not less than 20 m, though this distance depends ultimately on the magnitude of the difference. Note that not all the changes need to take place within that distance as the transition distance from one armour size to another actually depends on the length of the revetment face, i.e. the height from toe to crest.

A more elaborate example is shown in Figure 6.4, where it involves Core-loc units and rock armour with side slope and crest width changes. The

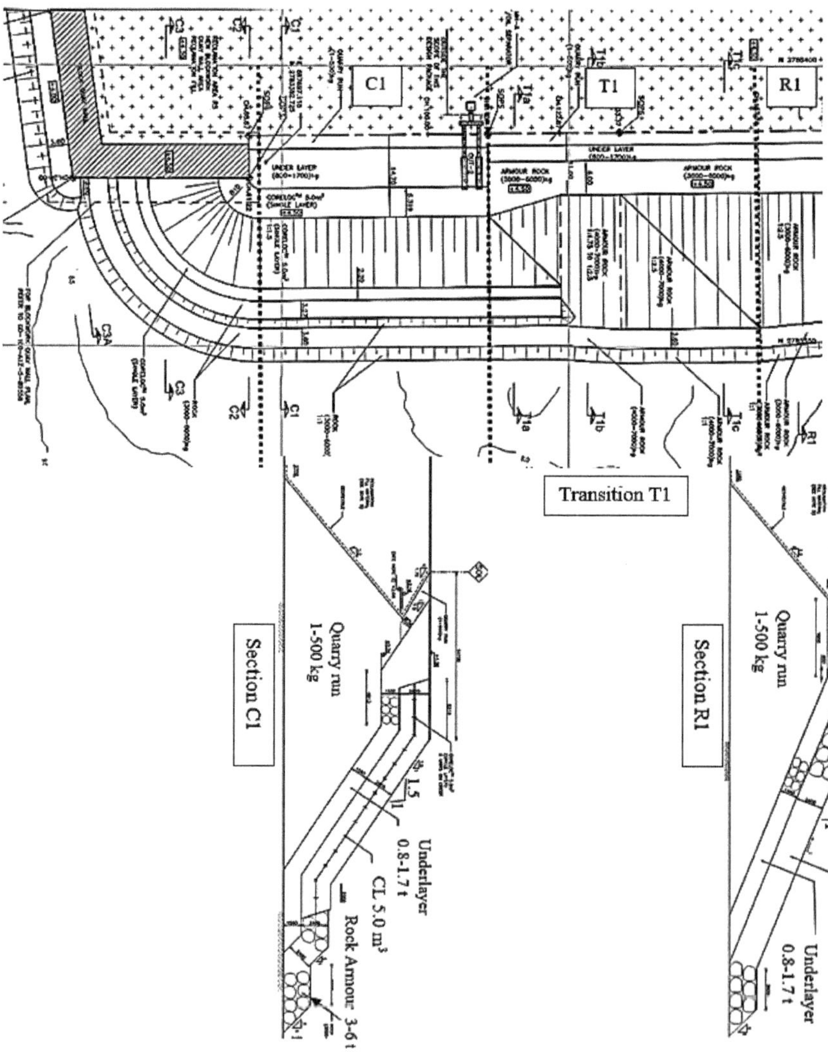

Figure 6.4. A transition from Core loc to rock armour.

Figure 6.5. The physical model for the Core-loc — Rock transition of Figure 6.4.

transition needs to consider an armour change (from 5.0 m^3 Core-loc to rock armour of 3–6 t class) and there will be a change of side slope from 1:1.5 for the Core-loc to 1:2.5 for the rocks. The most exposed stretch is protected by Core-loc 5.0 m^3, as in Section C1, going inshore the required protection will be rock armour of 3–6 ton in a slope of 1:2.5 as per Section R1. However, joining these two slopes presents a problem as neither the Core-loc slope could be made any flatter nor the rock armour slope could be made any steeper, as both types of armour will be unstable at the flatter/steeper slopes, respectively. The side slope at the transition zone needs to change gradually from 1:1.5 to 1:2.5, and the transition zone will therefore be protected by a third type of armour, namely a 4–7 t rock class. This rock class is slightly heavier than the 3–6 ton and is stable in a 1:1.5 side slope. The physical model for this critical transition is shown in Figure 6.5.

A different transition case is given in Figure 6.6, between Core-loc 0.7 m^3 with a slope of 1:1.33 at the rear slope of a breakwater and 2–5 t armour rock protecting the slope under an open quay. This transition is inside the port basin, i.e. not exposed to wave loading, however each type of armour is under different source of hydraulic loading. The Core-loc is to be stable under downrush loading of the wave overtopping, whereas the under quay slope shall be stable against vessel propeller/thruster hydraulic loading, with the loading almost perpendicular to the section in both cases. Therefore, and after slightly oversizing the rock size for the slope protection, the transition line need not be at 45° and could be made perpendicular to the SOL, as shown in the plan view in Detail A.

It is emphatically important to indicate the level of all horizontal flat surfaces on the plan view, as in all above examples, as this will help the designer visualise the transition and how to configure it. This is normally applicable in most crests and berms but not in toes and aprons, as quite

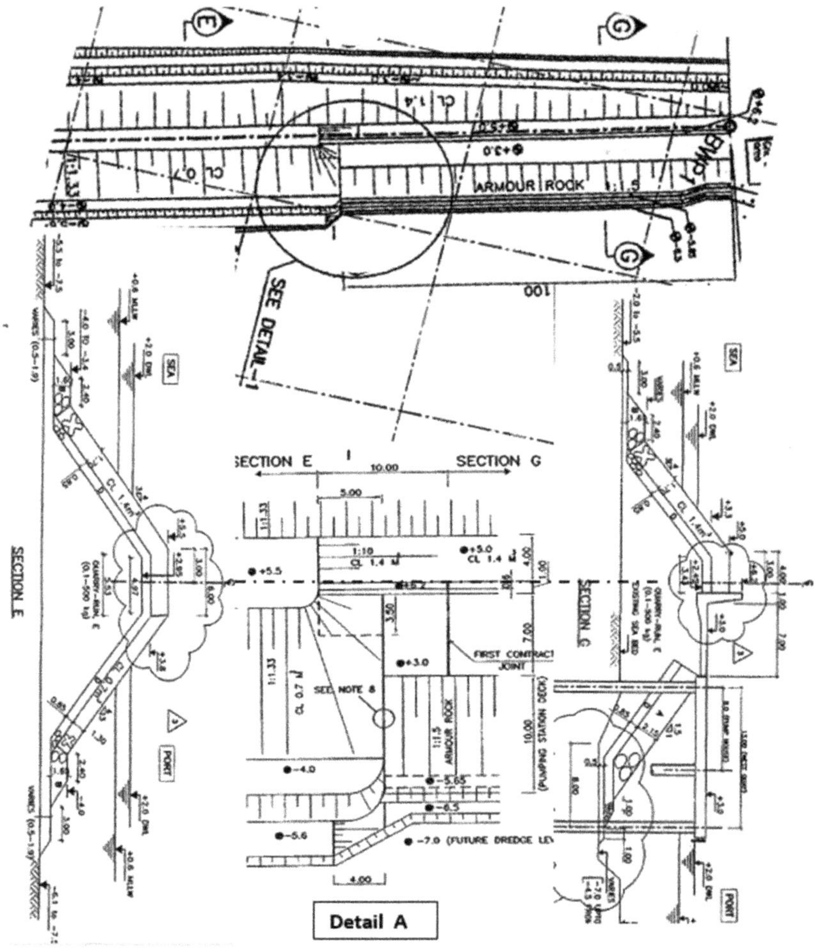

Figure 6.6. A transition from a breakwater section to an under-slope protection of an open piled berth.

frequently these are made to follow the seabed level and a layer thickness is specified for these elements instead of the level.

The transition from one armour size/type to another may also be horizontal, i.e. parallel to the SOL, as shown in Figure 6.7. The slope facing the sea waves understandably requires the heavy armour, however for the crest (if high enough) and the rear side of the breakwater a lighter armour could be sufficient. In some cases, though not frequent, the crest level of the breakwater may be relatively high enough above sea level where the

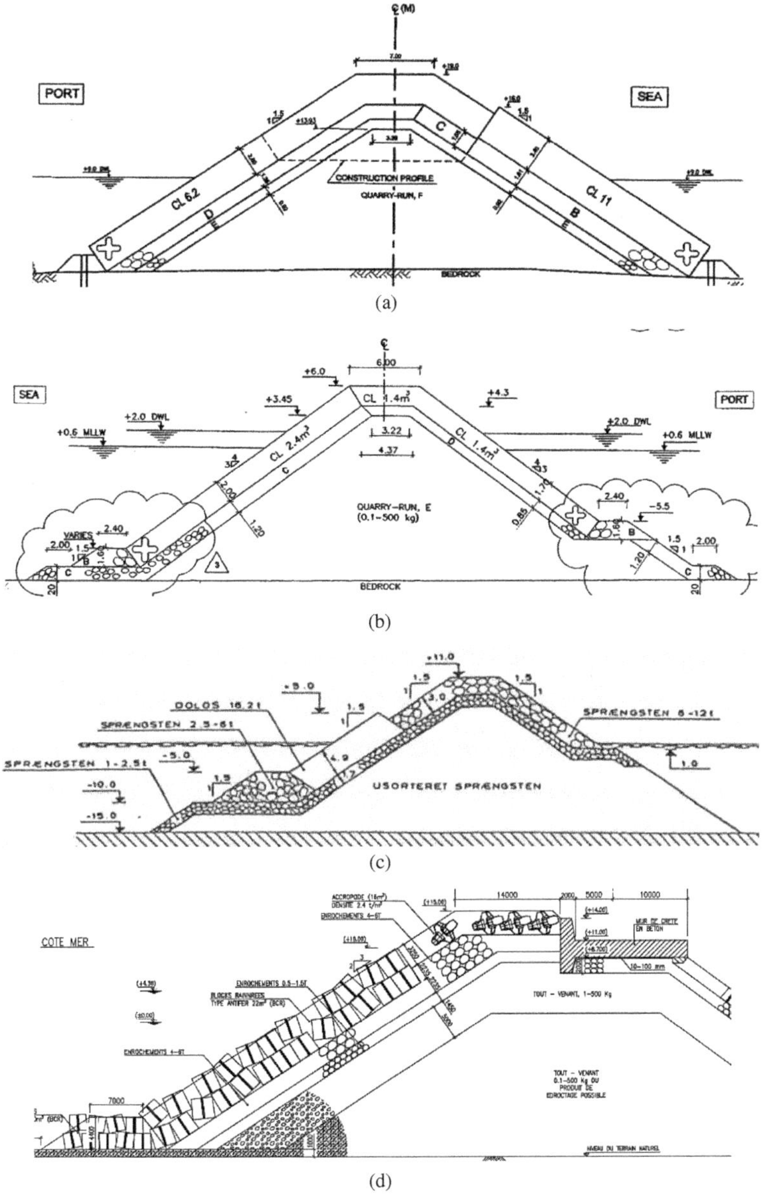

Figure 6.7. Horizontal transitions between different armour units (parallel to SOL): (a) Core-loc 11.0 m^3 to Core-loc 6.2 m^3 at the seaside face; (b) Core-loc 2.4 m^3 to 1.4 m^3 at the crest; (c) Dolos to rock armour at the seaside face (Jensen, 1984); (d) Cubes to Accropodes.

Figure 6.8. Selected transition photos between different CAUs or CAU to rock armour: (a) Accropode — Rock; (b) Antifer blocks — Rock; (c) Stabits — Rock; (d) Shed — Rock; (e) Antifer blocks — Tetrapods; (f) Xbloc — Rock; (g) Core-loc — Antifer.

top part of the face is not exposed to the full design wave loading, as in example (a), as the full wave loading is considered to be applicable in the zone of ±Hs from the sea level. Accordingly, a reduction in the armour unit size, together with the underlayer rock size, could be cost-effective. In other cases, it may be preferable not to have an armour unit change on the seaside face but to have it on the crest, as shown in example (b). Different types of horizontal transitions are shown in (c and d), namely between two different armour types. The first, as shown in (c), is between Dolos armour units and rock armour, which is possible with the two-layer CAU but highly not recommended for the single layer CAU. The second as shown in (d) is between two different CAUs, namely Cubes and Accropodes.

Photographs from different projects are shown in Figure 6.8 for transitions between different Concrete Armour Units (CAU) or CAU and rock armour along the longitudinal axis of the structure, i.e. at 45°.

6.3. Tie-ups

The details of a tie-up of a new structure to an existing one or at the termination/start point at the shoreline need careful consideration and must not be left to the construction team to patch this zone up. The tie-up zone if not carefully evaluated and designed could be a weakness point in the structure with possible unwanted consequences.

Tie-ups could come in different shapes and arrangements; some common examples shall be presented in this section. The termination/start point at the shoreline, i.e. the root of the breakwater/revetment and the connection of two breakwaters/revetments at an angle. The tie-up of a new revetment to an existing one shall also be presented.

Figure 6.9 is given as a guide to the extent of wave/current action limit, and accordingly the required extent and degree of the protection to the structure which is most likely to start beyond the coast limit.

Figure 6.10 shows two different examples for the arrangement at the root of a breakwater. In the first example in (a), with no shore erosion expected, the breakwater has been extended inshore with an embankment of quarry run until it reaches a level well above sea level. The second example in (b) shows two flanks, one on each side of the root of the breakwater, to provide protection to the breakwater root against shore erosion. The key issue of this type of tie up is to fully study the shoreline nature

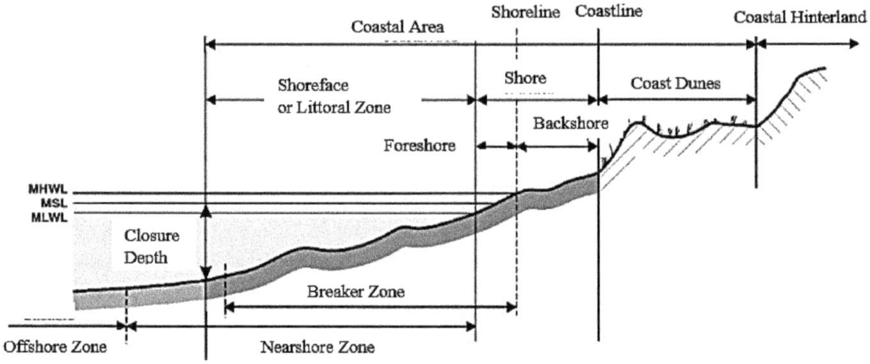

Figure 6.9. Definition of coastal terms, (SPM 1984).

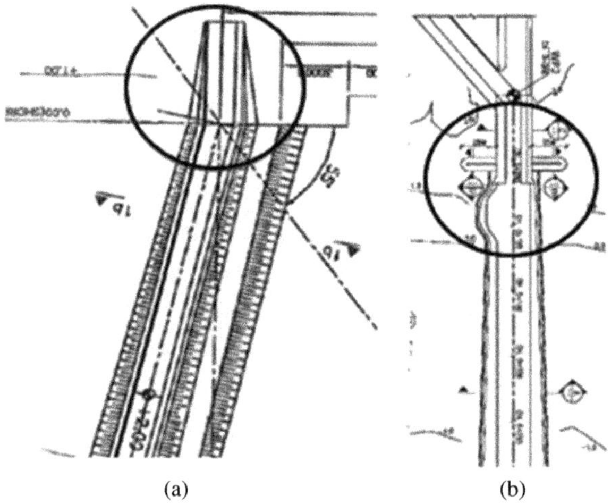

(a) (b)

Figure 6.10. Breakwater root protection design examples.

and behaviour in view of the wave climate in the area and the nature of the shoreline. In Figure 6.11, an example is given for a tie-up of a revetment to the shoreline, the causeway embankment is protected up to contour +4.0 mCD to ensure its integrity against possible erosion.

Figure 6.12 presents more photo examples (a, b, c, d and e) illustrating various breakwater tie-up arrangements to the shoreline, including short/long tie ups to sandy/cliff shoreline. However, in all these examples sufficient protection is provided to the root of the structure, which extends either inland or along the shoreline.

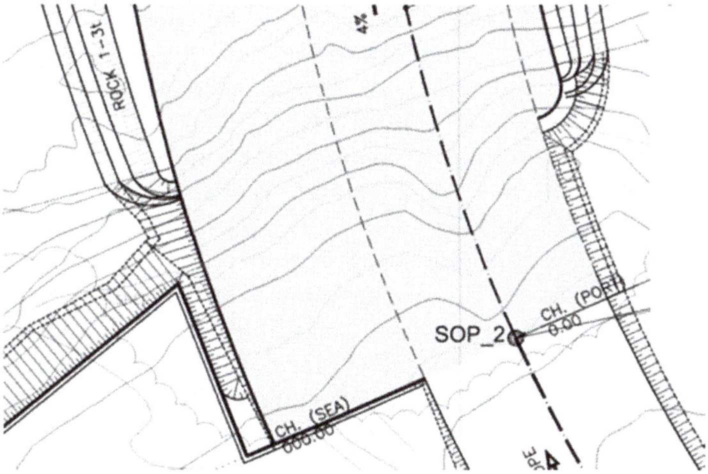

Figure 6.11. Tie-up of a shoreline to a reclaimed causeway.

(a) (b) (c) (d) (e)

Figure 6.12. Breakwater root tie-up to shoreline.

A second type of tie-ups is where two breakwaters meet at an angle, probably an existing and a new breakwater, or a main breakwater and a spur breakwater as in Figure 6.13. The connection could be close to a sharp angle if the structure is in relatively calm waters, however it is always recommended to have a small curvature at the intersection of the side slope faces avoiding sharp corners. A more detailed example is shown in Figure 6.14 for a tie-up of a groyne/spur breakwater to a revetment. A general plan is shown in (a), and the relevant cross-sections are shown in (b, c, d and e). The starting point is to finalise the plane views of the groyne and the revetment and make an initial connection as shown in (f), then start slowly and carefully to refine it to its final form as shown in (g). With such a complicated tie-up and transitions, few iterations could be required. What is shown in (f) is in fact not an initial connection trial but actually the final tie-up plan as drawn by a junior designer. This designer made many mistakes, which contributed to the lack of ability to visualise the connection in a 3-D form. The first of these mistakes is the missing, or not clearly indicated, levels of the plane surfaces on the plan, as indicated by 1a, 1b, 1c. This has led to having two horizontal surfaces (as in 1a, b), or a horizontal surface meeting a slope (1c), with abrupt change of level as if it is a vertical wall. As for 1a in particular, which is the transition from the groyne crest level at +7.0 to the revetment crest at +8.5, it is not just a smooth transition that is to be provided, but there should be a mild slope to allow for service vehicle access. The second mistake is not being accurate or consistent in using the hatch (as in 2a, 2b, 2c), where 2a should be similar to 2b, and in 2c the same hatch has been used for 2–5t rocks and for quarry run. The third type of mistakes is to

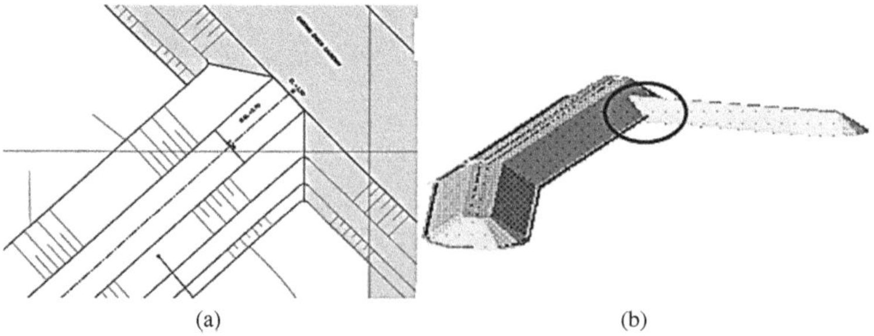

(a) (b)

Figure 6.13. Spur breakwater connection to a main breakwater (grey-shaded part is the existing/main breakwater).

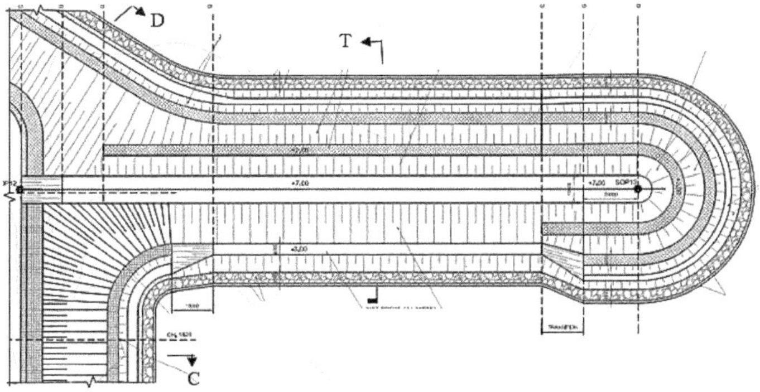

(a)

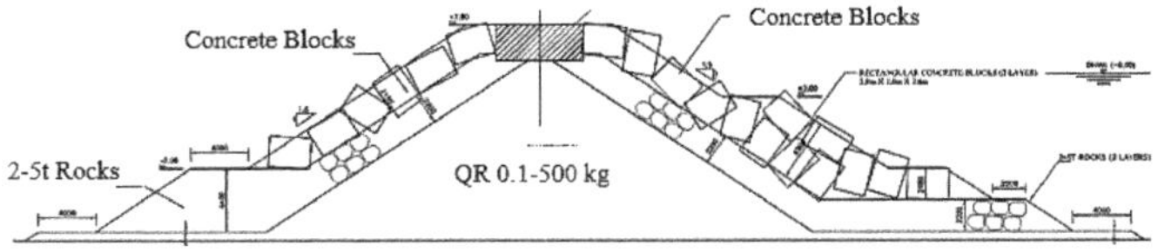

(b)

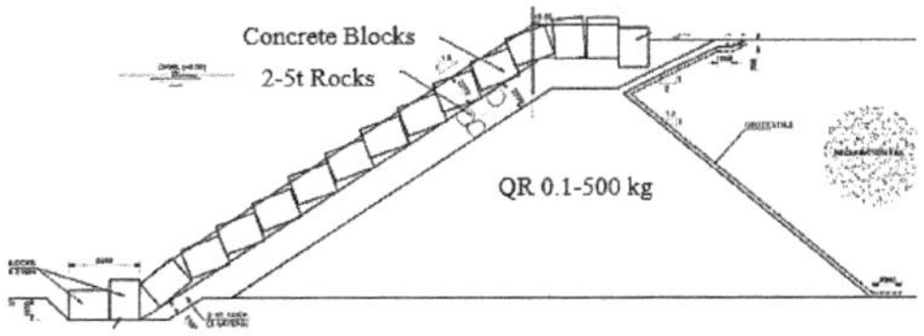

(c)

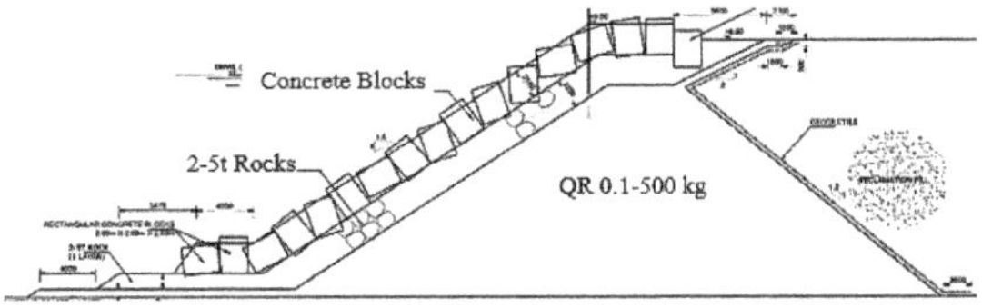

(d)

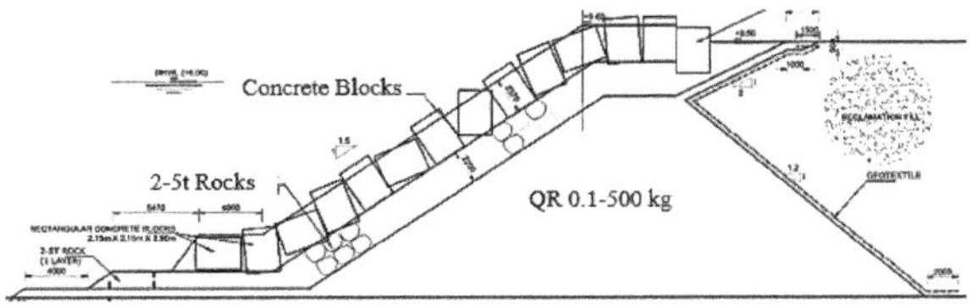

(e)

Figure 6.14. Tie-up of a groyne to a revetment: (a) General plan view; (b) Cross-section T-T; (c) Cross-section B-B; (d) Cross-section C-C; (e) Cross-section D-D; (f) Initial tie-up plan; (g) Final tie-up plan.

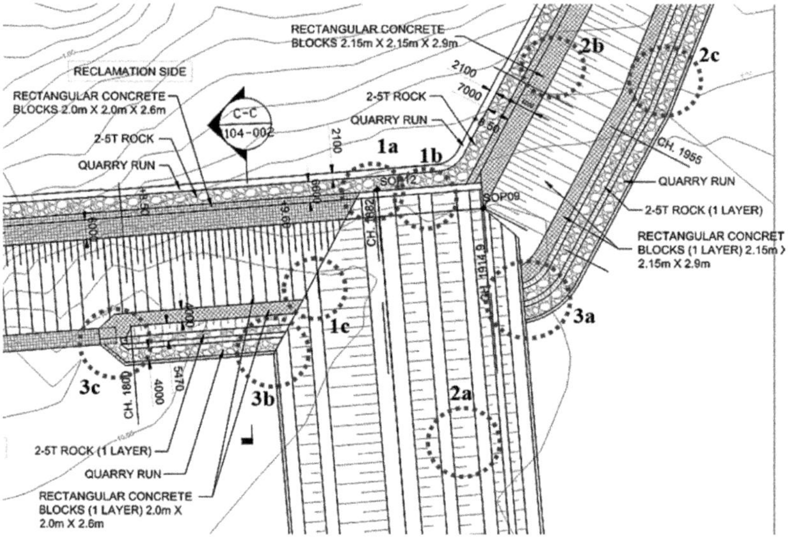

(f)

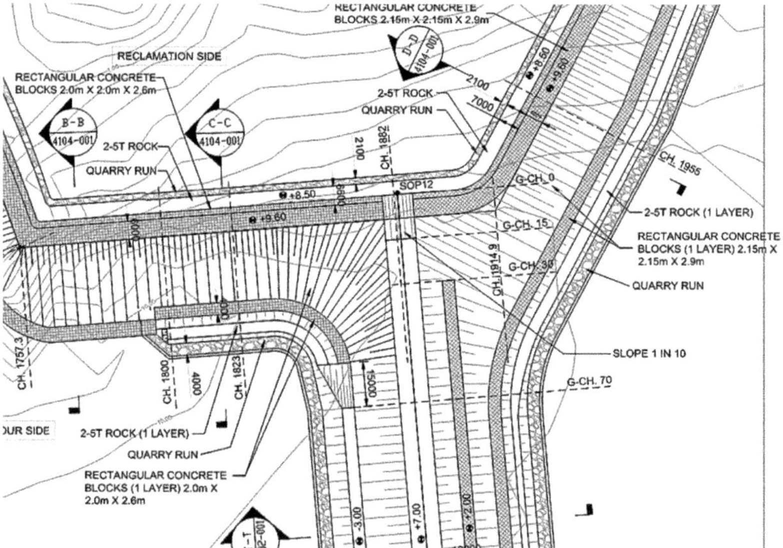

(g)

Figure 6.14. (*Continued*)

have a discontinuity in the toe (as in 3a, b), as this is only expected to take place if there is a change from trenched to non-trenched toe as 3c. All the above-mentioned mistakes have been rectified in (f).

Figure 6.15 shows a tie-up between a new and an existing revetment. It is not advisable, as mentioned before, to have a sharp connection angle at such a tie-up as in (a), especially where there is high wave action. Such a connection will result in weak interlocking between the CAU if this is to be the armour units or weak support and connection between the amour rocks. Therefore, it is highly recommended that the connection have some curvature, and the transition between the new and the existing armour be smooth and not abrupt, as shown in (b).

A third type of tie-ups is presented in this section between two different types of armours, the existing armour is Stabits (CAU) and the new one is a rock armour. This example is for an existing corroded sheet pile wall extending between two revetments with Stabits armours, as shown in

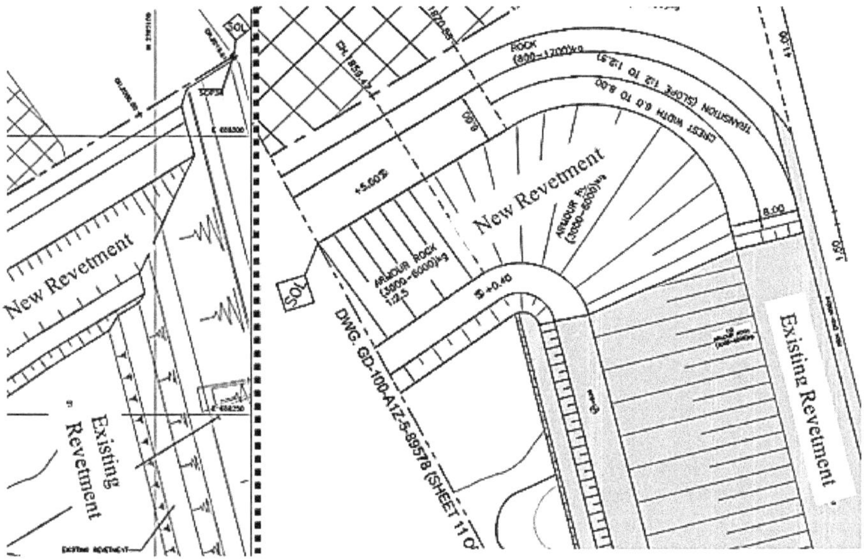

Figure 6.15. Corner tie-up (grey-shaded part is the existing revetment): (a) Incorrect; (b) Correct.

Figure 6.16. Sheet pile wall with Stabits revetment at the end connection.

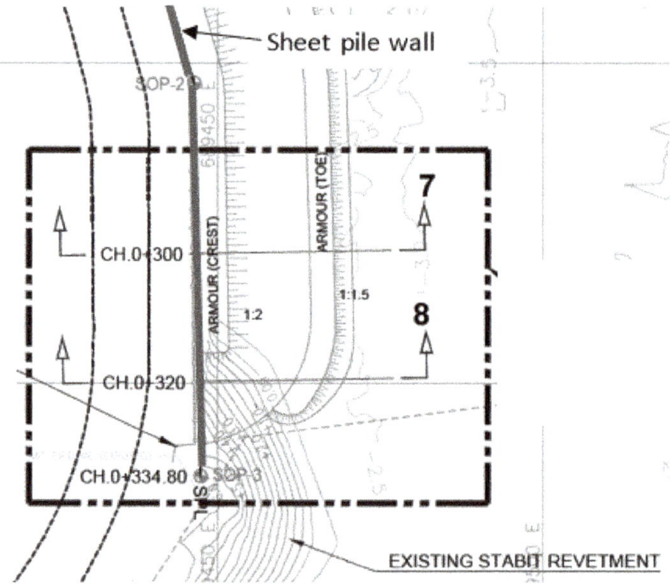

Figure 6.17. A plan view of the new revetment in front of the sheet pile wall.

Figure 6.16 . It is required to eliminate the functional requirement for the sheet pile as a retaining wall by constructing a rubble mound revetment with rock armour in front of the wall. The crest level of the new revetment needs to be at the same level as the existing Stabits revetments, i.e. at +6.0 m. Figure 6.17 shows the plan view of one side of the wall with the new revetment and the existing Stabits revetment. Figure 6.18 gives two cross-sections in the new revetment, cross-section 7 outside the tie-up zone and cross-section 8 in the tie-up zone. It is however to indicate that in such a tie-up connection it is not enough for the designer to give a plan and two cross-sections, as in Figures 6.17 and 6.18, for the construction team to achieve the intended design. More details need to be provided for the extent and levels of each of the construction rock layers in the new revetment, as shown in Figure 6.19. The construction plan views of all layers are given in (a, b and c) for the quarry-run core material, the under-layer and the armour rock layer, respectively.

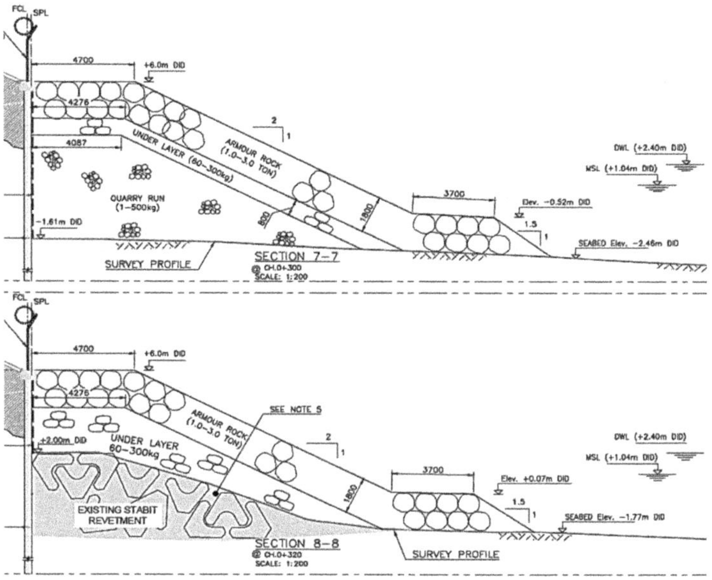

Figure 6.18. Cross-sections 7-7 and 8-8 as indicated in Figure 6.15.

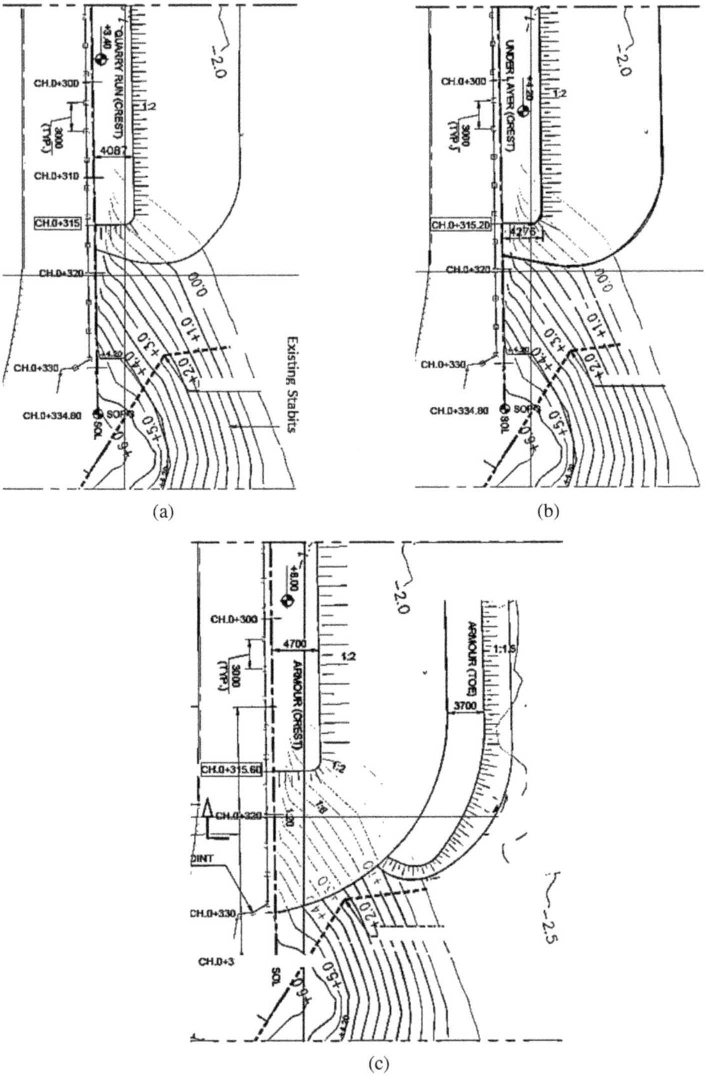

Figure 6.19. Plan view of the quarry run, underlayer and rock armour: (a) Quarry run plan view; (b) Underlayer plan view; (c) Armour rock plan view.

6.4. Junctions

The junction is when one type of structure ends, and another different type starts. Many different examples are given in what follows to illustrate this connection.

The connection between a slopping revetment and a bulkhead (i.e. vertical wall) is a clear example of structural discontinuity, as shown in Figures 6.5, 6.16 and 6.20 (under construction). The design layout plans for the last two junctions are shown in Figures 5.2 and 6.17. The main issue to be considered is the integrity of the revetment, in particular the armour layer, which overlaps with the wall. Therefore, it is preferred to start the placement of the armour blocks (CAU/rocks) in such joints from the junction end towards the revetment stretch. As the core and the underlayer are already in place, proper placement to the armour blocks could be achieved. Otherwise, if the construction of the revetment is to end at the junction, then there is a risk that the armour blocks could be cramped with some blocks protruding above the slope profile. However, the revetment construction will often be concluded at the junction as in Figure 6.20. Therefore, great care should be taken in the placement of the armour blocks to ensure that they are fully embedded and supported/interlocked, as unsupported and protruding blocks could easily be extracted out in adverse conditions, and this will lead to a progressive failure of the revetment.

In relatively sheltered waters, for example inside a port/marina basin, the armour placement could be carried with more ease. Figure 6.21 is an example for such a junction, with land and sea views of the junction.

When stairs need to be erected on a revetment slope to access, for example, a navigation light at the crest, as shown in Figure 6.22, the CAU armour placement pattern will be abruptly discontinued and this needs to be carefully considered, or if the armour is rock, then the placement should start from the stairs to ensure all rocks are properly supported by the surrounding rocks and the side of the stairs.

A bridge abutment, as shown in Figure 6.23, could also be considered as an example for a junction transition. It could be as a revetment-vertical

Figure 6.20. Rock revetment with block quay wall — exposed condition.

(a) (b)

Figure 6.21. Rock revetment with block quay wall — sheltered conditions: (a) View from landside; (b) View from seaside.

Figure 6.22. Stairs on the revetment slope.

Figure 6.23. Rock revetment with bridge abutment.

wall case or the revetment could partially encircle the abutment. Both cases should be designed with consideration for the expected high currents under the bridge.

Chapter 7

Interfaces

7.1. Introduction

The longitudinal transition, i.e. along the SOL, between two adjacent cross-sections was discussed in Chapter 6. In this chapter, however, the interface which is between two adjacent/overlapping layers of different materials and perpendicular to the SOL shall be discussed with the help of numerous examples.

7.2. Filters — Geotextile Fabric and Granular Filter

In rubble mound revetments, there is normally a geotextile fabric at the interface between the rock material and the fine fill material. In order to ensure that the fabric is, and will be, doing its job for the project design life, few points need to be considered and/or carefully assessed. These points are as follows:

(a) Anchorage: The top and bottom ends of the geotextile fabric must be anchored well. At the bottom, pegs (though not recommended) or sandbags could be used, and the top is to be embedded in the adjacent layer (see Figure 7.2).
(b) Quarry run: The geotextile can well be placed directly on the quarry run rock if properly designed for such use, see Figures 7.2 and 7.5. However, it is recommended to specify a trimmed profile for the quarry run.

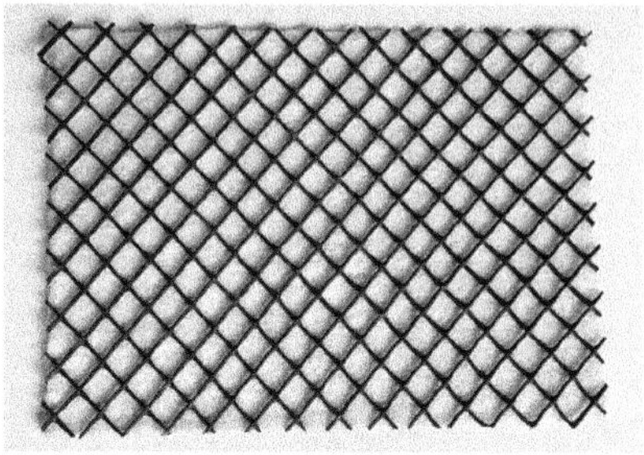

Figure 7.1. Geocomposite (Geotextile + Geonet).

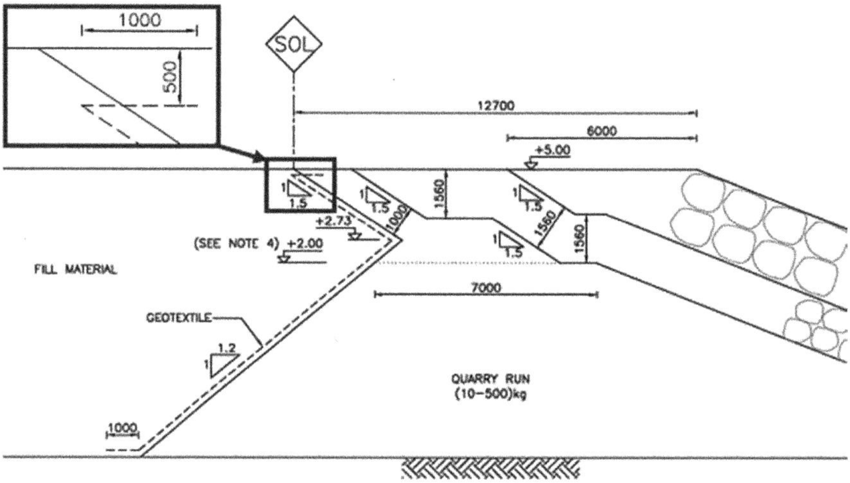

Figure 7.2. Geotextile anchorage example.

(c) Blinding layer: The use of a layer of small rock sizes could be required to separate the geotextile fabric from the quarry run, see Figure 7.3. One possible cause for this requirement is to avoid exerting excessive pressure on the geotextile due to the use of vibro-compaction or other soil improvement methods for the fill material. Other reasons could be the difficulty in trimming the quarry run profile surface due to the limited reach of the available equipment or the configuration and depth of the section or it could simply be an enhancement to the geotextile safety against puncture for a longer design life.

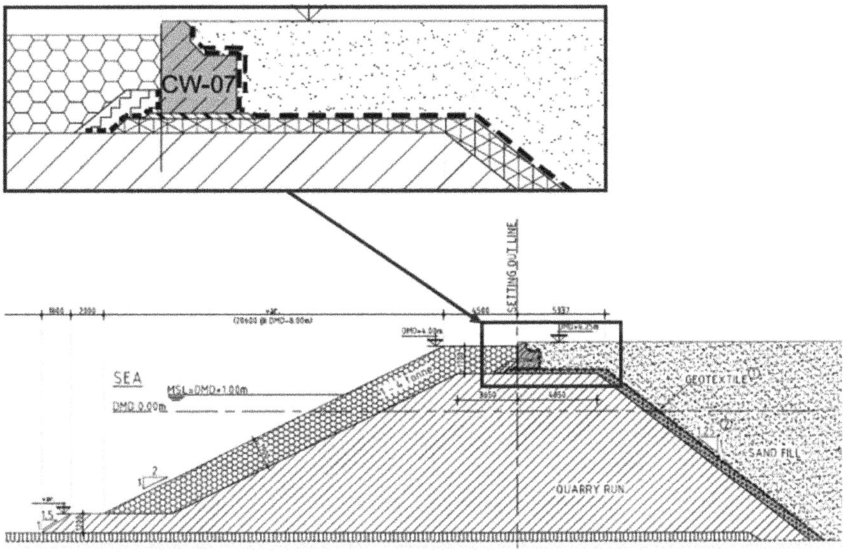

Figure 7.3. Geotextile with blinding layer.

(d) Filter layer: One or more granular filter layers could be used instead of the geotextile in certain situations. Figure 7.4 is an example where a filter layer has been used as an interface between the fill material and the quarry run. The difficulty of geotextile placement, cost or other construction methodology or constraints could be the reason for this replacement. It is however of note that the granular filter could be a close or open filter, depending on the wave action and the hydraulic gradient.

(e) As shown in Figure 7.4, the crown wall is normally to be placed on a fine granular bedding layer. The interface between this layer and the armour rock on the seaside of the crown wall does not usually fulfil the filter requirement, accordingly, a geotextile fabric needs to be used as an interface layer in this location.

(f) It is not recommended to place narrow graded larger rock, i.e. underlayer or armour rocks, directly in contact with a geotextile fabric, as shown in Figure 7.5. This is because the geotextile will not be fully covered and protected as in (a), and therefore it will be prone to deteriorate at a faster rate. It is also advisable to avoid exposing the geotextile to possible concentrated pressure points from sharp edges in

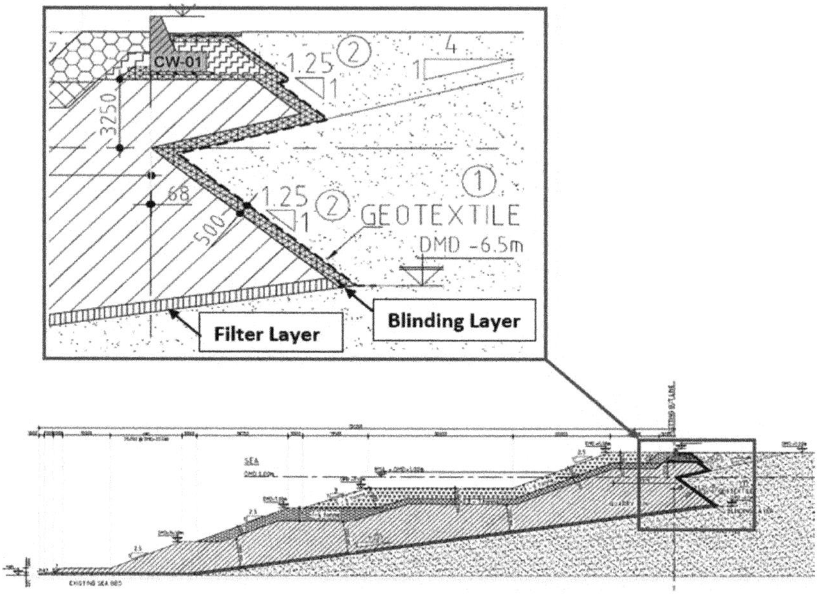

Figure 7.4. Use of geotextile for the upper part of the section.

the rocks. However, some designers are open to using geotextiles against underlayer rock sizes if they are well designed to withstand possible concentrated pressures.

(g) Further to points (c) and (d) above, quarry run may be placed directly on sand fill without a geotextile fabric at the interface or granular filter layers, see Figure 7.6. However, it is important to note that this is acceptable where the hydraulic gradient is relatively low.

Figure 7.6 also illustrates the interface between the blocks of a quay wall and the quarry run backfill. In sheltered situations where there is very little wave action, e.g. inside a port or marina basin, the quarry run could be placed directly against the blocks, as is the case in Figure 7.6. If the block quay is exposed to relatively high wave action, then it is recommended to use a geotextile fabric as an interface between the blocks and the quarry run. This is because of the wave-induced suction effect that is created between the front and the lee side of the wall, which will suck the fine material out. This will undoubtedly induce some subsidence to the surface.

(h) The example shown in both Figures 6.15 and 7.7 will require a geotextile to be placed against the corroded sheet pile to fulfil the following requirements:

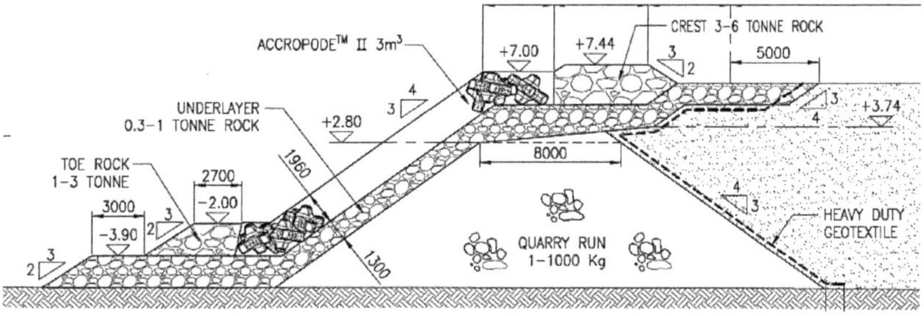

(a)

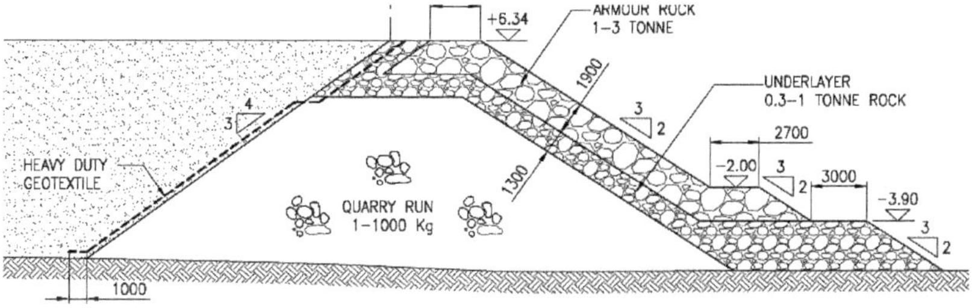

(b)

Figure 7.5. Geotextile and large rocks.

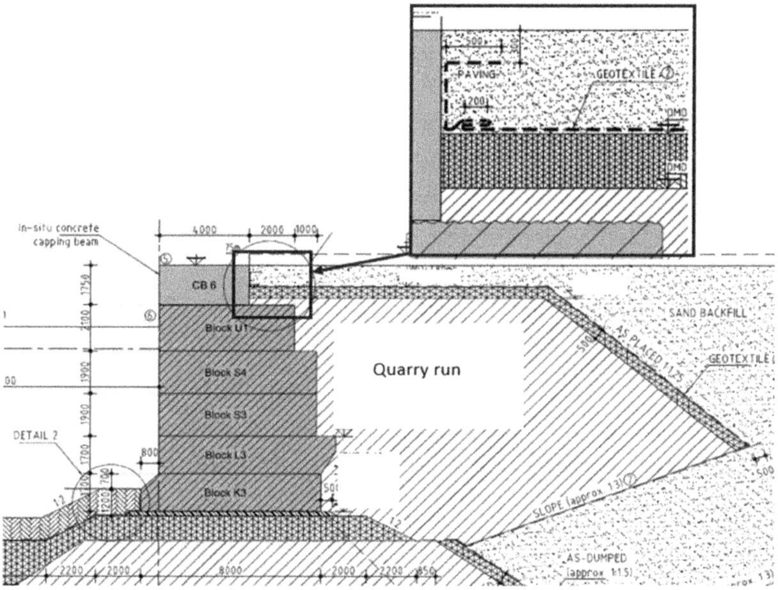

Figure 7.6. Partial use of the blinding layer.

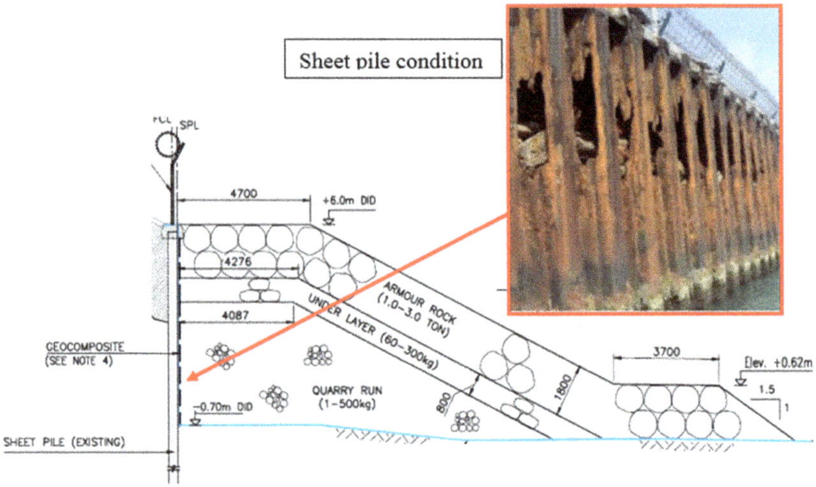

Figure 7.7. Geocomposite on a sheet pile.

- Retention of backfill material behind the sheet pile.
- Filtration, i.e. allowing the movement of water.
- High tear resistance.

The sheet piles need to be cleaned from any sharp protruding edges before placing the geotextile as these might puncture and tear the geotextile. However, future formation of such edges due to the continuous degradation process of the steel sheet piles is a possibility. Therefore, a second separation layer is required between the nonwoven fabric and the sheet piles to further minimise this risk. It is therefore recommended to use a Geocomposite, combining a nonwoven geotextile fabric with a geonet which will provide the required separation and it will also provide the additional tear resistance. Such a Geocomposite, see Figure 7.1, will have the hydraulic properties of the nonwoven geotextile with enhanced mechanical properties.

7.3. Surface Interface Between Fill and Rock

The interface between the fill material (normally sand) and the revetment rock material is always a difficult area to deal with, therefore a variety of solutions were developed by the designers covering a variety of situations.

Different approaches are followed on how to finalise such an interface, some are better than the others. Examples are given in this section to present what is recommended and what is incorrect.

An unacceptable interface arrangement is shown in Figure 7.8, where the geotextile has been used as an interface between the fill material and the underlayer rocks as in (a), or the armour rocks as in (b), and without anchoring the geotextile at the top. Another unacceptable example is given in Figure 7.9 with an interface between the fill material and quarry run. Using a geotextile as an interface in this example could be acceptable, but the finish at the surface is poor. In these examples. there will definitely be a loss of fill material at the surface through the rocks.

A better arrangement for the surface interface is shown in Figure 7.10, where the interface between the fill material and the core material is made of two layers. The granular filter layer (20–75mm) and a geotextile both will make a good and durable interface. However, the used size of the granular filter is slightly larger than what is required to prevent the sand fill from passing through, hence there will be after some time a slight

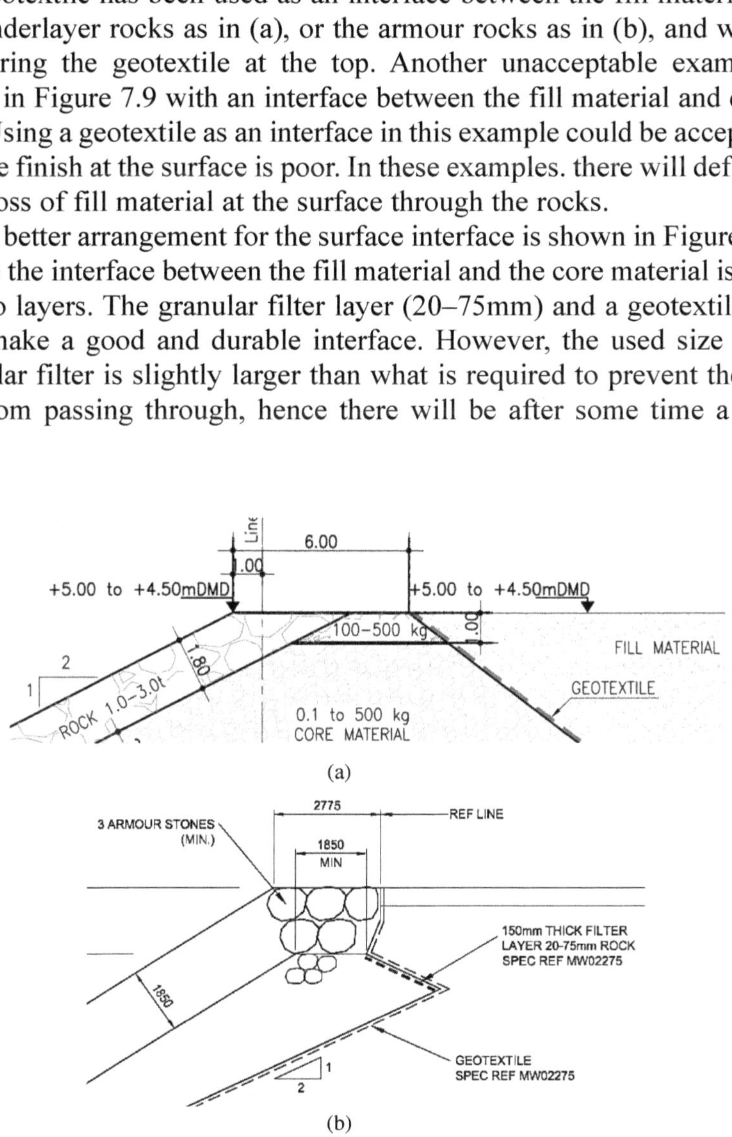

Figure 7.8. Interface with larger rocks and without special treatment.

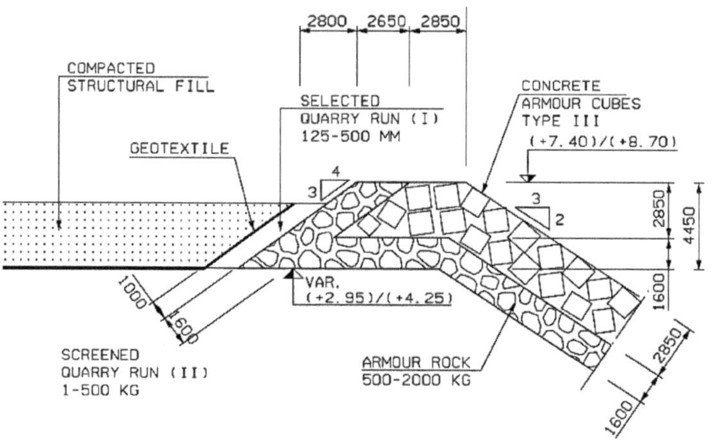

Figure 7.9. Interface with quarry run without special treatment.

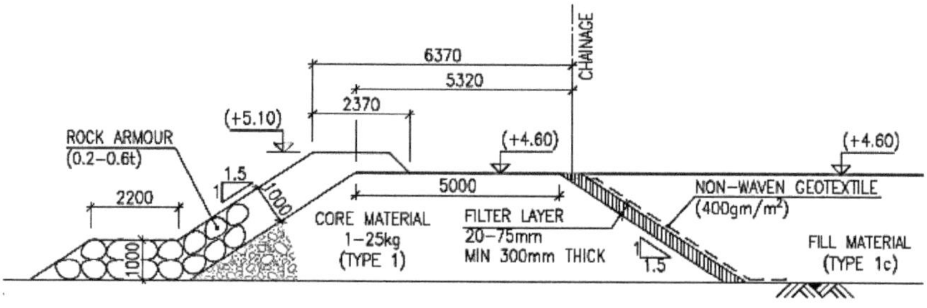

Figure 7.10. Interface with a filter layer.

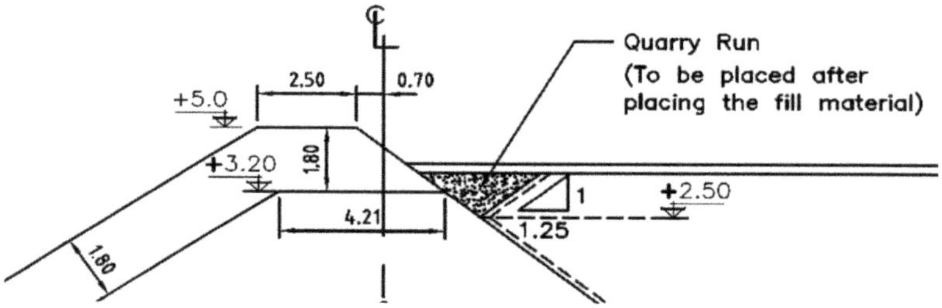

Figure 7.11. Interface with a quarry run wedge.

depression at the interface surface. In order to cancel out this surface depression, a wedge of coarser material could be used at the surface, as shown in Figure 7.11 with quarry run and in Figure 7.12 with sub-base material.

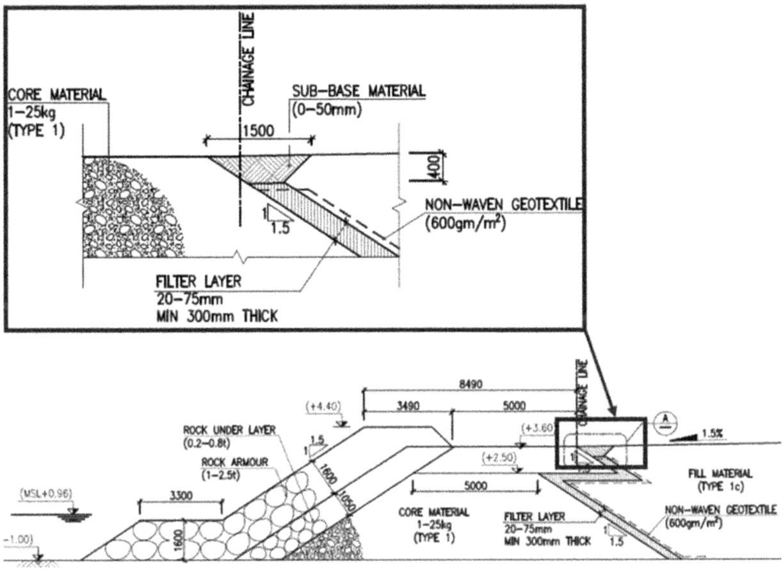

Figure 7.12. Interface with a sub-base material wedge.

Chapter 8

Roundheads, Bends and Corners

8.1. Introduction

The offshore end of a breakwater, normally called the head or roundhead, is usually exposed to more severe hydraulic loading than the trunk of the breakwater. This is because it normally reaches deeper water, hence larger waves than those at the trunk. However, the main reason is attributed to its curvature as the waves are forced to break like a water jet over a cone-shaped structure which leads to large velocities. The most affected part of the roundhead is found to be at the water level and located between 120°–150° from the incident wave direction (Jensen, 1984), see Figure 8.1. Therefore, the damage is expected first to take place on the inner quadrants of the roundhead.

Further to the intense hydraulic loading, and because of the curvature, the armour blocks (CAU/rocks) are less supported by each other or less interlocked. Therefore, special attention should be given to the placement of the CAU or rocks on the roundhead.

It is however important to note that bends and corners in breakwaters or revetments must also be carefully designed as these resemble, to some extent, the breakwater roundhead. What they have in common is the curved surface and the possibility of having an area exposed to high currents if the bend/corner has an angle greater than 90°.

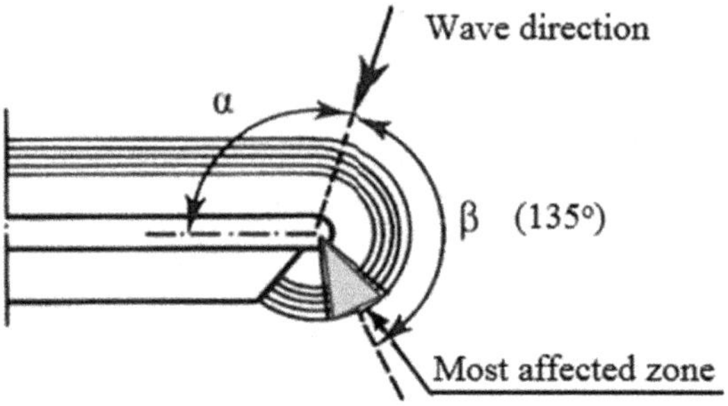

Figure 8.1. Roundhead most affected zone (Jensen, 1984).

8.2. Roundhead Common Configurations

In order to ensure adequate stability of the roundhead to counteract the curvature-induced increase in hydrodynamic forces, certain configurations are commonly used that are associated with certain measures depending on the breakwater's primary armour type. The required stability enhancement measures for the roundhead which are to be considered could be summarised as follows:

- Using larger size armour block (CAU/rocks) than the trunk armour. However, in some cases the trunk armour block can still be used at the roundhead if it has sufficient reserve stability, i.e. it has been slightly over-designed for the trunk.
- Increasing the density of the CAU by adding heavy aggregates like iron ore, although this is rarely used.
- Using a different and heavier type of CAU at the roundhead than that at the trunk.
- Using a flatter slope for the roundhead than that at the trunk.

After implementing one or more points of the above, the resulting roundhead arrangement could be categorized into two groups, as shown in Figure 8.2, and as listed in what follows:

Category 1: Using larger size armour and no change to the side slope, Figure 8.2(a–c). This category is normally associated with Concrete Armour Units (CAU) as the side slopes for the trunk and the roundhead are the same or nearly the same. The recommended slope range starts

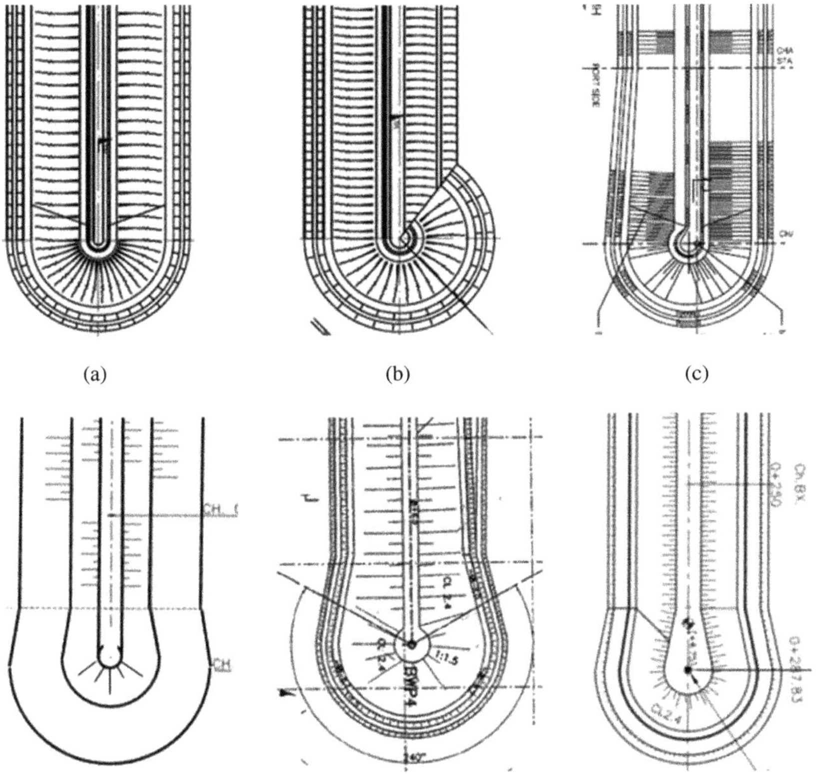

(a) (b) (c)

(d) (e) (f)

Figure 8.2. Common Roundhead configurations category 1 (a, b, c) category 2 (d, e, f).

from 1:1.33 for a single armour layer, reaching 1:1.5–1:2.0 for single and double armour layers, respectively (see Table 10.5), hence, larger units are to be used at the roundhead of the same type of CAU or a different type. *Category 2*: Using flatter side slope with/without heavier armour, as in Figure 8.2(d–f). This category is normally associated with rock armours as slopes flatter than 1:2.0 are very common in roundheads to avoid using very large rocks, which could be very difficult/impossible to procure locally.

Four photo examples of different roundhead arrangements are shown in Figure 8.3.

In all of the roundhead configurations, it is important to keep the transition to the trunk as smooth as possible, especially if it is exposed to some

(a)

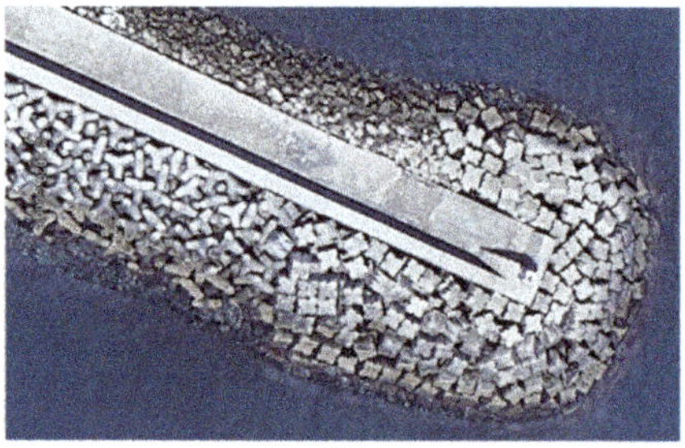

(b)

(c)

(d)

Figure 8.3. Examples of different roundhead arrangements: (a) Using larger Core-loc unit size at the roundhead; (b) Using a different and heavier CAU at the roundhead; (c) Using a flatter slope at the roundhead; (d) Using an enlarged roundhead.

degree of hydraulic loading. However, if this is not possible, or if it is required due to construction preference, then the transition must not be exposed to the incoming waves. Examples of a simple configuration frequently used with CAU are shown in Figure 8.4, as this type (as shown

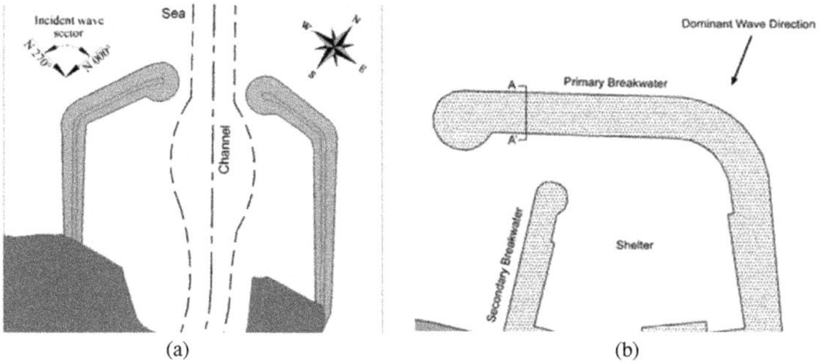

Figure 8.4. Typical examples of primary/secondary breakwaters' roundhead layout: (a) Two primary breakwaters; (b) Primary and secondary breakwaters.

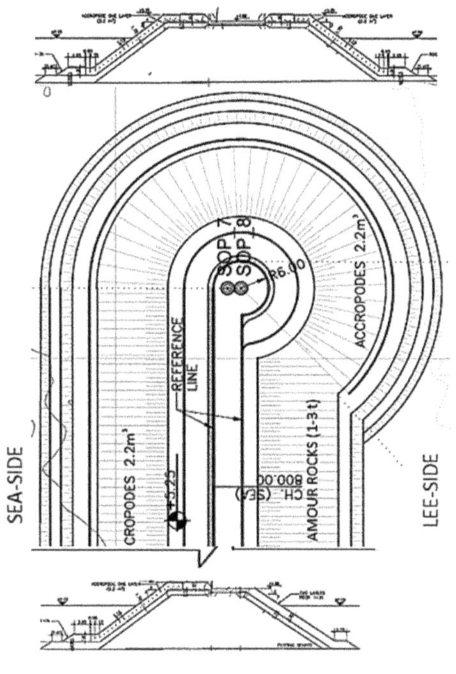

Figure 8.5. Breakwater roundhead, category 1-b with CAU (plan and cross-sections at the trunk and the roundhead).

in Figure 8.2 as category 1-(b) is preferred due to ease of construction. A plan view and two cross-sections for the trunk and the roundhead are shown in Figure 8.5, these are extracted from detailed design drawings for a project with this type of a roundhead and Accropodes armour. Other roundhead configurations can of course still be used with CAU, probably with some project specific adjustments. An example is shown in Figure 8.6 for category 2-(e), however this is with Core-loc units and not

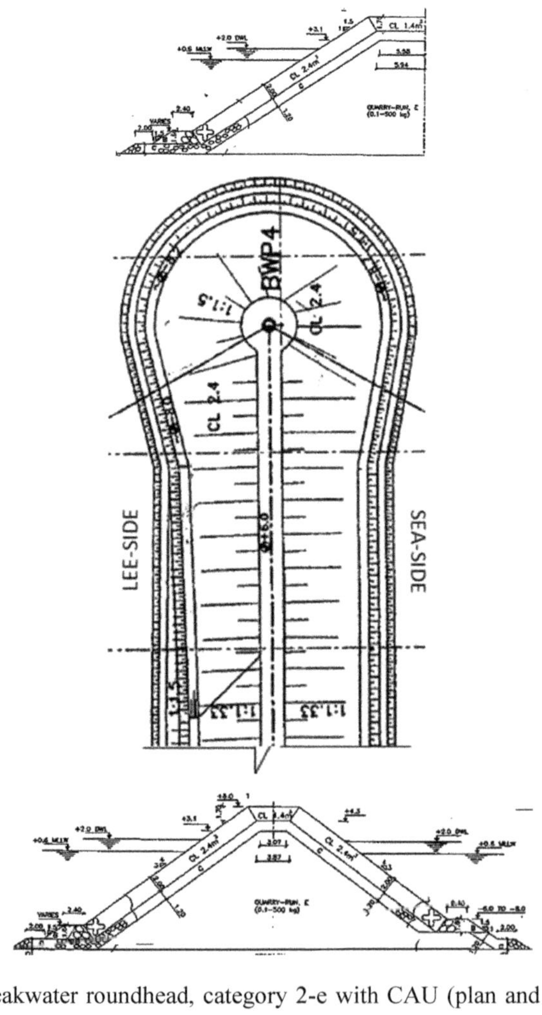

Figure 8.6. Breakwater roundhead, category 2-e with CAU (plan and cross-sections at the trunk and the roundhead "with slightly enlarged scale").

rock armour. The trunk is protected by 2.4 m^3 units with a 1:1.33 slope (with 14.5% reserve capacity for higher waves), the same unit is used at the roundhead, however with a slightly flatter slope of 1:1.5 while still having some reserve capacity of 9% for higher waves. By adopting this arrangement, the transition between the trunk and the roundhead has been practically avoided.

8.3. Roundheads, Stability and Function

Concrete Armour Units (CAU) are typically placed following a grid, specially the single layer armour, to ensure the desirable interlocking between the adjacent units. However, at the roundhead the placement pattern is expected to deviate from the standard trunk placement pattern. The placement at the roundhead is characterized by varying distances between the adjacent units, varying packing density and most certainly larger gaps between the units than in the trunk. The convex shape of the roundhead surface will further add to the reduction of the interlocking. Accordingly, design guidelines for various CAU (CLI, 2012; Xbloc, 2014) advise to have the roundhead radius not less than 2.5Hs (with Hs being the design significant wave height) at high water level (HWL), see Figure 8.7. This is to minimise the curvature effect and ensure having sufficient

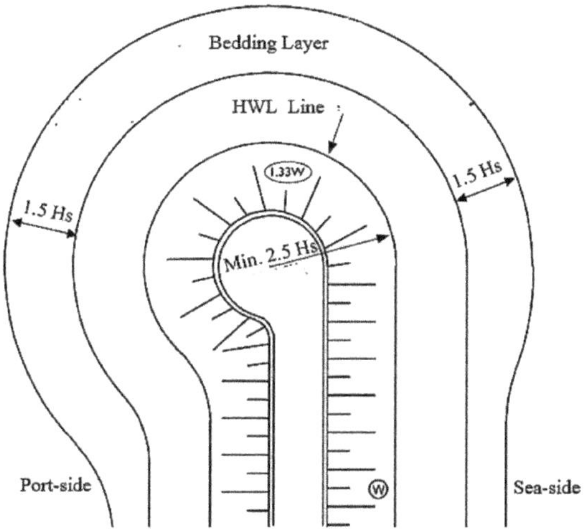

Figure 8.7. *Recommended Roundhead dimensions for Core-Loc armour units* (CLI, 2012).

interlocking between the units, which can maintain the fabric of the single layer armour units intact. The slope of the roundhead is to be kept the same as that of the trunk while increasing the weight of the armour unit by 33%.

No such radius limitations were suggested in literature for breakwaters with rock armour, however designers are always inclined to follow the same recommendations for the CAU roundheads. Moreover, roundheads with rock armour are normally designed with a flatter side slope than that of the CAU roundhead, which minimises the effect of the radius. Two examples are shown in Figure 8.8 to demonstrate the front view of a CAU and rock armour roundheads.

The drawback of having large roundheads is the undesirable consequence of a wide entrance opening to the port/marina basin, which means large wave energy entering the basin. This has often led to difficulties in meeting the design wave disturbance criteria inside the basin. A typical example is shown in Figure 8.9, where (a) indicates the navigational requirement to be satisfied (clear width of 65.0 m at –7.5 mCD), and (b)

(a)

(b)

Figure 8.8. Front view of roundheads with CAU and rock armour: (a) With Core-Loc armour units — sharp slope; (b) With rock amour — flat slope.

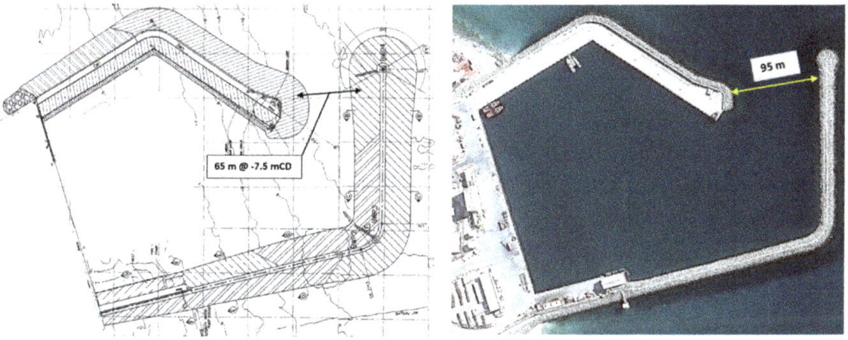

(a) (b)

Figure 8.9. Port entrance width with rubble mound roundheads: (a) Port breakwaters — design layout plan; (b) Port breakwaters — after construction.

shows the final open entrance width at approximately MSL with a top width of 95.0 m. Note that the roundheads on both sides are with Core-loc as the primary armour, and a side slope of 1:1.5. This shows that if the roundheads are of rock armour with a side slope of, say, 1:2.5, it will lead to an even worse situation.

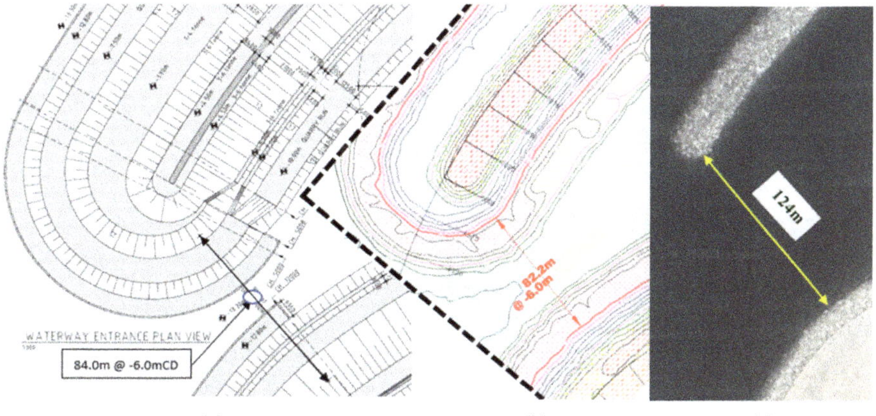

(a) (b) (c)

Figure 8.10. Marina entrance width — design (84m @ –6.0 mCD), after construction survey (82.2m @ –6.0 mCD) and at water level (124m): (a) Design; (b) Survey; (c) Ariel photograph.

It is important to note that the design clear width between two structures, as per the requirement for safe navigation of the design vessel, must have sufficient tolerance to accommodate all construction tolerances so that the final constructed width can still satisfy the design criteria. Figure 8.10 shows an example of a marina entrance design layout and after construction survey, both sides are with armour rocks. The entrance is between a breakwater and a curved revetment, the entrance with design width of 84.0 m at –6.0 mCD, as shown in (a). After construction survey, as in (b), it has been revealed that the actual width is 82.2 m at –6.0 mCD, which is 2% less, however this is within the acceptable tolerance limits. The top width is shown in (c) to be 124.0 m at approximately MSL.

8.4. Alternatives to Rubble Mound Roundheads

It is always desirable to minimise the difference between the design width for navigational requirement and the width at the water level in order to reduce wave penetration inside the port/marina basin. Therefore, using CAU armour with relatively steep side slopes is beneficial in comparison with rock armour with normally flatter side slopes. The use of concrete caissons is adopted in many cases as an alternative to the rubble mound roundheads, as shown in Figure 8.11. Note that the use of caissons may be required in some ports where safety regulations impose the need for a vertical reference at the entrance.

The use of caissons does have other advantages also in dismissing the need for larger armour blocks at the roundhead, and accordingly eliminating the risk of damage to the breakwater head. However, it is a different

(a) (b)

Figure 8.11. Examples of concrete caissons used as a breakwater head.

(c)

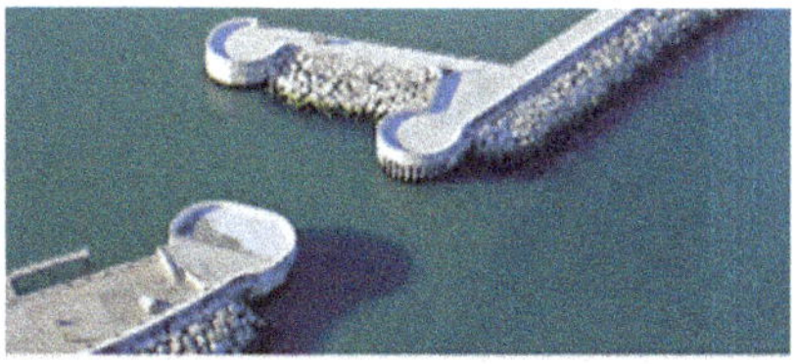

(d)

(e)

(f)

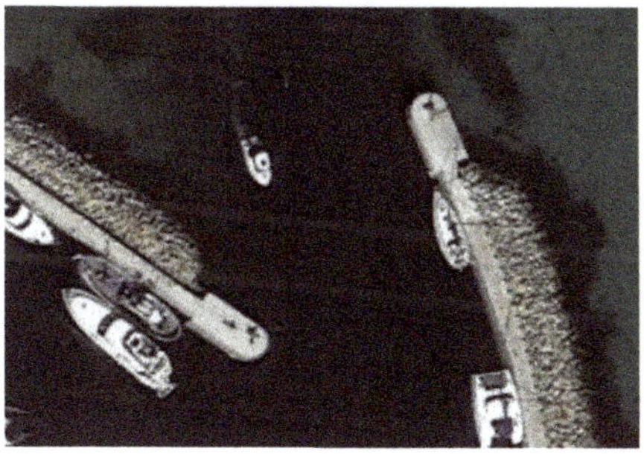

(g)

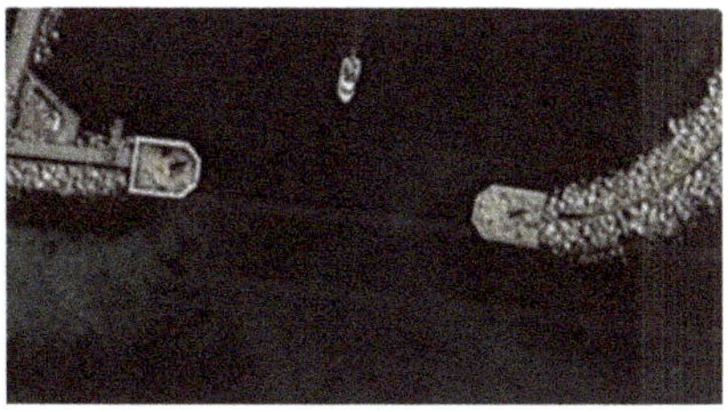

(h)

Figure 8.11. *(Continued)*

structure than a rubble mound and this requires different construction equipment and methodology.

Caissons may be constructed in different shapes, sizes and materials, as illustrated in Figures 8.11 and 8.12. The size of the caisson should be large enough to allow for the cone-shaped termination of the armour layer meeting the caisson.

(a)

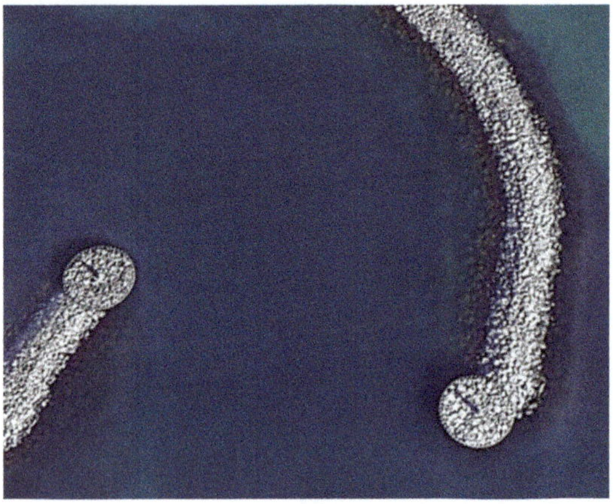

(b)

Figure 8.12. Timber crips filled with rock: (a) Timber crips during construction (timber crips of tropical wood timber held together by steel rings); (b) After being filled with rock.

8.5. Bends and Corners

Most of the discussed issues in Section 8.3 need to be carefully considered before being adopted for the bends and corners. These are mainly regarding the radius of curvature, the requirement to increase the armour size/use a flatter slope and the applicability of the roundhead most affected zone as illustrated in Figure 8.1. All of the aforementioned issues are actually influenced by the geometry of the bend/corner and the wave direction.

The same guideline for the minimum radius of the roundhead, as shown in Figure 8.7, is still applicable for the bend/corner to minimise the negative effect of the curvature on the armour layer integrity. The arc's central angle (ɵ), for most bends/corners, is ≤90°, as illustrated in Figure 8.13. This means that the curvature will end before reaching the

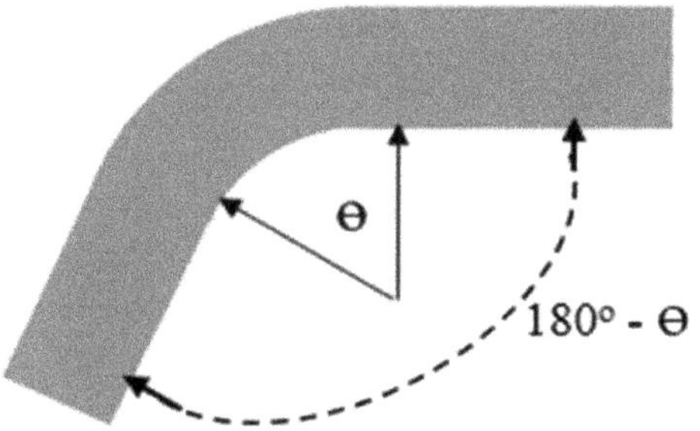

Figure 8.13. Illustrative sketch for the central angle of the bend/corner arch.

critical angle range of 120°–150° for the most affected zone on the curved section (depending on the principal wave direction). Therefore, it is frequently neither required to increase the armour block size nor use a flatter slope, especially when the trunk armour has some reserve capacity. However, some designers prefer to use one, or both, of these measures for extra safety, albeit it may not be needed. Such a precaution is in fact recommended if part of the arc makes 135° with the wave direction, or in case of high bends, i.e. in deep water, where the upper part of the bend cone is more affected than the lower parts due to wave reflections, as observed in many model tests (Jensen, 1984), or if the armour block size is large where the gaps between the blocks will relatively be larger in comparison with smaller blocks.

Practical examples are presented in this section to demonstrate the different approaches by different designers to breakwater bends (three examples) and to revetment corners (three examples).

Figure 8.14 shows a 90° bend in a breakwater, the plan is given in (a), and two cross-sections, the first is for the straight trunk stretch in (b) and the second for the bend in (c). Although the same armour unit of 12.5 m^3 Tetrapods has been used in both, the designer opted to give the bend section a slightly flatter slope, i.e. providing extra safety. The designer obviously thinks that the bend is sharp and in deep waters, hence needing an extra safety measure, such as the use of a flatter slope. This concern could possibly be addressed without altering the section, which is something not preferred by the construction team. The curvature of the bend could be reduced either by increasing the radius of the bend, as shown by the dotted line, or using two 45° bends with a short straight stretch between them, similar to the breakwater in Figure 8.16.

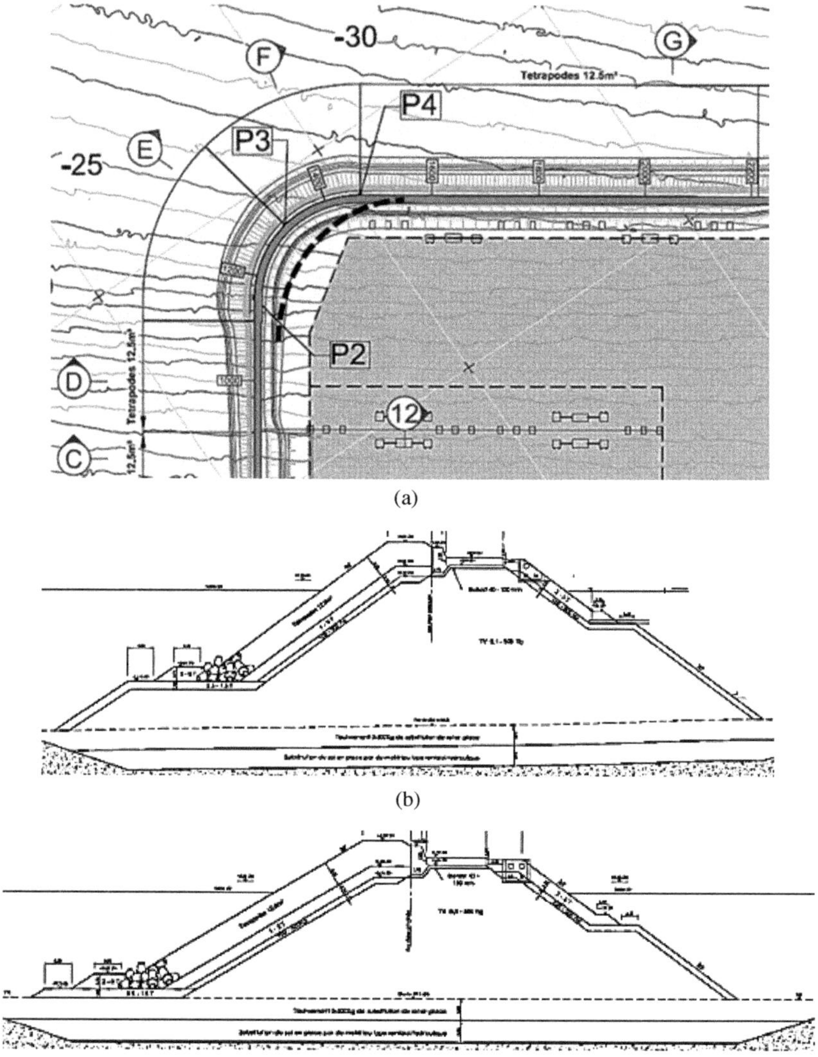

Figure 8.14. A 90° bend in a breakwater: (a) 90° bend — plan; (b) Cross-section G (12.5 m^3 Tetrapods at 3:4 slope); (c) Cross-section E (12.5 m^3 Tetrapods at 3:5 slope).

In Figure 8.15, the breakwater has two bends, 90° and 45°, the same armour unit of 6.0 m^3 Accropodes and the same slope is used along the straight trunk stretch and the two bends of the breakwater. It is however obvious in this example that the designer has no concerns about the

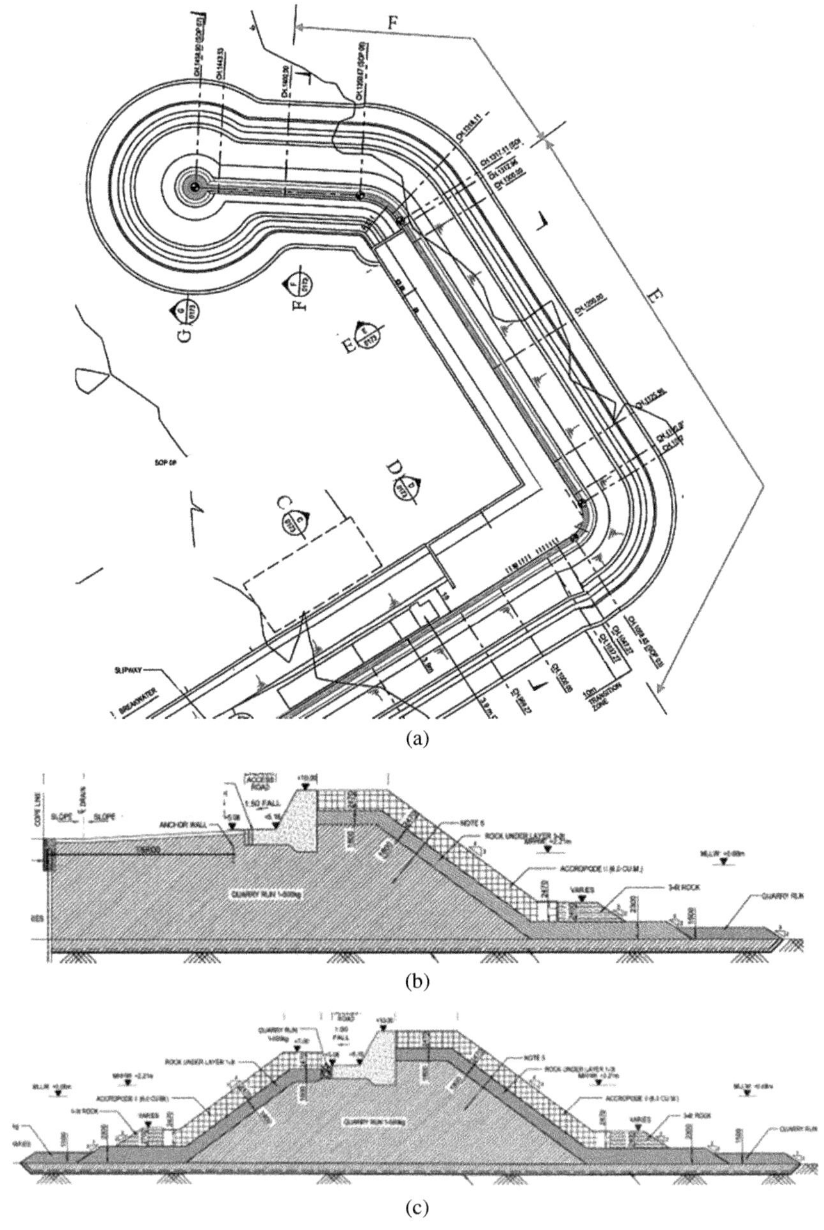

(a)

(b)

(c)

Figure 8.15. CAU breakwater with 90° and 45° bends: (a) 90° and 45° bends — plan; (b) Cross-section E (6.0 m^3 Accropodes with 3:4 slope); (c) Cross-section F (6.0 m^3 Accropodes with 3:4 slope).

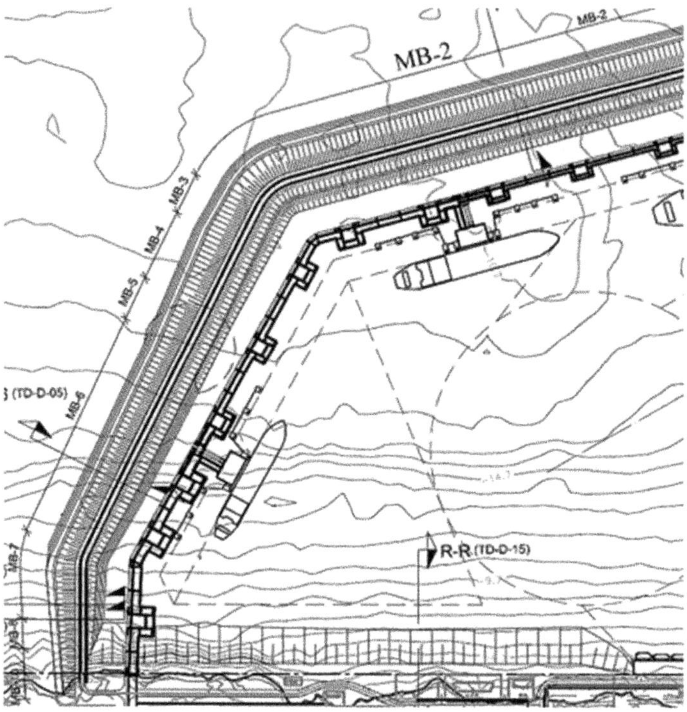

(a)

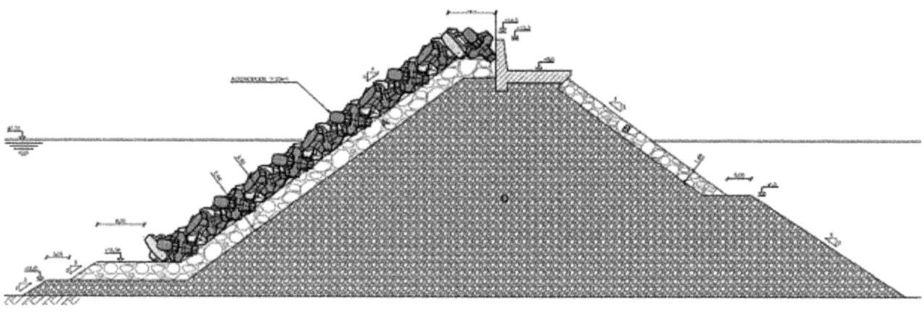

(b)

Figure 8.16. CAU breakwater with two 135° bends: (a) Two 45° bends; (b) Cross-section MB-2 (20 m^3 Accropodes with 3:4 slope).

stability of the armour units on the bends, and accordingly no extra measures were taken.

In the third breakwater example in Figure 8.16, the designer has not used a 90° bend but two 45° instead. Therefore, it is with more certainty the same armour unit of 20 m^3 Accropodes and the same slope used along the straight trunk stretch and the offshore 45° bend.

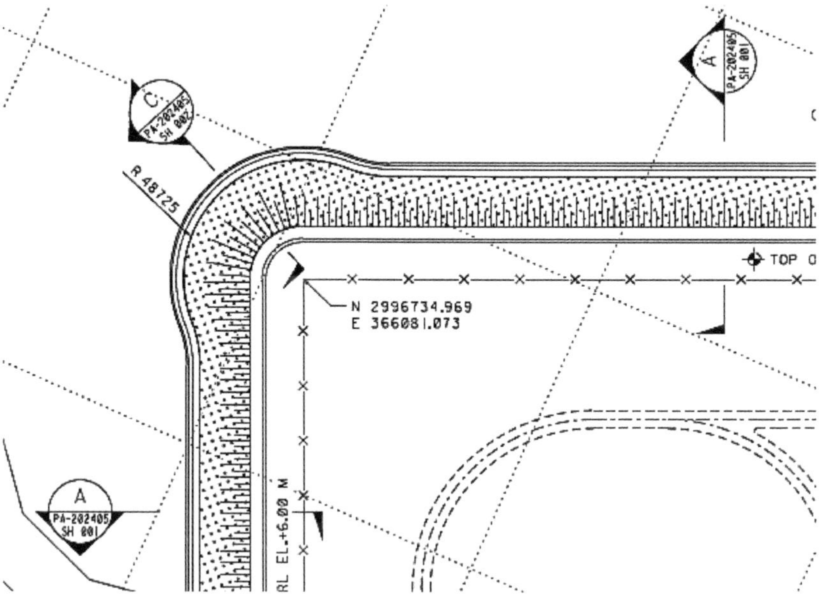

(a)

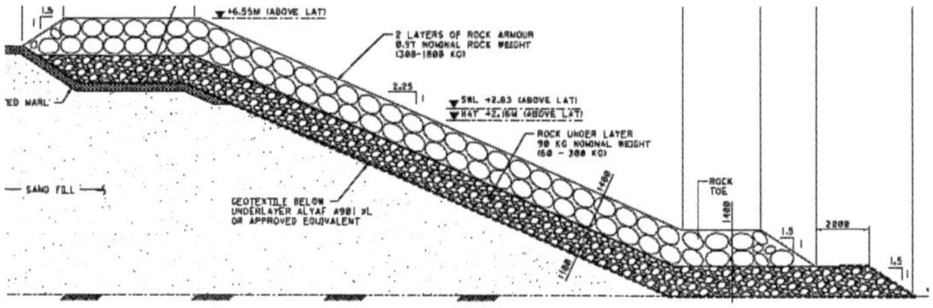

(b)

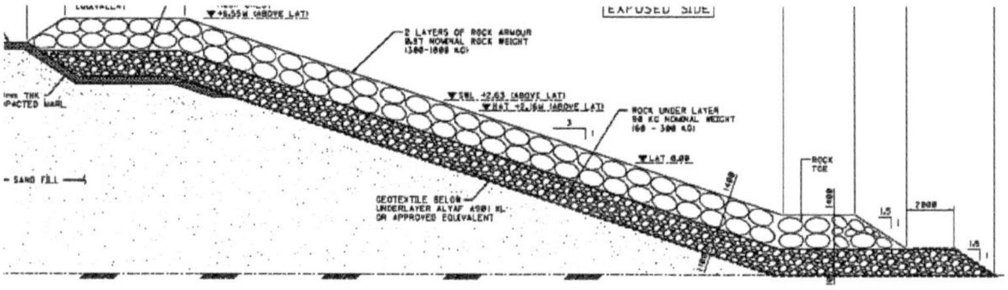

(c)

Figure 8.17. Rock armour revetment corner with flat slope: (a) Revetment corner — plan; (b) Cross-section A (1-3 t rocks with 1:2.0 slope); (c) Cross-section C (1-3 t rocks with 1:3.0 slope).

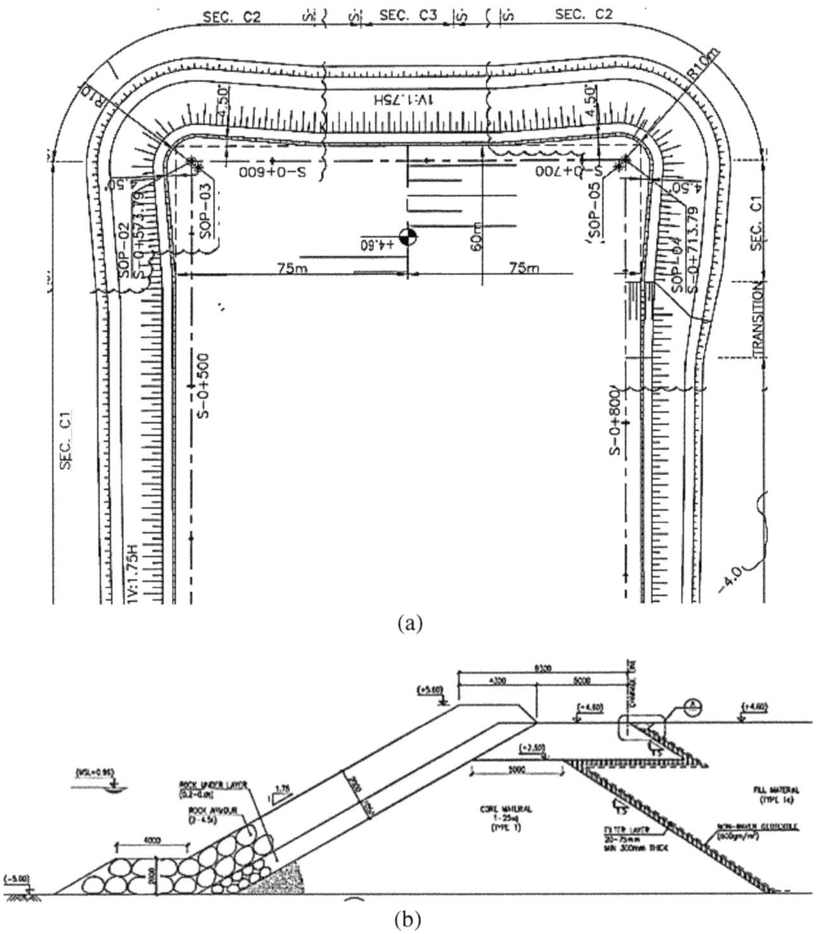

(a)

(b)

Figure 8.18. Rock armour revetment corner with enlarged curvature radius (same cross-section for the straight and curved stretches): (a) Revetment corners — plan; (b) Cross-sections C1, C2 and C3 (2–4.5 t rocks with 1:1.75 slope).

Note: The three cross-sections are different with the crest level only.

The first example of corners in revetments is shown in Figure 8.17 where the designer has used the armour rock class of 1–3 *t* for both the straight and curved sections, however with a flatter slope at the corner.

In Figure 8.18, the same armour rock and the same slope are used along the straight stretches and at the two corners. However, it is clear that the designer was a little concerned about the tight curves at the corners.

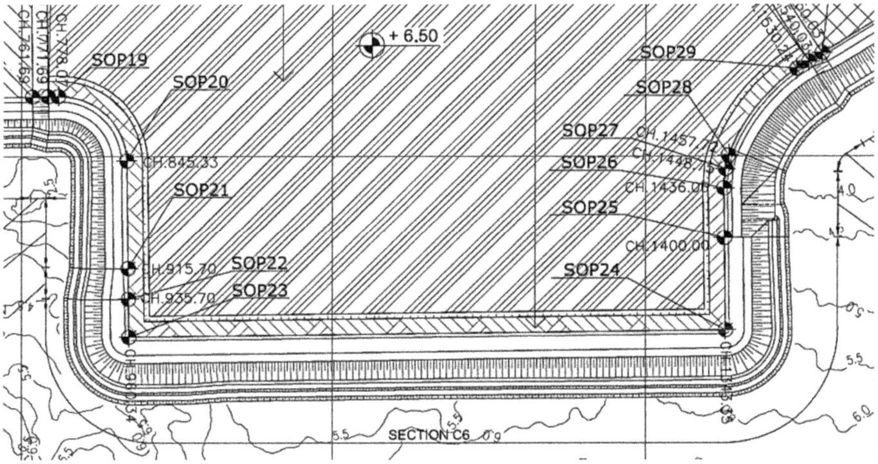

(a)

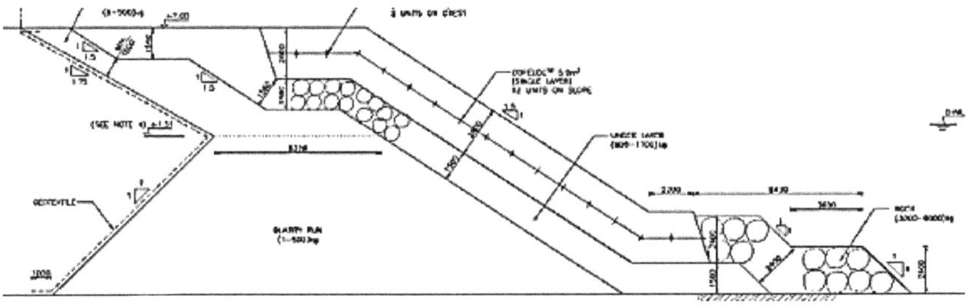

(b)

Figure 8.19. CAU revetment corners (same cross-section for the straight and curved stretches): (a) Revetment corners — plan; (b) m Cross-section C6 (5.0 m^3 Core-loc with 1:1.5 slope).

Thus, instead of altering the section, by changing the rock size or the slope, the designer has enlarged the radius of the corners to reduce the curvature and improve the armour integrity.

The third example on revetment corners is shown in Figure 8.19 where the armour unit is 5.0 m^3 Core-loc at 1:1.5 slope. It is clear that the designer in this example has no concerns about the integrity of the corners, and as such the same section was used throughout the straight stretch and the two corners of the revetment.

Chapter 9

Rock Armour

9.1. Introduction

Rock armour, which in most cases is produced by blasting in quarries, is frequently used in coastal structures for protection against waves and currents. Armouring the shoreline structures may also include the protection against scouring in front of or near these structures. Rock armour is normally the most common type of armour used in shoreline management to construct groynes, nearshore detached breakwaters and rock headlands.

The merits of using rock armour in coastal structures are many but can be summarised as follows:

- More economical to use in comparison with other armouring solutions.
- Ease of construction and maintenance with no specially trained staff required.
- Environmentally friendly.
- Forms a flexible armour layer which could easily absorb small local settlements and tolerate minor damages.
- Long-term performance.

9.2. Rock Suitability Assessment

The following points are to be considered when assessing the suitability of using rock armour in the design of coastal structures:

(a) Rock size and grading
   - Resistance to wave action
   - Effect of run up and overtopping
   - Likely effect on erodible foreshore
   - Availability
   - Amenity consideration

(b) Necessity for filtration and rock armour foundation stability (Van der Meer, 1988; Kik *et al.* 2012)
   - Underlayer specification
   - Core material specification
   - The use of a geotextile fabric

(c) Quality of rock material (Ch. 3 of CIRIA, CUR, CETMEF, 2007)
   - Durability
   - Availability
   - Cost
   - Method of delivery

9.3. Rock Armour Gradings

EN 13383 (2002, and the updated version of 2013) divides armour stones into three categories, which are also presented in CIRIA, CUR, CETMEF (2007), each category has a set of standard classes and each class is defined by some passing requirements, as follows:

- **Heavy gradings** – the classes under this category are appropriate, and normally used, for armour rock layers.
  10 000–15 000 kg
  6000–10 000 kg
  3000–6000 kg
  1000–3000 kg
  300–1000 kg
  The associated defining limits for these gradings are as follows:
  - ELL (Extreme Lower Limit) — the mass below which not more than 5% passing
  - NLL (Nominal Lower Limit) — the mass below which not more than 10% passing
  - NUL (Nominal Upper Limit) — the mass below which not less than 70% passing

- EUL (Extreme Upper Limit) — the mass below which not less than 97% passing

Each of the above classes comes with an acceptable range for M_{em}, defined as "effective mean mass, which is the average mass of the sample without the fragments, i.e. all stones below ELL".

It is worth noting that EN 13383 designates "Category A" to the class grading specification if the fragments are excluded and M_{em} is used, which could be converted to M_{50} as per Table 9.1. The class grading specification would otherwise be of "Category B" if the fragments are included and M_{50} could directly be used. Note that this type of rock grading specification, i.e. "Category B", is normally used in most of the projects.

- **Light gradings** — The classes under this category are appropriate for underlayers and filter layers.
  60–300 kg
  10–60 kg
  40–200 kg
  5–40 kg
  15–300 kg
  The associated defining limits for these gradings are similar to those of the heavy gradings apart from the ELL, where it becomes 2% instead of the 5%.
- **Coarse grading** — The classes under this category are often used for filter and bedding layers.
  45/125 mm
  63/180 mm
  90/250 mm
  45/180 mm
  90/180 mm
  The associated defining limits for these gradings are 5%, 15%, 90% and 98% for the ELL, NLL, NUL and EUL, respectively.

The heavy grading is the one of interest in this chapter, however, light and coarse gradings are given for clarity and completeness.

The grading limits and standard rock classes grading curves for the heavy gradings are illustrated graphically in Figures 9.1 and 9.2, respectively.

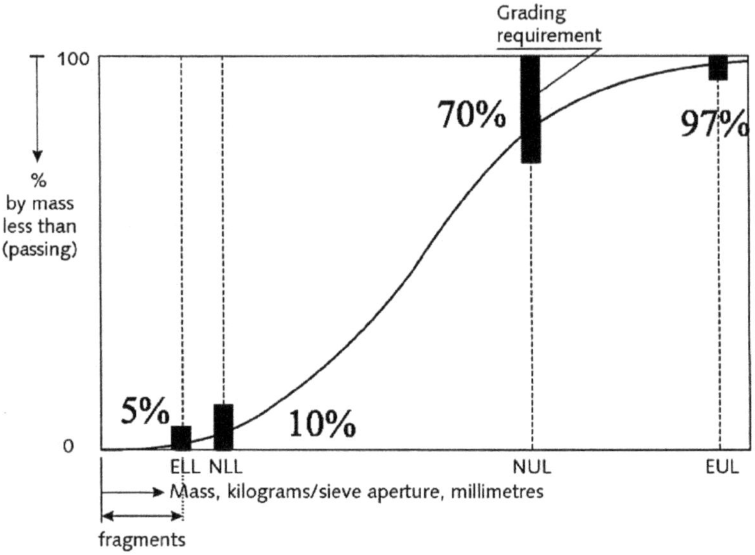

Figure 9.1. System for limits of EU standard gradings (Based on EN 13383 and CIRIA-CUR-CETMEF, 2007).

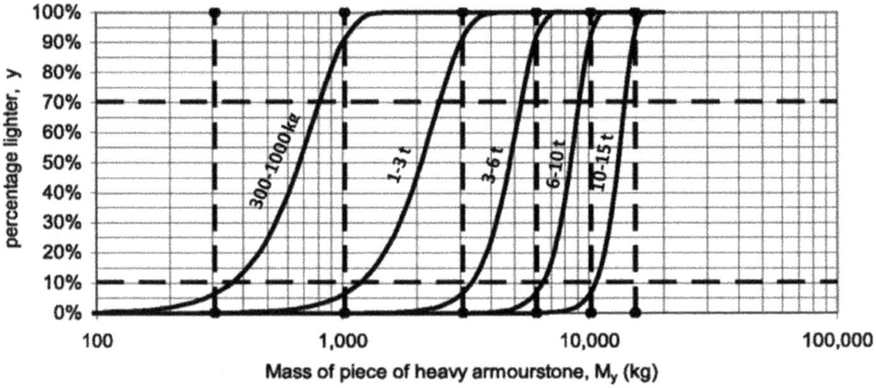

Figure 9.2. Heavy grading rock classes as per EN 13383.

It is important to note that the standard heavy gradings of EN 13383 were designed to become systematically wider, i.e. larger NUL/NLL, with decreasing M_{50}. This is demonstrated by the flatter S-curves in Figure 9.2 (which is slightly exaggerated by the log-scale) and in Table 9.2.

Table 9.1. M_{50} calculations for the heavy gradings standard classes (based on Table 3.6 of CIRIA, CUR, CETMEF, 2007).

a	b	c	d
Rock Class (kg)	M_{50}/M_{em} (=1.61 $M_{50}^{-0.05}$)	M_{50min} = M_{emll} x (b) (kg)	M_{50max} = M_{emul} x (b) (kg)
10,000–15,000	1.002	12,000	13,000
6,000–10,000	1.024	7,680	8,710
3,000–6,000	1.054	4,430	5,060
1,000–3,000	1.099	1,870	2,310
300–1,000	1.163	628	802

It is also observed from the M_{50min} and M_{50max} values given in Table 9.1 that the $M_{50,}$ i.e. the nominal value, deviates towards the $M_{50max.}$ Starting with null deviation at the heaviest class, it peaks at 3–6 t to approximately 5%, then reduces again. This complex behaviour of the relation between the average M_{50}, i.e. 0.5(NUL+NLL), and M_{50}, i.e. 0.5(M_{50max} + M_{50min}), is better demonstrated in Table 9.2.

The data given in Table 9.2 could be translated mathematically to calculate M_{50}, i.e. the nominal M_{50}, by Equations (9.1) and (9.2). The calculation of the acceptable range for the nominal M_{50}, i.e. between M_{50min} and M_{50max}, is given by Equation (9.3). Note that these equations are applicable for standard and non-standard classes.

Table 9.2. Relationship between the average M_{50} and the M_{50}. (the table is based on Table 9.1).

NLL (t)	NUL (t)	NUL/NLL	M_{50avg} (t)	M_{50min} (t)	M_{50max} (t)	M_{50} (t)	Acceptable M_{50} range ±(%)	$M_{50} - M_{50avg}$ (t)
10.0	15.0	1.50	12.50	12.0	13.0	12.5	4.0	0.00
6.0	10.0	1.67	8.0	7.68	8.71	8.20	6.3	0.20
3.0	6.0	2.0	4.50	4.43	5.06	4.75	6.6	0.24
1.0	3.0	3.0	2.0	1.87	2.31	2.09	10.5	0.09
0.3	1.0	3.33	0.65	0.63	0.80	0.72	11.9	0.07

$$M_{50} = 0.5(NUL + NLL) + E \tag{9.1}$$

$$E = 0.305\,(NUL/NLL)^3 - 2.44\,(NUL/NLL)^2 + 6.21(NUL/NLL) - 4.83 \tag{9.2}$$

$$M_{50max/50min} = M_{50} \pm M_{50}\,[0.039\,(NUL/NLL) - 0.011] \tag{9.3}$$

As sieving is practically impossible for larger rocks, the D_{50} is replaced by D_{n50}, which is often called the median diameter, and accordingly the M_{50} is the "median mass". But it is in fact incorrect to use the term "median" for both, i.e. M_{50} and D_{n50} (Verhagen and Jansen, 2014). Statistically the term "median" will result in a different value than the theoretical rock mass for which half of the mass of the sample is lighter.

9.4. Rock Armour Class Selection

The governing parameter in representing the stability of rock armour is the M_{50}, and this is defined as 50% by mass of rock passing. This mass gives the nominal diameter, D_{n50}, which is used in the stability number $H_s/\Delta D_{n50}$, and also in defining the layer thickness. There are however many cases where the required M_{50} from design calculations is either slightly above or below that of a standard grading class. The designer therefore is to decide on adopting one of the following approaches:

(a) Move one step up in the standard gradings and select a heavier class than what is actually required. This option will result in an over-designed structure with more rock quantities and an increased cost.
(b) Move one step down in the standard gradings and select a lighter class than what is actually required. This option will result in an under-designed structure with less rock quantities and a cost saving.
(c) Select a non-standard rock class that could best satisfy the design from the engineering and economical sides.

For example, if the required M_{50} is 3.0 t, which lies in between the two standard classes 3–6 t and 1–3 t, with M_{50} of 4.75 t and 2.1 t, respectively.

The grading limits for the rock class with M_{50} equal to 3.0 t could be obtained in two ways (see Section 3.4.3.9 of CIRIA-CUR-CETMEF, 2007) by a simple or a detailed approach as explained in the following sub-sections.

9.4.1. *Simple approach for mass distribution similar to standard gradings*

- Set the NUL and NLL of the required rock class to fulfil the following:
  $M_{50} \approx 0.5$ (NUL+NLL)
  Setting NUL=5.0 t and NLL=2.0 t will result in M_{50} approximately equal to 3.5 t, which is acceptable.
- Set ELL = 0.7 NLL and EUL = 1.5 NUL.
  Leading to: ELL = 1.4 t and EUL = 7.5 t.
- The M_{em} rang could be defined as follows:
  $M_{emll} = 0.8\,(NLL + NUL)/2 \Rightarrow 2.8$ t.
  $M_{emul} = (NLL + NUL)/2 \Rightarrow 3.5$ t.

The above method, as presented in CIRIA, ends short of defining the M_{50min} and $M_{50max,}$ which are practically required in specifying the grading curve limits. However, it is possible to resolve this issue by using Equations (9.1)–(9.3) as presented in Section 9.3.

The nominal M_{50} is calculated as 3.71 t, with an acceptable range between:

$$M_{50min} = 3.40 \text{ t to } M_{50max} = 4.05 \text{ t}$$

The final grading curve and limits for the non-standard rock class 2–5 t are shown in Figure 9.3.

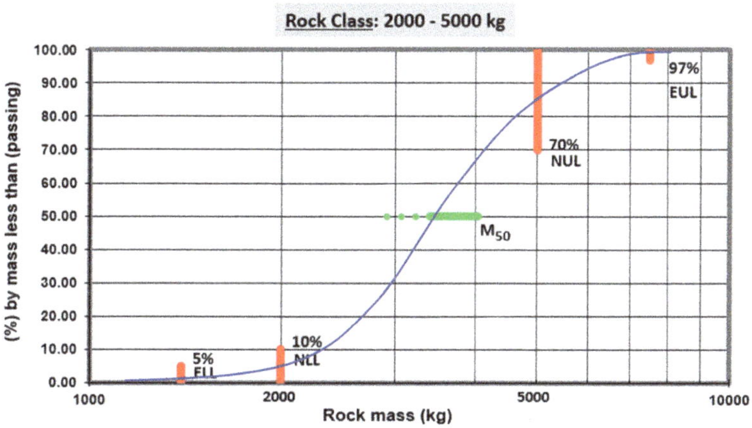

Figure 9.3. Grading curve for the non-standard class 2–5 t.

By selecting in the first step an average M_{50}, i.e. 0.5(NUL+NLL), slightly larger than the required M_{50} by design, a more practical and achievable grading curve is obtained. This is specially appreciated by quarries and construction teams (in avoiding rock supply rejections) by slightly easing the stringent requirement for the acceptable M_{50}, i.e. between M_{50min} and M_{50max}. In Figure 9.3, the solid green line represents the allowable "narrow" range of M_{50} according and similar to CIRIA methodology. This is often difficult to meet in practice and needs to be relaxed if possible. This relaxation could be made as EN13383 and CIRIA allow for the M_{50} to be lower than the actual M_{50} by approximately 4–11% (see Table 9.2). This means it is possible to be lower than the design M_{50}, i.e. the 3,000 kg, by say 5%, to a value of 2,850 kg and the rock grading would still satisfy the design requirement. This is shown in Figure 9.3 by the dotted green line extension. This makes the allowable range for M_{50} extend between 2,850 kg, as M_{50min}, to 4,050 kg, as M_{50max}.

9.4.2. *Detailed approach for mass distribution similar to standard gradings for Category A specification*

CIRIA introduced this approach to derive the specification grading limit masses, Y (kg), from the required M_{50} (kg) by using Equation (9.4) and Table 9.3.

$$Y = A\, M_{50}^{B} \quad \text{(where A \& B are coefficients given in Table 9.3)} \tag{9.4}$$

Table 9.3. Coefficients for non-standard grading requirements for the specification of gradings with similar properties to standard gradings (based on CIRIA-CUR-CETMEF, 2007).

		Class limits						$M_{15,\,50,\,85}$			
		ELL	NLL	NUL	EUL	M_{emll}	M_{emul}	M_{15}	M_{85}	$M_{50\,min}$	$M_{50\,max}$
Factor A		0.027	0.156	2.520	5.160	0.449	1.060	0.160	3.340	0.663	1.515
Factor B		1.32	1.160	0.920	0.880	1.080	0.990	1.171	0.918	1.041	0.950
Example for a non-standard grading similar to a standard grading (all mass values in kg)											
M_{50} = 3,000	calculated	1050	1685	3984	5923	2556	2935	1893	5180	2768	3279
	rounded	1000	1700	4000	6000	2500	3000	1900	5200	2750	3250

Table 9.4. Comparison between simple and detailed approaches for grading design examples.

Example gradings	ELL < 5% (t)	NLL **<10% (t)**	NUL **>70% (t)**	EUL **>97% (t)**	M_{50min} (t)	M_{50max} (t)	M_{50act} (t)	Grading width (NUL/NLL)
Example from Table 9.3	1.0	1.7	4.0	6.0	2.75	3.25	3.0	2.4
Example from the simple approach	1.4	2.0	5.0	7.5	2.85	4.05	3.5	2.5

The given example in Table 9.3 is to illustrate the *Detailed approach* method using the same M_{50}, i.e. 3,000 kg, of the *Simple approach* example.

A summary for the calculated grading limits and properties by the two methods, for comparison purposes, are given in Table 9.4.

Based on the results presented in Table 9.3, the following few remarks could be made:

- The grading width, NUL/NLL, is almost the same.
- The required rock quantities for the Simple approach are more by 4.5% to cover an area with double layer armour.
- The rocks, in general, are slightly larger in the Simple approach.
- The acceptable M_{50} range is $3.0 \pm 8.3\%$ t for the Detailed approach, and 3.5 (+15.7%)/(−18.6%) t for the Simple approach, which indicates a higher flexibility. This issue, i.e. the acceptable M_{50} limit, is actually very sensitive where the construction team is nearly always requesting the designer to specify the required grading limits as shown in Figure 9.1 and not as in Figure 9.3, i.e. not showing the M_{50} range. The rock supply grading curves are then to be presented as complying with the ELL, NLL, NUL and EUL limits, with no other restrictions, and this in fact will still be in full compliance with EN 13383 and CIRIA requirements.
- Although the outcome from both approaches is a grading fulfilling the design requirement, the Simple approach does provide a little extra safety.

It is however important to emphasise that the above examples are given for illustration only and it is for the designer to follow whichever approach suits the project best, the rock production and the construction team.

9.5. Rock Armour Placement

Rocks of heavy grading classes used as armour shall be placed individually, however, the way in which these rocks are placed affects the layer packing density (i.e. tightness), the layer thickness and the stability of the rock slope. A degree of permeability always exists in rubble mound structures. This permeability helps in absorbing the wave energy (and consequently enhances the rock armour stability), reduces wave run-up (and consequently the overtopping) and decreases wave reflection (which is specially important in port and marina basins).

The placement method of individual rocks could generally be described, or divided, into the following three main types (CIRIA-CUR-CETMEF, 2007):

1. *Random placement*: This placement is without control of orientation and without paying too much attention to contacts with adjacent rocks or voids between the rocks.

   This type of placement is normally performed by dumping the rocks in place and profiled to the required levels. It is used for the smaller sizes of the heavy grade classes, i.e. 0.3–1.0 t or similar, and specially if used in locations not exposed to waves from outer seas as shown in Figure 9.4.
2. *Standard placement*: This is, as the name indicates, the most common placement method where rock orientation control is applied during the placement, as shown in Figure 9.5 for a slope, and in Figure 9.6 for a crest. Adherence to the three points of contact between the rocks within the layer must be observed, this means the rock might be required to be rotated to achieve this requirement. However, creating large voids between the rocks, in the order of D_{n30} or larger, needs to be avoided. The crane operator or the supervisor tends, unfortunately quite often, to fill the larger voids with smaller rocks to avoid rearranging the armour rocks, as shown in Figure 9.7. This practice is definitely not recommended as these small rocks could easily be

Figure 9.4. Random placement example for 0.3–1.0 t armour layer.

Figure 9.5. Standard placement method on a slope.

Figure 9.6. Standard placement method on a crest.

Figure 9.7. Filling voids with small rock.

pulled out by the wave action and this could lead to further damage in the rock slope. One other issue regarding the voids is the deep voids, as shown in Figure 9.8, which are to be avoided. The voids in the lower layer if coincided with those in the upper layer will render the underlayer to be somewhat exposed to wave action.

Other points which are highly recommended to be considered:

— the longitudinal axis of each rock should be normal to the axis of the structure.
— The centre of gravity of the placed rock should be closer to the underlayer, i.e. in a more stable position.
— Rocks should not protrude excessively from the structure face to avoid excessive wave force on individual rocks.
— Cracked or fractured rocks found during placement are to be replaced, see Figure 9.9.

Figure 9.8. Deep voids exposing the underlayer.

Figure 9.9. Crackled and fractured armour rock.

3. *Dense placement*: The requirement for this type of placement is unfortunately misunderstood by a large number of crane operators and work supervisors. The objective of this placement method is to achieve minimum voids, and as such it involves the rotation of armour rocks, and removal and replacement of some rocks if necessary. The resulting rock surface should still have some roughness, and not smooth as shown in Figure 9.10. This type of surface will perform very badly in hindering the wave run up, which will result in more-than-expected overtopping discharge.

Orderly placed or pitched rocks, as shown in Figure 9.11, are normally placed in a single layer with a smooth surface. Such rock placement

(a)

(b)

Figure 9.10. Dense placement of armour rocks: (a) Revetment slope; (b) Revetment crest.

could be used for its aesthetics or to improve the stability of the rocks (Nurmohamed *et al.,* 2006). It is normally used in inner canals and lagoons, or in locations with low wave action. With this type of placement, quite often a grout is used, normally of cement or asphalt-based material, to fill the gaps between the rocks, as shown in Figure 9.11(b).

9.6. Porosity (p) and Layer Thickness (k_t)

The degree of rock material packing is defined either by the packing density (ϕ) or the porosity (p). The packing for bulk placed rock material is highly dependent on the slope of the grading curve, whereas this property for the individually placed rocks shall depend on the grading width of the rock material, as given in Table 9.5, and on the placement methodology.

(a)

(b)

Figure 9.11. Pitched rock slope: (a) Pitched rocks without grout; (b) Pitched rocks with grout.

Table 9.5. Rock armour grading width related to the uniformity (expressed in size and mass ratios) (based on CIRIA, CUR, CETMEF, 2007).

Grading width	D_{85}/D_{15}	M_{85}/M_{15}
Narrow or single-sized gradation (*normally used rock armour for coastal protection*)	<1.5	1.7–2.7
Wide gradation (*rip-rap for estuarial and river bank application*)	1.5–2.5	2.7–16
Very wide or quarry run gradation	2.5–5.0	$16–125^{+}$

Table 9.6. General guidance for layer thickness coefficient and porosity for various placement methods (based on Table 3.9 of CIRIA, CUR, CETMEF, 2007).

		Reference ($0.5D_{n50}$ spherical probe) survey method**	
Placement type	Parameter	Blocky rock eg mean BLc* = 65%	Irregular rock eg mean BLc* = 50%
Double Layer — Standard	k_t	0.91	0.87
Double Layer — Standard	p	32%	35%
Double Layer — Dense	k_t	0.91	0.87
Double Layer — Dense	p	31%	32%

Note: *BLc = [M/(*XYZ* · ρ_r) · 100], Blockiness is defined as the volume of a stone divided by the volume of the enclosing XYZ orthogonal box with a minimum volume.
**Guide values in the table are based upon assumptions of mean Length to Thickness ratio LT = 2.0, and deduced from research study of Stewart *et al.* (2003).

In view of the numerous variables involved in determining the volumetric porosity (n_v) and the layer thickness coefficient (*kt*), CIRIA compiled many results, from laboratory, full scale models and field studies, to present a set of recommended values, see Table 9.6. Many designers are however simply specifying the double layer thickness as $2D_{n50}$. This means the quantity surveyor will slightly overestimate the required armour rock quantity. This could be tolerated as it provides some extra reserve material against rock breakage during transportation. However, this could present a problem as the as-built layer thickness will be less than the design thickness, which has led sometimes to the construction supervisor accusing the construction contractor of using less of the rock material than required by design. Therefore, it is recommended to either use $kt = 0.9$ or add a note to the design drawing stating that the layer thickness could be less, up to 10%, than the design layer thickness.

9.7. Stability Considerations

The geotechnical stability of the rock structure in resisting the influence of gravity and applied static, hydraulic and seismic loads is to be investigated as part of the required design procedure. This is usually achieved by searching for the most critical slip circles and/or deep sliding pattern, see

Figure 9.12, by using limit equilibrium approaches that apply conventional soil mechanics techniques or a numerical modelling approach.

The above-described geotechnical check does not, or actually cannot, address the stability of the armour layer on its own. This point is better explained by drawing a comparison with the concrete armour units (CAU), as given in Chapter 10, to better demonstrate what issues need to be considered in this regard.

All design guidelines for the single layer CAU advise a maximum number of rows along the slope, therefore the slope length is proportional to the CAU size, the larger the unit, the longer the allowable slope length. A similar concept cannot be directly copied and applied to rock armoured slopes for the following reasons:

(a) The armour rock face is held in place by utilizing the following forces: resting on the lower row, anchoring to the rocks on the sides and underneath, and friction with the rock underneath. These forces are actually the same that are acting on the CAU face.
(b) The rock armour layer is more versatile to accommodate slight armour settlement (i.e. nesting), if any, or even loosing few rocks from the slope face.
(c) Armour rocks are normally placed on flatter slopes than those usually adopted for the single layer CAU. This means the critical weight component parallel to the slope gets smaller as the slope gets flatter.

Based on the above it seems that the factors which restrict the length of the CAU slope should not affect the length of a rock face slope. A set of sketches produced by Burcharth (1993), which were also cited in CEM

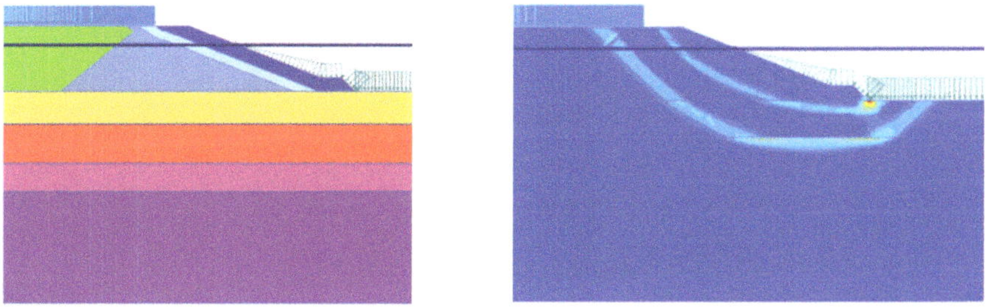

Figure 9.12. Numerical model example for a critical slip circle failure.

(2003), illustrate the acting forces on the armour rocks in different loading scenarios, as shown in Figure 9.13, and the possible failure modes as shown in Figure 9.14.

One of the possible failure modes is mode (d) in Figure 9.14, which is influenced not only by the wave action but also, and could be significantly, by the slope length, i.e. the longer it is, the less stable it becomes.

In order to minimise the risk of a sliding failure in an armour layer, the slope length is to be limited. Safe limits could not be made like that of the single layer CAU by limiting the number of rows (see Section 10.4) due to some fundamental differences. In single layer CAU, the armour units, after being exposed to few storms, tend to get consolidated at the bottom of the slope (a behaviour termed as nesting or settlement) and open up at the top rows. This movement mechanism will take place before the risk of having the units sliding along the slope is materialised. Rock armour slopes can usually absorb any settlement in the armour layer as the rocks will adjust themselves without leaving any gaps at the top, therefore the risk of armour sliding could materialise before having any noticeable deformation on the rock face. It is however important to note that the sliding stability of the rock face is not affected only by the number of

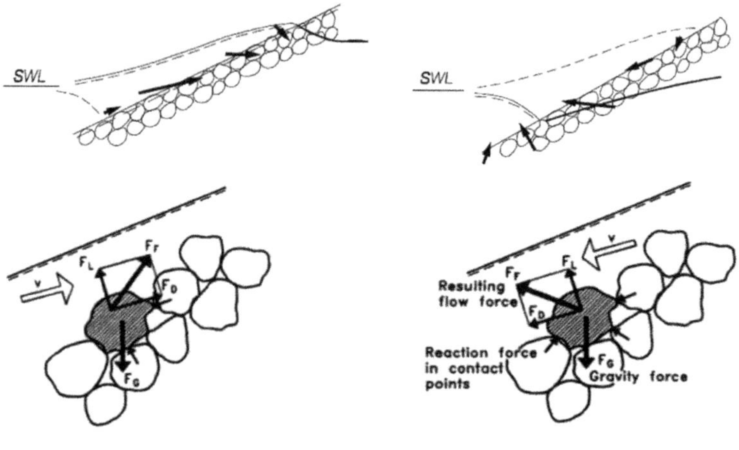

Wave up-rush & down-rush on permeable slope

Cyclic acting force $\overline{F_F}$ (= $\overline{F_D}$ + $\overline{F_L}$ + $\overline{F_I}$)

Figure 9.13. Acting forces on a rock armour block during wave up/down rush, (Burcharth, 1993).

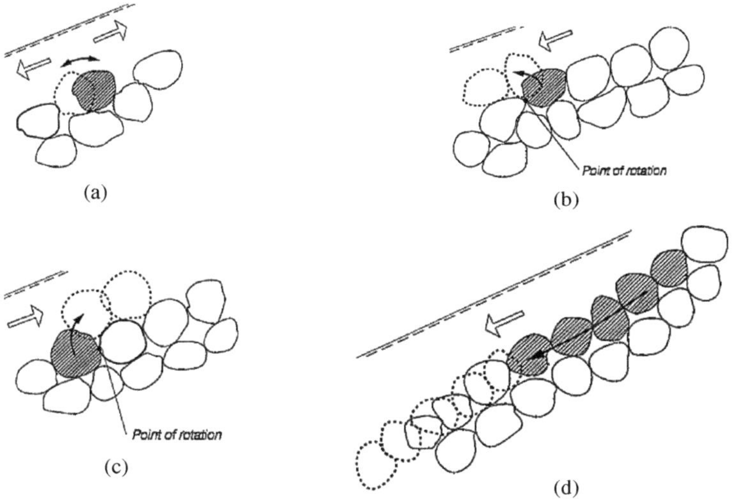

Figure 9.14. Typical rock armour layer failure modes (Burcharth, 1993): (a) Rocking of block during up/down rush; (b) Rotation and subsequent down-slope displacement during down rush; (c) Rotation and subsequent up-slope displacement of rock during up-rush; (d) Sliding of several armour rocks during down-rush.

rows (and consequently the rock size), but also the slope inclination. Other factors are related to the placement workmanship, exposure to waves, underlayer rock class and permeability of the core material. Conservative guidance could only be inferred from constructed projects.

Different examples of long rock armour layer slopes with different side slope angles are presented in Figure 9.15. Whereas schematic representations of slopes with different angels are shown in Figure 9.16 to illustrate that it is not the slope length which matters but the height from toe to crest, which is the controlling factor. This is because the total force component parallel to the slope is the same for all slope angles if the height from crest to toe is the same.

Based on the representative examples presented herein a maximum height from toe to crest of 14.0 m could be recommended for designing a straight slope of rock armour layer. The designer is advised to take into consideration the degree of exposure of the structure as this will consequently mean large or small armour rocks and the existence of wave-induced forces. Therefore, the recommended height is better to be considered applicable for inner, or moderately exposed, slopes.

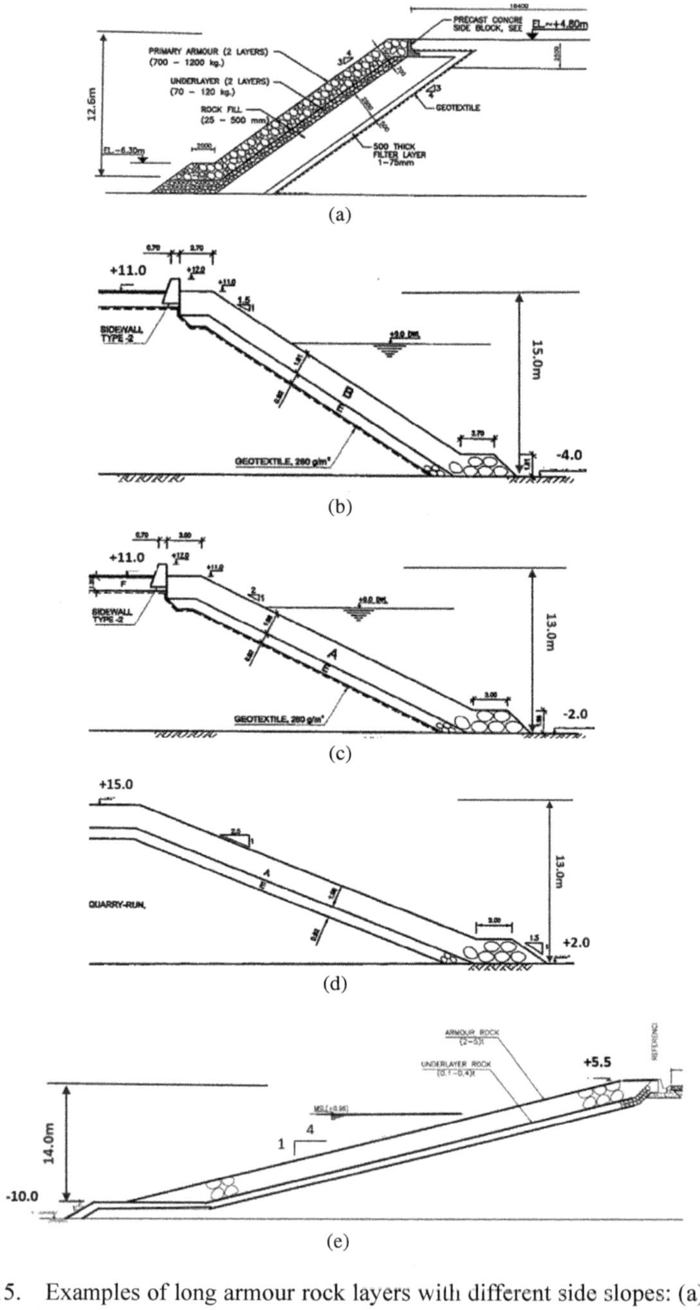

Figure 9.15. Examples of long armour rock layers with different side slopes: (a) 1:1.33; (b) 1:1.5; (c) 1:2.0; (d) 1:2.5; (e) 1:4.0.

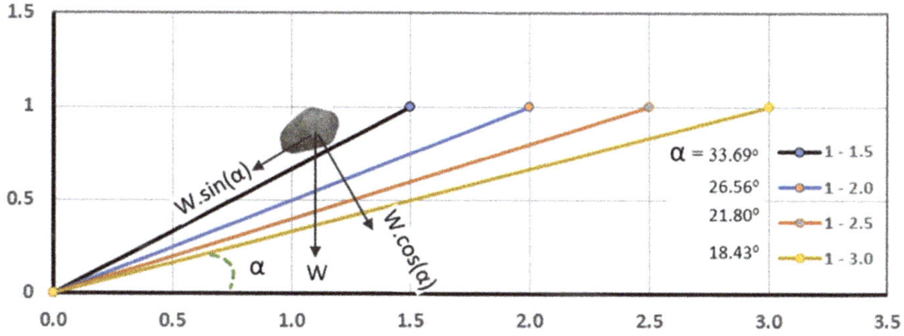

Figure 9.16. Rock weight components on a slope.

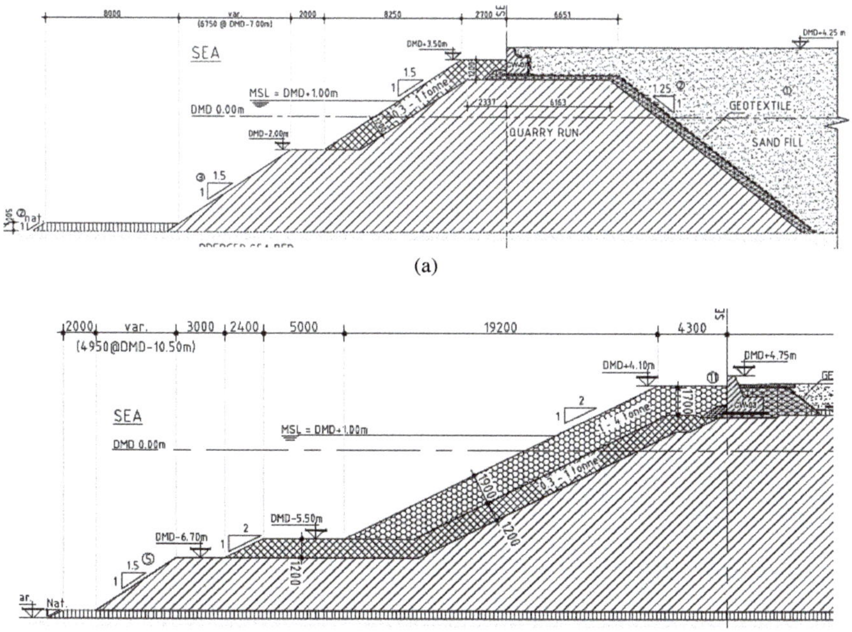

(a)

(b)

Figure 9.17. Sections with toe berm of core (quarry run) material.

The designer however has to keep in mind that the construction team will always push to have a one slope structure as it is easier to construct, faster and marginally less costly.

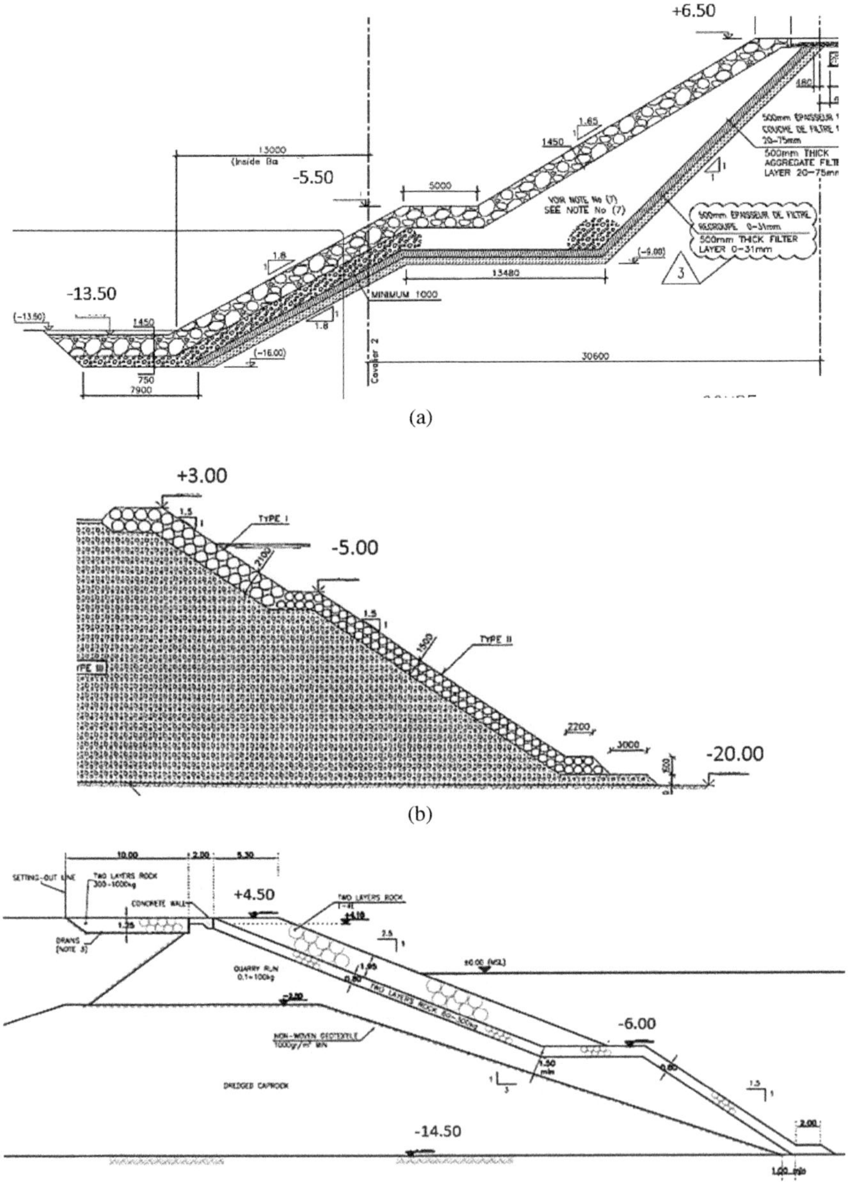

Figure 9.18. Long slopes divided into two separate slopes.

There are many deep water locations where the height from crest to seabed exceeds the suggested recommendation of 14.0 m limit, in such cases there are basically two approaches:

(a) To elevate the toe from the seabed level by including a berm from the core (quarry run) material, as show in Figure 9.17.
(b) To break the long slope into two slopes with a small step/berm, as shown in Figure 9.18. Note that the primary armour need not be continued down to the seabed, as shown in Figure 9.18(a), as the full wave loading is applied mainly in the zone of ±Hs from the water level, as mentioned previously in Section 6.2. Accordingly, a lighter armour rock is normally used for the lower slope, as shown in Figure 9.18(b), or further to that a steeper slope could also be used, as shown in Figure 9.18(c).

Chapter 10

Concrete Armour Units (CAU)

10.1. Introduction

The growing demand to expand existing ports or construct new ones frequently involves more exposure to higher waves. This could be due to encroaching into deeper waters or developing a coastline with no, or limited, natural shelter. Consequently, coastal structures will need to be protected against more adverse conditions, i.e. larger waves. Hence, there is an increased requirement for heavier armour units, and since rocks above 8.0 t are very limited in many parts of the world, artificial concrete armouring units were invented to cater to this new market.

Apart from the rock size issue, it is widely accepted that for small waves a conventional rock armour will be more cost-efficient, provided that rocks are available nearby. This is because the concrete armour units, with their required mould fabrications, casting and placement processes will not be economical.

A large variety of plain Concrete Armour Units (CAU) have been developed in the last century, see Table 10.1. The aim was to improve the interlocking, robustness and stability of the units, increase the armour layer porosity and reduce the concrete volume, thus optimizing the cost. The evolutionary journey of the CAU has gone through three distinct generations:

The **1st generation** is characterised by a massive unit with a simple shape, typical examples are cubes, parallelepipeds and antifer cubes. The stability relies on the unit weight with limited interlocking. Therefore, each cube must be sufficiently heavy to withstand the wave action.

Table 10.1. Overview of most commonly used CAUs.

	Interlocking			Friction	
Massive	Bulky	Slender		Bulky (hollow)	Slender (multi-hole)
1 - 2 Layers	1 Layer	1 Layer	2 Layers	1 Layer	1 Layer
Cube 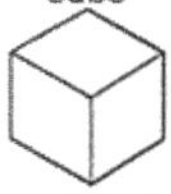	Accropode I 	Core loc 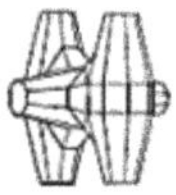	Tetrapod 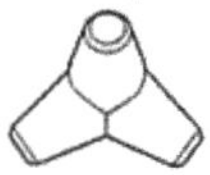	Haro 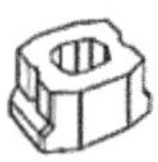	Shed 
Parallelepiped block 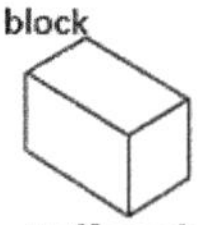	Accropode II 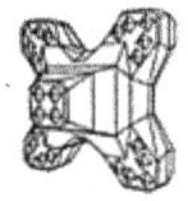		Stabit 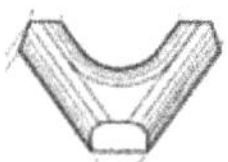	Seabee 	Cob 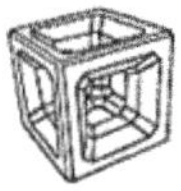
Antifer cube 	Xbloc 		Dolos 		
Cubipod 					

The placement of the units is random and in two layers, however, there are also few structures constructed with one layer.

The **2nd generation** was developed in the 1950s and 1960s and is characterised by a complex shaped unit to be placed in two layers, typical examples are the Tetrapod, Dolos and Stabits (though the Stabits could be placed in a single layer). These units derive their stability mostly from their weight and partially from interlocking with neighbouring units. The placement of the units is either random or in pattern, and in two layers.

Apart from the enhanced interlocking, another feature was also increased in comparison to the cubes, and that is the porosity of the armour layer. This is highly effective in contributing to the absorption of the wave energy and reducing the wave run-up.

Reinforcement (rebars or fibre) has been used in some units of this generation especially in projects where exceptionally large units were required.

The **3rd generation** is characterised by specially shaped units to enhance the interlocking which enabled them to be placed in one layer only.

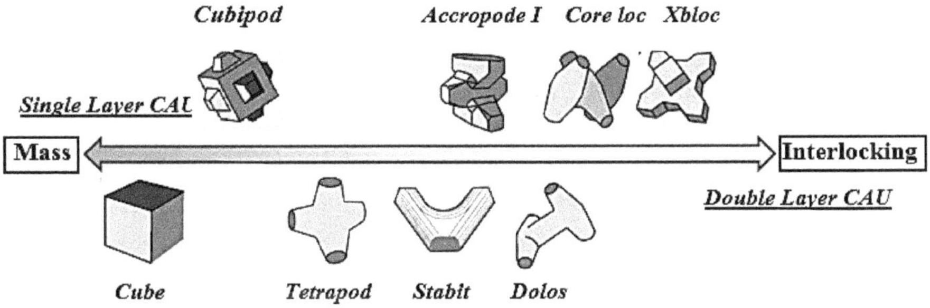

Figure 10.1. Qualitative presentation of CAU stability sources.

Typical examples are the Accropode I & II, Core-loc and Xbloc. The stability of the unit relies equally on the unit weight and the interlocking between the units.

Part of this generation is a different family of concrete armour units consisting of Hollow cubes, with a porosity of approximately 60%. The concept of one layer is still to be followed, however with uniformly placed units, with the Shed, Cob, Haro and Seabee as typical examples. These units gain their stability not from their weight or interlocking but from lateral support and friction between the surrounding units.

The “Massive” and “Interlocking” categories could be presented in a simplified line diagram for illustration purposes only, see Figure 10.1. Note that the Cubes and Cubipod units are presented as a double layer and single layer, respectively, following their common use, but they both could be used either in a single or a double layer.

10.2. CAU Suitability Assessment

10.2.1. *General*

Although the CAU has gone through many development phases, which have been characterised into three generations, this doesn't mean that the 3rd generation is always better to use than the 2nd or the 1st generation. Units from all generations are still in use today as each unit has its merits and the selection of the suitable CAU is project specific. The selection of the proper CAU should be made after a careful study and analysis of the following: project location, the foreshore configuration, wave climate, wave characteristics and its breaking mechanism, the preliminary design section of the structure, locally available rock material, logistics and transportation, cost, construction-related issues including damage risk during the construction, failure risk and consequences, maintenance and repair of armour layer, and most importantly design basis and design criteria (as these might stipulate some requirements which limit the choice of the CAU). Many of these issues have been discussed in various parts of this book but some important issues related to the CAU and its characteristics which could have some bearing on the CAU assessment and selection process will be discussed in the following sections of this chapter.

A summary of some defining properties of the main types of CAUs are given in Table 10.2, which should be taken into consideration in the CAU selection and assessment procedure.

The Accropode I, Core-loc, Accropode II and Xbloc have become the first choice for alternative armours to rocks. It is however worth noting that all these units are being patented with royalties and can only be used after having a license agreement with the developer, as follows:

Accropode™: CLI (Concrete Layer Innovations), France [earlier: ARTELIA, France]
Core-loc™: CLI (Concrete Layer Innovations), France [earlier: US Army Corps of Engineers (USACE), U.S.A.
Xbloc®: BAM Infraconsult by (Delta Marine Consultants), Netherlands.

In return for the royalties, technical services are to be provided for the design consultant and the construction company, including the placement pattern and training of the crane operator regarding the handling and orientation of the units.

An illustrative comparison table between these four units is presented in Table 10.3 demonstrating the qualities of each of these units. Note that the recommended K_D values in the table are according to the patentees.

It is however worth noting that there is a size limitation, i.e. maximum unit size, to each of these units as given in their relevant design table (see www.concretelayer.com and www.xbloc.com), and as summarised as follows, together with their weights (for a concrete density of 2.4 t/m^3):

Accropode I & II:	28 m^3	(67.2 t)
Core-loc:	11 m^3	(26.4 t)
Xbloc:	20 m^3	(48.0 t)

Table 10.2. Summary of the Pros and Cons of selected CAUs.

CAU Type	Pros	Cons
Cube[a] (2-Layers)	• Easy to manufacture, stack and handle • High structural strength • Low risk of progressive failure • Low maintenance cost	• Low stability coefficient (K_D = 6–8)[b] • Relatively smooth surface leading to high overtopping • Low friction with filter layer • High construction cost • Long construction time
2-Layers CAU	• Low maintenance cost • Relatively rough surface leading to low overtopping	• Low stability coefficient (K_D = 6–8 *for Tetrapods,* K_D = 10–12 *for Dolos & Stabits*)[b,c] • High construction cost • Long construction time • Risk of progressive failure • Risk of unit breakage
1-Layer CAU[d]	• High stability coefficient (K_D = 15–16 *for Accropode, Core loc & Xbloc*)[c,e] • Relatively rough surface leading to low overtopping • Low cost • Short construction time	• High maintenance cost • High risk of progressive failure • Risk of unit breakage (some units)

Notes: [a]Cubes have been presented as a separate category only because each mainly depends on its mass with no, or little contribution from interlocking.
[b]K_D values are for double-layer trunk section, (SPM 1984 and BS 6349-7). The waves and foreshore characteristics are to be taken into consideration which could reduce the K_D value.
[c]Most of the given K_D values are based on the patentees recommendations, however, in practice quite often lower values of K_D are used and due to different reasons.
[d]The Accropode I & II, Core-loc and Xbloc, excluding hollow units and Cubipods.
[e]K_D values are for a trunk section (concretlayer.com for Core loc & Accropod, and xbloc.com for Xbloc). The waves and foreshore characteristics are to be taken into consideration, which could reduce the K_D value.

The Dolos, although it is a double-layer armour unit, has a maximum unit size comparable with the above single-layer units. The Dolos maximum size is 20 m^3 with a weight of 48.0 ton (Dolos Technical Design Specifications).

Table 10.3. A comparison table between the commonly used single-layer CAU (Jensen, 2013).

Comparison of Monolayer Armour Units (data for sea bed slope 1:100)					
		Accropode I	Accropode II	Core-locs	Xbloc
K_D recommended, trunk	—	15	16	16	16
K_D recommended, roundhead	—	11.5	12.3	13	13
Example, 10 m³ unit	m³	10	10	10	10
H_s max, for design	m	7.80	7.95	7.95	7.95
Block height	m	3.09	3.25	3.56	3.11
Weight (2.4 t/m³)	t	24	24	24	24
Armour layer thickness	m	2.78	2.93	3.27	3.00
Number of units per unit area	units/100m²	13.70	13.40	13.30	12.50
Porosity	%	50.83	54.30	59.41	58.7
Max. Number of rows up slope	—	20	20	20	20
Max. Size of unit in design table	m³	28	28	11	20

Other units of the double-layer category, i.e. cube, tetrapod and Dolos, have been used with considerably larger weights. Kasai *et al.* (2006) reported the use of 80 ton Tetrapods and 50 ton unreinforced Dolosse in Japan. The weight of the Dolos could however reach up to 90 ton if reinforced (Treadwell *et al.*, 2006). Rectangular blocks of 82 ton have been reported to be used in the Port of Bilbao breakwater (Jensen, 2013).

A recent development to the basic Cube unit has been made by introducing the Cubipod, which improves on the Cube as an armour unit by enhancing its friction with the underlayer rocks, provides better interaction between the cubes, ensures high porosity to the armour layer and provides roughness to the layer surface. With all these added benefits the Cubipod has increased the K_D factor of the Cube by almost 50% (Medina, 2016).

The Cubipod® is patented by SATO-OHL, Spain.

Further to footnote c of Table 10.2, it is worth noting that many designers, with the consent of the construction company, are inclined to use lower K_D factors in the design of the interlocking CAU. This will consequently result in slightly larger units than what is theoretically required, which could also be proved to be sufficient in a model test. There are many advantages, and some disadvantages, to this practice which could be summarised by the following:

- The required number of units will be less, and this will reduce the placing activities and save time.
- The units will be more robust with less risk of breakage or settlement.
- The increase in the armour layer thickness, i.e. concrete consumption, is not significant.
- The rock sizes for the underlayer will be larger, i.e. more rock material.

10.2.2. *Unit breakage*

The CAU development was basically driven by the notion of reducing the weight of the units and enhancing the interlocking between them. This led to the production of units with slender limbs like Tetrapods, Dolosse, Core-locs …etc. The design of the slender units aims to achieve a balance between the structural integrity of the unit and its hydraulic stability (resistance to displacement). Not only this, as these units also increase the layer porosity, a highly desirable and beneficial property of the armour layer.

The slender unit shape however turned out to be vulnerable to breakage, especially for larger size units. This is because the units are not reinforced (for avoiding corrosion problems), thus the structural integrity depends largely on the tensile strength of the concrete. When the unit size increases linearly, the cross-sectional area of the limbs increases to the second power whereas the forces increase to the third power. This means that the tensile stresses will increase linearly as the units get larger, ($TS = F/A$, where TS is the tensile stress, A = cross-sectional area and F = applied dynamic and static forces). Since the concrete tensile strength is constant, a larger unit will be more vulnerable to structural damage if the applied stresses exceed the available tensile strength.

Units categorised as "Interlocking/Slender" in Table 10.1 are statistically more susceptible to breakage than others. Two examples of CAU breakages are shown in Figures 10.2 and 10.3. When a CAU is not well-interlocked with adjacent units, it will experience a rocking movement, i.e. a slight cyclic movement of a unit, when exposed to a hydraulic loading, i.e. waves. This type of movement will often be the cause of the unit breakage.

Figure 10.2. Core-loc fluke breakage example.

Figure 10.3. Dolos trunk breakage example.

The importance of the CAU breakage has gained much attention after some failure incidents: Tetrapods (48 ton) in Arzew El Djedid breakwater (Algeria-1981), Dolosse (42 ton) in Sines breakwater (Portugal-1978), Dolosse in San Ciprian breakwater (Spain-1980) Port. It is worth noting that other CAU breakwater failure incidents did occur not because of a

CAU breakage but because of other causes, like the Bilbao breakwater with an armour of rectangular blocks, (82 ton), (Spain-1976) (Jensen, 2013; Maddrell, 2005).

The failure of the Sines breakwater instigated many surveys (REMR, 1994), research and studies of the strength of the Dolosse and other slender CAUs and whether they should be reinforced (see for example Burchart *et al.*, 2000). Many breakwaters in Japan (Treadwell *et al.*, 2006) have been successfully constructed with reinforced Dolosse weighing up to 90 ton.

The breakage of a CAU in a severe storm event will have negative consequences in two ways: reducing its weight, and loss of interlocking with the adjacent units. Both of these factors could lead to the extraction of some of the broken units. Once the interlocking between adjacent units is lost, progressive extraction of units will follow almost immediately or in subsequent storms, and ultimately the situation will deteriorate ending in the failure of the structure.

Based on the above review, it is justifiable to introduce some additional safety measures in the design of CAU structures. The reasoning for the need of this extra safety has been well formulated by Jensen (Jensen, 2013) as follows: "It became evident that extreme care should be exercised in the interpretation of results from small scale model tests where the fragility of the units could not be modelled and where the armour layer in the model would be intact after the tests. Minor rocking of a number of units and settlements of the armour layer looked innocent in the model, but in nature it would mean breakage of units and possible failure of the breakwater".

For more details on the armour layer damage definitions and categorisations, see section VI-5-3 of CEM 2003 (Part VI-2011, the latest version).

In order to assist the designer in translating the above caution from qualitative to quantitative expression, it is proposed to reduce the recommended K_D factor for all types of interlocking slender CAUs, see Table 10.1, by 2.5% for each 1.0 m^3 above the 5.0 m^3 unit size. That is by applying the relevant recommended K_D factor first, and if the required CAU size is above 5.0 m^3, repeat the calculations with the reduced K_D factor. Noting that this added safety measure is intended to neutralise the breakage issue only.

Further to the above discussion regarding the unit breakage, it is important to note that the breakage could also be influenced by causes other than the CAU shape and size. Not adhering to the construction and placement specification and recommendations (e.g. position tolerance, dropping distance, orientation … etc.) is one cause, while other causes revolve around concrete manufacturing not being to specifications as this will accelerate the CAU deterioration mechanisms and shorten its life cycle. A summary of the external and internal agents (i.e. forces and deterioration sources) as well as the possible effects on the concrete are presented in Table 10.4.

Table 10.4. Overview of agents and effects on CAU concrete (Dupray *et al.*, 2009).

External	Internal	CAU concrete
Seawater ^{C,S}	Temperature ^{P,F}	Sulphate attack ^{F,C,S}
Ambient temperature ^{P,F}	Alkali content ^{P,F}	Cracking ^{F,C,S}
Atmospheric CO_2 ^{S}	Sulphate content ^{P,F}	Spalling ^{F,C,S}
Dynamic loads ^{F,C,S}	Chloride content ^{P,F}	Scaling ^{F,C,S}
Static loads ^{F,C,S}	Other chemical content ^{P,F}	Section loss ^{C,S}
Abrasion ^{S}		Strength loss ^{F,C,S}
Biological attack ^{S}		Alkali-aggregate-reaction ^{C,S}
Freeze and thaw ^{F,S}		Expansion ^{C,S}
Ice ^{S}		
Forces (accident) ^{F,C,S}		
Deterioration (accident) ^{F,C,S}		

Note: ^{P} relevant for Preparation, ^{F} relevant for Fabrication, ^{C} relevant for Construction, ^{S} relevant for Service.

10.3. Placement of CAU

10.3.1. *Placement styles*

CAUs could be placed in many different styles and patterns ranging from simple to complex, the simple being "random", and the complex being in a "pattern with special unit orientation". However, no matter what the placement style is, a minimum placement density, i.e. number of armour units per unit surface area, needs to be achieved.

There is no clear agreement between the designers on the categorisations of the different placement styles and their exact interpretation [see for example Allsop (1988), Burcharth (2007), Dupray *et al.* (2009), Pedro *et al.* (2012) and Van Gent *et al.* (2013)]. Random, regular, irregular, pattern and uniform styles are used as descriptions.

CAU are frequently placed using cranes equipped with pressure clamps or slings to place the units according to the coordinates, if applicable, given by the placement grid. The term "Random placement" is commonly used by many designers if the crane operator is not to place the unit in specific orientation in relation to its neighbouring units. Therefore, Allsop (1988) used instead the term "Random orientation" placement.

The author, however, is not in favour of the above confusing terminology and prefers to adopt the following style names:

Random: This is to include all units which are to be placed randomly, either with two degrees of freedom (position and orientation), or one degree of freedom (position). This means that randomness is considered to be basically related to position. Some general flexible guidelines and recommendations regarding the unit position or orientation may exist and these are preferably to be observed during placement. Example units are: Cubes, Cubipods and Tetrapods.

Pattern: Units are to be placed according to a placement grid following a certain pattern. The orientation of the units is to be observed following the relevant guidelines and rules to ensure proper interlocking, and these could be different for different units. For example, the requirements for the Accropode I & II, Core-loc and Xbloc normally include: adjacent units have different attitude, each unit is to be keyed into two units of the row below and limited number of units shall have the plane face parallel to the slope.

Example units are: Tetrapods (this unit has different placement styles), Stabits, Accropodes, Core-locs, Xblocs and Cubipods.

It is worth noting that most single-layer CAUs are considered by a number of designers to be under the "random placement" category, the randomness here does not refer to the placement grid but to the unit orientation (although the orientation is not actually completely at random as there are some guidelines to be observed by the patentees).

Uniform: Units are to be placed uniformly in a grid, e.g. Haro, Seabee, Shed, Cob, Cube (one layer) and Cube (two/one-layer pavement).

10.3.2. *Armour layer porosity and packing density*

Porosity (p) is the voids fraction in a certain volume, and the packing density (φ) is the number of placed units per square nominal diameter (D_n). These two properties, which are very much inversely linked together, are the direct outcome of the type of the CAU and its placemat style. The porosity is essential for absorbing the wave energy and reducing the wave run-up, while the packing density is crucial for having sufficient interlocking between the units to resist the wave loading as a netted fabric and not as individual units.

The packing density, layer porosity and other related geometric terms are given in Equations (10.1)–(10.7). In Table 10.5, a summary of placement styles and key properties relevant to those styles are given for all main CAUs.

Table 10.5. CAUs key properties related to its placement.

CAU	n	Placement	k_s	K_t	p	φ	cot α
Cube	2	Random	1.0	1.10	0.37	1.17[a]	1.5 to 3.0
	1	Pattern	1.0	1.00	0.28[b]	0.72[b]	1.5 to 2.0
Cubipod	1	Pattern	1.0	1.00	0.41[c]	0.59[c]	1.5 to 2.0
	2	Pattern	1.0	1.00	0.41[d]	1.17[c]	1.5 to 2.0
Tetrapod	2	Random	0.28	1.02	0.50	1.04[a]	4/3 to 2.0
Dolos[d]	2	Random	0.16	0.94	0.56	0.83[a]	4/3 to 2.0
Stabit[e]	2	Pattern	0.30	1.10	0.52	1.06	4/3 to 2.0
Accropode I	1	Pattern	0.34	1.29	(0.50–0.51)[f]	(0.64–0.62)[f]	4/3 to 1.5
Accropode II	1	Pattern	0.29	1.36	(0.53–0.55)[f]	(0.63–0.61)[f]	4/3 to 1.5
Core-loc	1	Pattern	0.22	1.52	(0.58–0.59)[f]	(0.64–0.62)[f]	4/3 to 1.5
Xbloc	1	Pattern	0.33	1.40	(0.587)[f]	(0.58)[f]	4/3 to 1.5

Notes: [a]Recommended packing density is given as it could vary as the placement style is random (SPM, 1984).

[b]Van der Lem *et al.* (2016).

[c]Cubipod (2016); Medina *et al.* (2016).

[d]For units with $r = 0.32$ (waist to height ratio).

[e]For double-layer placing method.

[f]The porosity and the packing density are increasing and decreasing, respectively, as the CAU size increases (www.concretelayer.com and www.xbloc.com). Noting that almost all single layer CAUs reduced the layer porosity which was recommended at the launch of the unit. This will result in a more compact armour face with a reduced armour settlement down the face due to the continuous wave loading

$V = M/\rho_c = D_n^3 = k_S D^3$	CAU volume, nominal diameter and characteristic length	(10.1)
$A_a = n/N$	Area covered by one armour unit	(10.2)
$t_a = n\, k_t\, D_n$	Armour layer thickness	(10.3)
$p = 1 - [(n \cdot V)/(A \cdot t_a)]$	Armour layer porosity	(10.4)
$\varphi = n\, k_t\,(1 - p)$	Packing density coefficient	(10.5)
$N = \varphi/D_n^2 = \varphi/V^{2/3} = t_a\,(1 - p)/V$	Number of units per unit area	(10.6)
$V_c = NV = t_a\,(1 - p)$	Concrete volume per unit area	(10.7)

10.3.3. *Single-layer CAUs*

This type of CAU is to be placed primarily in a uniform or well-defined pattern to ensure a proper restraining of the units against movement, either by interlocking, friction or being restricted by neighbouring units. This type of restraining will improve the unit stability with less reliance on the unit weight, however in different degrees, depending on the unit type. The orientation of the units will either be specified or follow some general guidelines during the placement.

The armour layer, in general, shall not have large gaps between the units as that could expose the underlayer to strong wave action, and it could also weaken the support and/or interlocking between adjacent units. However, an increased porosity of the armour layer is always welcomed to increase the saving in concrete and also increase the wave energy absorption.

CAUs used as a single armour layer need to be designed with sufficient reserve capacity, normally 20% for interlocking units. This is due to the fact that the failure of these structures progresses very rapidly if one unit fails. Therefore, the physical model testing of the structural integrity is to satisfy 0% damage criteria for an overload test with $1.2H_s$ (H_s is the significant wave height of the design storm).

The Interlocking Units (Accropode I & II, Core-loc and Xbloc) placement grid is to be prepared for the crane operator to follow, identifying each unit position by coordinates (Eastings and Northings) which are to be uploaded to the crane computer. The positioning could tolerate some deviation, typically 10% of the unit height.

An illustrative sketch is given in Figure 10.4 for the Core-loc placement grid. Similar grids are produced for Accropode I & II, Xbloc and Cubipod.

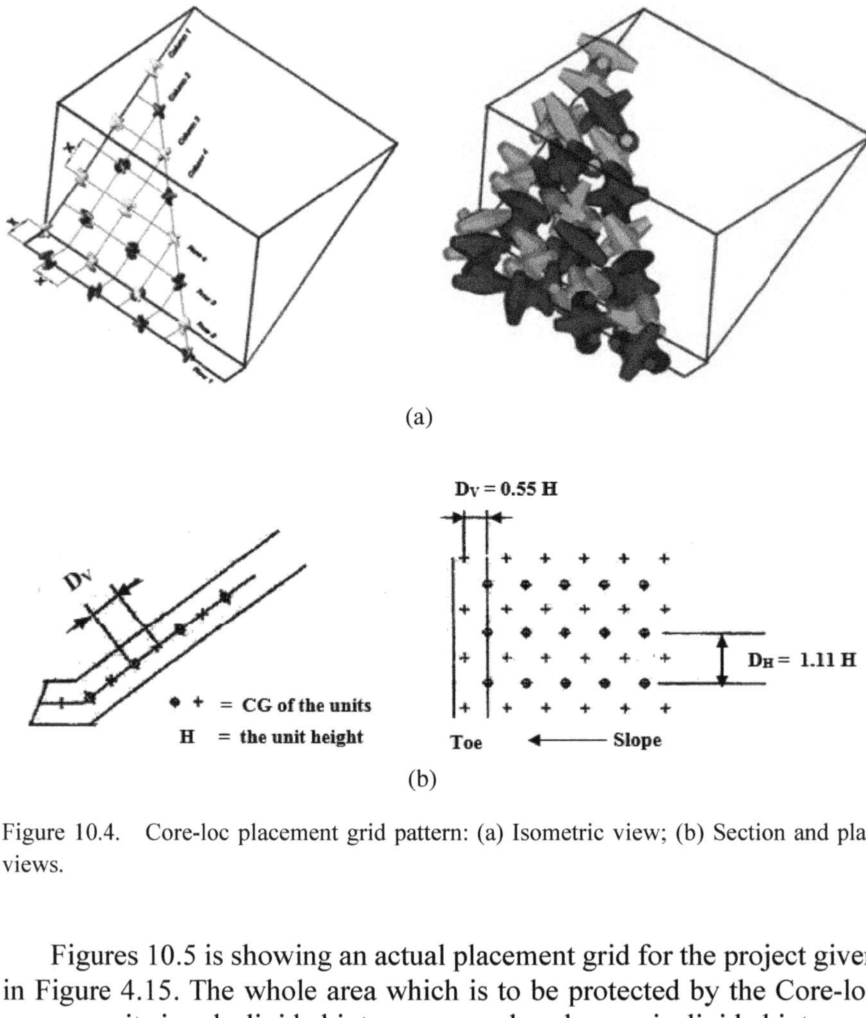

Figure 10.4. Core-loc placement grid pattern: (a) Isometric view; (b) Section and plan views.

Figures 10.5 is showing an actual placement grid for the project given in Figure 4.15. The whole area which is to be protected by the Core-loc armour units is sub-divided into zones, and each zone is divided into sectors numbered according to the placement sequence, i.e. starting triangle first in sector P1b, then P2b followed by P3b and so on. The width of these sectors depends on the placement crane capacity and reach.

The placement in each sector starts from the toe units up the slope with 45° inclination between the sectors. With all the units of the same size in this example, the inclination could go either way, but if there is a change in size, the smaller unit must always rest on the larger unit.

The placement grid can well tolerate a mild change in the seabed level, however not steeper than 1V:10H. This means that the structure side slope length, from crest to toe, could be variable, as is the case in Figure 10.5. In case steeper changes in the seabed level are encountered, then these need to be adjusted with the use of rock fill material or by dredging to bring the seabed to what could be tolerated, otherwise the units will not have the required interlocking.

Representative design sections for various CAUs are presented in Section 14.3, and example photos of constructed breakwaters are shown in Figure 10.6 for each of the four different interlocking single-layer CAUs.

The Hollow units are placed uniformly, as shown in Figure 10.7, but they are generally difficult to place underwater and only possible with extensive diver assistance. That is because these units need to be lowered to their predetermined position and to be very close to the adjacent units. The placement tolerances will include the distance from the adjacent units and the difference in level, as all units should attain almost the same face level (see Allsop, 1996). The hollow units could be placed on straight and curved sections; however, the curvature will result in small, tapered gaps opening up between adjacent units. It is therefore the units at the curvatures are placed very tightly together so that they contact one another at the base. The Hollow units, unlike the Accropode, Core-loc and Xbloc units, cannot tolerate any changes in toe level. All units at the first row needs to be placed at the same level to be able to place the subsequent rows.

The Cube armour unit, and all the units under the "Cube" category, being dependent on its mass and not on interlocking, have the flexibility to be placed in a single or double layer, with no limitation to the number rows as is the case with the interlocking single-layer units. The single layer could be placed either uniformly or randomly (Van Gent *et al.*, 2013), the terms used by Van Gent were "regular and irregular", but in both cases the cubes are placed with one side parallel to the slope. This will largely ease the placement and contribute positively to the stability of the individual cube.

The Cube in a single-layer placement style could be used on seaside face of a breakwater or revetment, see Figure 10.8. Where the pattern is shown in (a) with sufficient gaps between, the cubes are required to dissipate the internal water pressure, the design section is shown in (b) and photos during construction in (c) and (d). For the same design conditions, the cube as a single-layer armour unit will definitely be larger than an interlocking unit, however a simple formwork, simple placing pattern and no royalties to be paid are the gained advantages.

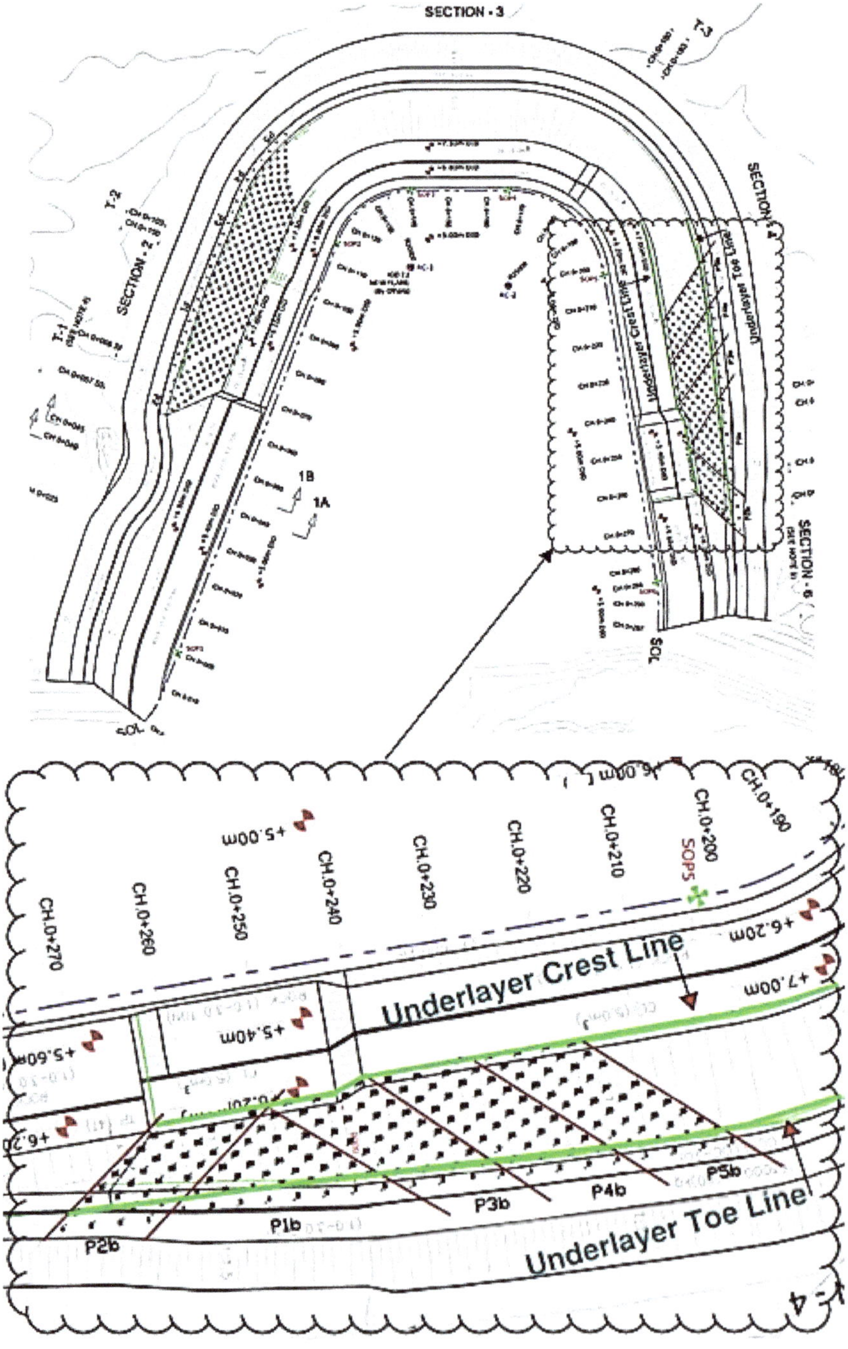

Figure 10.5. Core-loc placement grid — Part B of the project.

(a) (b) (c) (d)

Figure 10.6. Constructed examples for single-layer armour units: (a) Accropode II; (b) Core-loc; (c) Accropode I; (d) Xbloc.

Figure 10.7. Single-layer armour layer of Shed unit.

Single-layer blocks are frequently used on the rear side of breakwaters as they offer little resistance to the falling discharge of the wave overtopping and this means better stability. An example is shown in Figure 10.9.

The Cubipod is not an interlocking unit, it is a massive “randomly” placed unit with self-arranging tendency capable of maintaining the

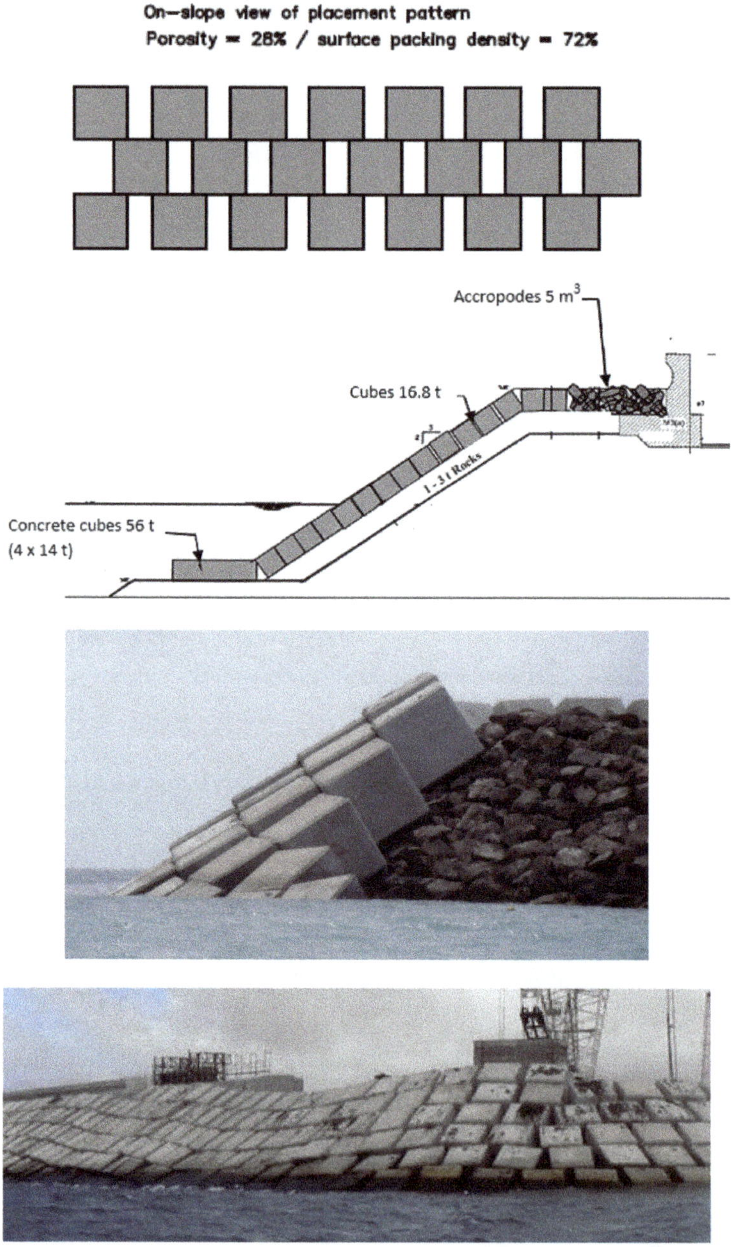

Figure 10.8. Single-layer Cube armour units on the seaside face of a breakwater: (a) Placement pattern (van der Lem *et al.*, 2016); (b) Design section (van der Lem *et al.*, 2016); (c) The breakwater roundhead during construction; (d) The breakwater during construction.

Figure 10.9. Single-layer antifer cube armour units on the rear side of a breakwater (Hellinga, 2016).

Figure 10.10. Single-layer Cubipod armour — Random (position & orientation) placement.

initial armour porosity, with few variations (Medina *et al.*, 2016). Figure 10.10 shows the Cubipod units in random placement, with randomness in position and orientation, whereas Figure 10.11 shows the unit placed with randomness in position while the orientation is fixed by having the face of the unit parallel to the slope.

Figure 10.11. Single-layer Cubipod armour — Random (position) placement.

Figure 10.12. Cubipod diamond-type placement pattern (Pardo *et al.*, 2014).

It is however worth noting that the recommended placement style is a "diamond-type" placement grid as shown in Figure 10.12. No limitations are imposed on the number of rows.

10.3.4. *Double-layer CAUs*

The placement style in double layer CAUs was generally stated as being random, but some designers are not fully satisfied with this option. It is always preferable to have the placement more controllable, i.e. in a pre-defined pattern, than to be left to the construction team judgment, as carrying out rectifications after an inspection is time-consuming and costly. Besides that, an improvement to the stability of the armour layer might also be achievable. However, placement in a pattern will require more time, but this could be more than compensated if the pattern placement would increase the unit stability and thus a reduction of unit size could be possible for the given design event.

Representative design sections for various CAUs are presented in Section 14.3, and examples of constructed breakwaters are shown in this section for the four main units in this category of double-layer CAU, namely the Cube, Tetrapod, Dolos and Stabit.

The CAU in a double layer, and unlike the single-layer CAU, could be designed for a slight damage at the design event, in the order of 0–2%. Reference could be made to Table VI-5-21 of the CEM, 2011, for more details.

The Cube was the first CAU to be developed, it was placed randomly as a double layer armour. Since then, many variations have been introduced to the shape of the cube and to its placement style to improve its stability and performance.

The different shapes of the cube included one or a combination of the following: simple cube, cuboid, parallelepiped, trapezium prism (by having four sides sloping and one of the two parallel faces slightly smaller than the other), grooved on the sides (antifer cube), and a hollow cube (to reduce thermal cracking). Most of these variations aim to enhance the interlocking between the blocks, and subsequently increase the stability of the armour layer, at the same time it will increase the slope surface roughness which will help in reducing the overtopping.

Various placement patterns were also developed throughout the years (Frens, 2007), with few examples shown in Figure 10.13. It is however worth noting that complete randomness (position & orientation) is difficult to achieve in practice, especially for the first layer as the cube units tend to rest on one face parallel to the slope and fit face-to-face with neighbouring cubes. Thus, some armour layers appear to be more randomly placed than others. Furthermore, adopting a pattern placement style is normally welcomed by the construction team to avoid any arguments with the inspection team which might lead to some rework and rearrangement of the cube units.

(a) (b) (c)

Figure 10.13. Illustrations of antifer cube different placement styles (Frens *et al.*, 2008): (a) Random; (b) Pattern; (c) Pattern.

Figure 10.14. Cube armour with a random placement.

Figures 10.14 and 10.15 show two breakwaters constructed with random and a uniform placement style, respectively.

The Tetrapod unit has been a widely used CAU in the past and still is the preferred option for many designers. The unit was developed and patented by Sogreah (Artelia now), France, in 1950, however this has expired, and the unit is free to users without a license. This unit was the first attempt to demonstrate that the stability factor, K_D, could be substantially improved by having an interlocking between the units.

The placement style of this unit, and similar to the cube, started as a random style, as shown in Figure 10.16. The total number of Tetrapod units, for each Tetrapod unit size, in a control area of 100 m^2 as shown in the figure, could be calculated as per Table 10.5. The Tetrapod armour layer is pervious enough to dissipate the internal water pressure, absorb wave energy and provides sufficient roughness to reduce the wave run-up and overtopping.

The 1st layer is placed with three legs resting on the underlayer rock surface and the fourth points upwards. This is not meant as a strict

Figure 10.15. Antifer cube with a pattern placement.

requirement but it will ease the placement work of the construction team by having the unit rest in its natural stable position. The distance between the units is preferred to be in the order of the diameter of the unit leg end, as in Figure 10.16.

The 2nd layer is then to be installed pointing, as much as possible, in the opposite direction to the 1st layer. Tetrapods in the 2nd layer have actually a natural tendency to settle in this manner. Each unit in the 2nd layer must at least extend over two units from the 1st layer ensuring a well-interlocked fabric of the units. When the second layer has been placed, it is difficult to add few more units to improve the packing density, as this would in any case be useless or even harmful (Danel, 1953).

Different placement styles with varied packing densities have been used and researched throughout the years, with some examples of the possible pattern placement styles shown in Figure 10.17 (Fabião, 2013). More placement examples are also given by Min (2015).

Figures 10.18 and 10.19 show two breakwaters constructed with random and a pattern placement styles, respectively.

***Dolos** (plural **Dolosse**)* represents the next step in the CAU development after the Tetrapod. The unit was developed by Eric Merrifield, an East London port engineer, South Africa, in 1963, and it was not patented.

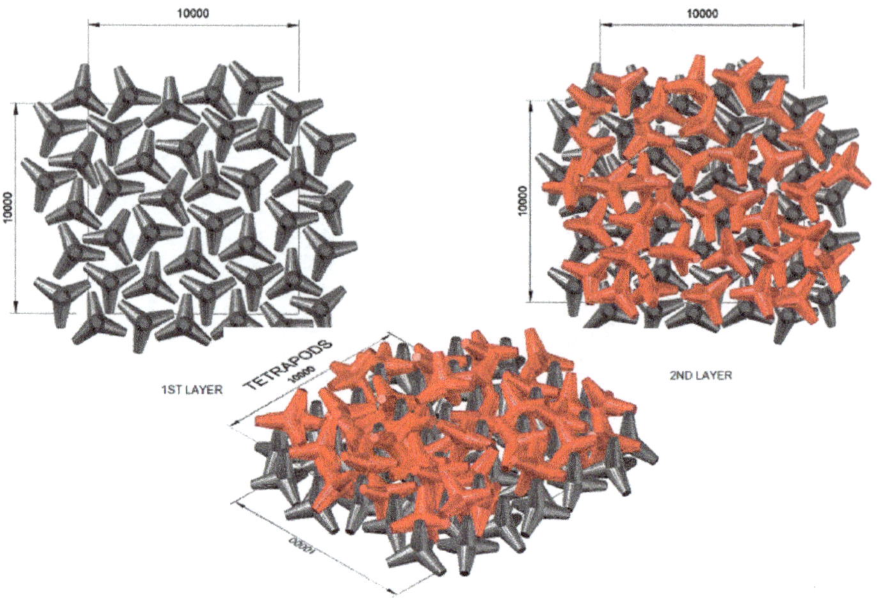

Figure 10.16. Tetrapod random placement style illustration (1st layer, 2nd layer and isometric view).

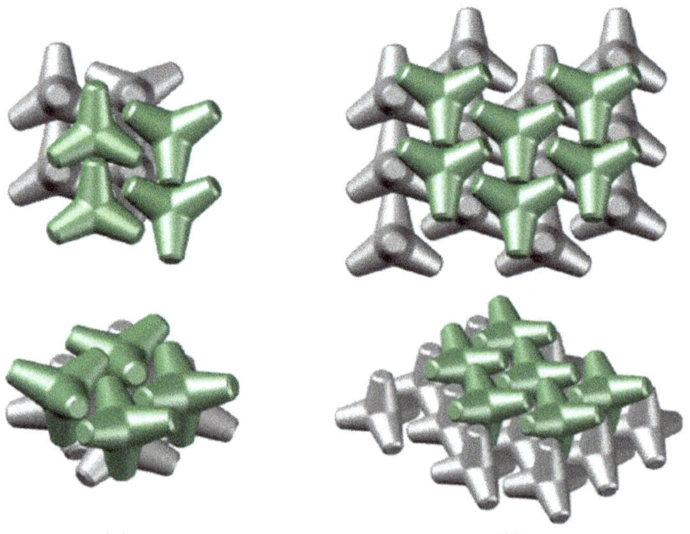

(a) (b)

Figure 10.17. Examples of Tetrapod placement pattern styles (Fabião, 2013): (a) Square mesh; (b) Triangular mesh.

Figure 10.18. Tetrapods with a random placement.

(a) (b)

Figure 10.19. Tetrapods with a pattern placement: (a) 1st layer; (b) 2nd layer.

The basic placement style of this unit is random placement. Attempts have been made to develop a possible placement pattern in the hope that it will improve the unit stability (Carver *et al*., 1977), with no decisive outcome.

As the placement style is random, then theoretically, a higher packing density for the armour layer gives more support and interlocking between the neighbouring units, and hence fewer displaced units, i.e. less damage. Therefore, Dolos stability coefficient (K_D) is expected to be directly proportional to the packing density (φ), however, this was shown by Burcharth *et al*. (1993(b)) to be only true within a range ($\varphi = 0.61$ to 1.0).

Figure 10.20. Dolos armour unit.

A higher packing density of ($\varphi > 1.0$) might actually reduce the interlocking and stability because of the limited available space for the legs to stick in between each other. Therefore, it is recommended to use an optimum packing density in the design (see Table 10.5) in order to achieve the optimum stability of the unit. An example is shown in Figure 10.20 for Dolos armour layer with random placement style. It is important to mention that today the use of Dolos is very limited after the disastrous breakwater failure in Sines, Portugal 1978, (Burcharth, 1987).

Stabit unit was developed in the UK in 1961, nearly at the same time with the Dolos, by Halcrow (presently part of Jacobs Engineering Group), however the patent has expired, and no license is required for its use. This unit, unlike the Dolos with its worldwide use, has predominately been used in projects designed by Halcrow in the UK and the middle east (the Gulf region). Few recent coastal projects in the Gulf region did use Stabit units as an armour, this is only because these projects were either upgrading or an extension of existing structures with Stabits armour.

Two different patterns are used for the placement of this unit as explained by Singh (1968) and Stabits (1979). These two methods are termed as: "Double Layer" and "Brickwall". The former being used for

Figure 10.21. Stabits armour unit.

slopes flatter than 1:2 and the latter for slopes of 1:2 and steeper. Thus, Stabits are to be placed using the "double layer" method for the horizontal toe berm, while the "Brickwall" method is used for armouring the relatively steep side slopes.

Double-layer method places the lower Stabits forming cradles (legs up), while the upper Stabits sits on them (legs down). The second method shall basically use the Stabits in one layer by closely following the principle of bonded brickwork construction. Stabits in the first row are placed side by side, in the next row up the slope, Stabits are placed staggered from those in the first row. Each succeeding row partly overlaps the lower row. Figure 10.21 shows an example of a breakwater with Stabits armour units.

10.4. Settlement of CAU Layers

Due to the wave action, the CAUs tend to slide down the slope face of the structure and this will consequently result in having the units become more compacted, i.e. reduced layer porosity, down the slope and less compacted or opened up, i.e. increased layer porosity, up the slope. This phenomenon is termed as "armour layer settlement".

The settlement of the units cannot be completely avoided as these units cannot be placed very tightly, and with the wave action dragging the units down the slope, a tighter packing density will result down the slope. As the settlement cannot be well modelled in the laboratory, it is difficult to be estimated or assessed accurately. Settlement could take place in all types of CAUs, and in single- or double-layer types, though it is more critical in the interlocking single-layer armour units.

There have been guidelines developed for the Core-loc and Accropode units which could generally be used to evaluate the single-layer unit's performance (Besley *et al.*, 2009) and as follows:

- No, or very minimal, unit movement during a "service event", which is normally with a return period of 1:5 or 1:10 years.
- Slight movement and consolidation of the armour can be accepted for the "design event". This could be with a return period of 1:100 or 1:200 year. The acceptable settlement is generally limited to less than 30% of the height of the unit.
- Slightly greater displacements are acceptable during an overload condition; however, no unit extraction is allowed.

Figure 10.22 shows an example for Core-loc settlement in a 2-D model test, and Figure 10.23 illustrates one possibility of how the armour settlement could be manifested by developing a gap between the units and the crest wall.

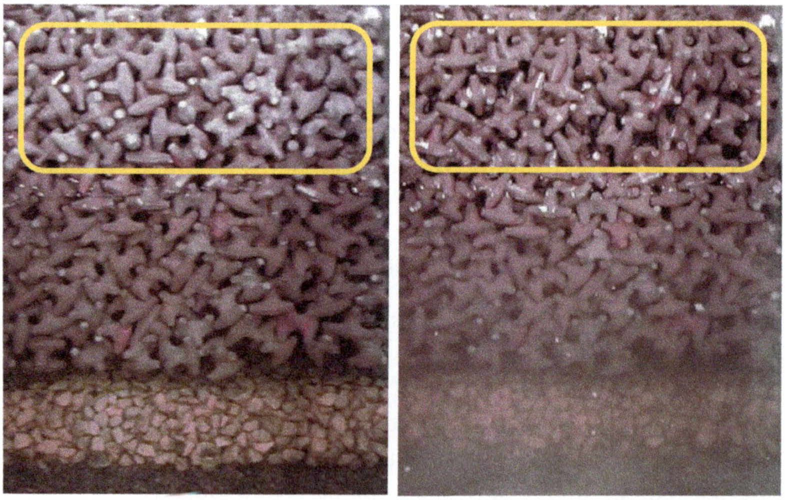

(a) (b)

Figure 10.22. Settlement of Core-loc unit in a 2-D model test with visible gaps above the water line level: (a) Before settlement; (b) After settlement.

Figure 10.23. A gap at a revetment crest due to Accropodes settlement.

CAU settlement is expected to be accelerated by the following contributing factors:

a. Higher layer porosity:
   This means less restraining of the unit sliding tendency.
b. Long slope:
   All CAU design guidelines for the single-layer interlocking CAU advise a maximum number of 20 rows along the slope, therefore the maximum slope length shall depend on the CAU size, the larger the unit size, the longer the slope. This is because the armour layer is held in place due to frictional forces with the underlayer and anchoring to the underlayer, accordingly a longer armour face requires more frictional and anchoring forces, especially for the lower layers.
c. Steeper slope:
   Resulting in higher driving force component parallel to the face and smaller stabilizing force component perpendicular to the face.
d. Smoother underlayer:
   This could result either from using a wide gradation with relatively small rock sizes, or the rocks are rounded in shape and not angular, both will not provide the CAU with enough grip (friction + anchoring).
e. structure vertical settlement:

Compaction of core material. The core material is typically composed of a very wide graded rock material, i.e. quarry run, with very limited settlement to be expected. However, there are numerous core construction possibilities utilising different materials that are prone to settlement, see Chapter 12.

f. Deformation of the supporting toe:
   This problem is more likely to be encountered in structures with an elevated toe-berm, see Figures 9.17 and 9.18. If the toe-berm is not properly designed, it could suffer damage and deformation, which in turn will affect the armour layer.
g. Stability of CAU toe units against rocking:
   It is imperative to have the toe units properly restrained against rocking otherwise it will greatly accelerate the settlement of the whole armour layer face (Mohammad and Jensen, 2002). Therefore, the units cannot be placed directly on the seabed if it is a bedrock, the designated underlayer for the CAU is preferably used under the toe units to provide proper anchoring for the units which will greatly reduce any rocking movement. If the units are to be placed in a trench in the bedrock, which will prevent the units from sliding but not from rocking, then additional measures should be considered, e.g. filling the trench with in-situ concrete.

The above is a general root cause listing which needs to be considered by the designer. However, when the design has progressed to the model test stage and failed due to settlement, then practically nothing can be improved apart from increasing the unit size, and this will normally be coupled with an increase in the underlayer rock size, which will significantly reduce the settlement.

Chapter 11
Underlayer

11.1. Introduction

A rubble mound structure is basically composed of core material (quarry-run rock material/sand/dredged material/Geotubes, etc.) covered by armour blocks (mainly quarry rocks or CAU). The interface between the armour blocks and the core normally consists of one or multiple under/filter layers, as required by filter rules and other stability issues. Each of these layers is to be of minimum two relevant rocks' size in thickness. The basic functions of the underlayer could be summarised as follows:

- to provide a foundation layer supporting and restraining the armour blocks.
- to act as a filter layer.
- to dissipate part of the wave energy due to its porosity as higher porosity means enhanced armour stability, a reduction in wave run up and a reduced wave reflection.
- to provide temporary protection to the core material during construction.

11.2. Underlayer for Rock Armour

The function of the underlayer is clearly to act as a filter between the armour layer and the core, or between the upper and lower layers in a multilayer structure. However, the stability of the armour blocks against sliding on the underlayer is affected by the friction and mechanical anchoring/gripping. A larger size ratio between the underlayer and the armour will provide a better surface for the armour blocks to have a grip on, but it will not increase the frictional forces as the armour weight and the material type are not changing. Therefore, the recommended rock class for the first underlayer is to satisfy the filter requirements and to use the largest possible size.

For rock armour layers, the recommended mass ratio (in many design manuals and guidelines, see Figure 11.1) for the first underlayer is

$$M_{50\text{-armour}}/M_{50\text{-underlayer}} < 10\text{–}15$$

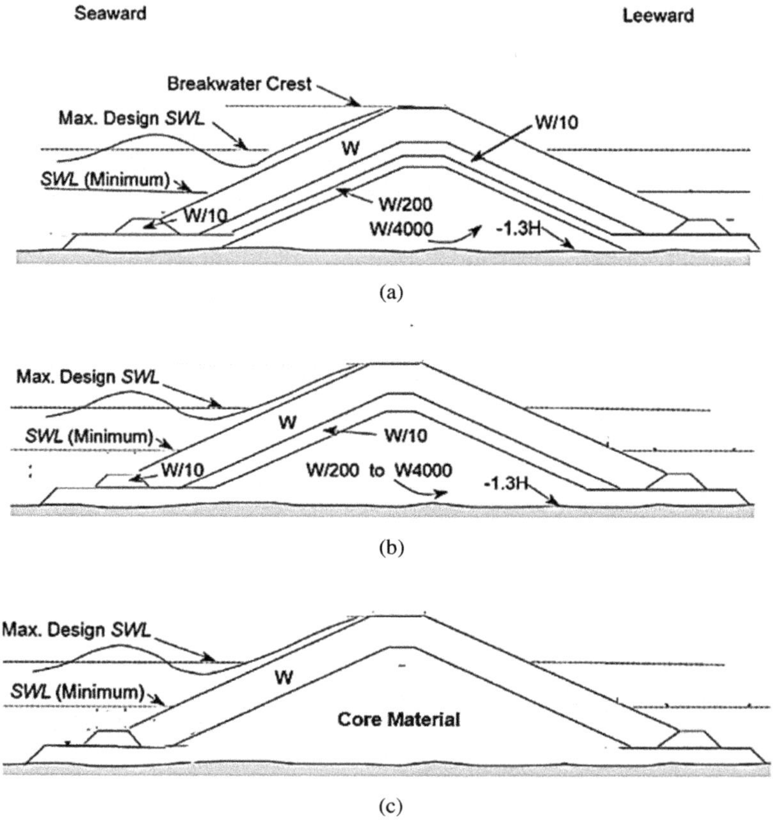

Figure 11.1. Basic configurations of multi-layered rubble mound breakwater, (a & b) rock weight ratio as recommended by "SPM 1984 and CEM 2011". (c) no underlayer is required.

which means a rock size ratio in the order of

$$D_{50\text{-armour}}/D_{50\text{-underlayer}} < 2.2\text{–}2.5$$

and this is on the larger side and very conservative than what is required for geometrically closed filter criteria, with a grading width ratio of $D_{85}/D_{15} \approx 2$, and stability requirement of $D_{15\text{-armour}}/d_{85\text{-underlayer}} < 5$, $D_{50\text{-armour}}/D_{50\text{-underlayer}} < 10$ (see Chapter 13).

After following the above-recommended sizing of the underlayer, the rock class may be too large to act as a filter layer for the core material, thus a second filter layer may well be required, see Figure 11.1(a). Jensen (1984) and others have shown that these original SPM 1984 recommendations are too strict for practical breakwater design as they frequently, and unnecessarily, lead to many under/filter layers.

It is however important to note that if the rock armour is relatively small and the quarry run rock material for the core is large enough, then no under/filter layer will be required, see Figure 11.1(c).

11.3. Underlayer for Concrete Armour Units

The need for an underlayer for the CAU, and unlike the rock armour, is mainly to provide a robust and rough foundation layer holding the CAUs in place and allowing for the placement of the units in their allocated

Table 11.1. Recommended underlayer rock sizes for Concrete Armour Units.

CAU (of wight W)	NLL (Underlayer)	NUL (Underlayer)
Cube	W/20[e]	W/10[e]
Tetrapod	W/20[a]	W/10[a]
Dolos	W/10[a]	W/5[a]
Stabits	W/10[a]	W/5[a]
Accropode	W/15[b]	W/7.5[b]
Core-loc	W/15[b]	W/7.5[b]
Xbloc	W/15[c]	W/7[c]
Cubipod	W/20[d]	W/10[d]

Note: [a]BS 6349-7, [b]CLI (concretelayer.com), [c]Xbloc (xbloc.com), [d]Cubipod Manual (Medina *et al.*, 2016), [e]usually the given ratio is used up to 1–3 t rock class, however for heavier cube category armour the same rock class of 1–3 t could also be used.

places and directions. This is of course in addition to its prime function as a filter for the underlying rock material.

The recommended underlayer rock sizes for each of the CAU types, as given in Table 11.1, have been derived from model tests and from

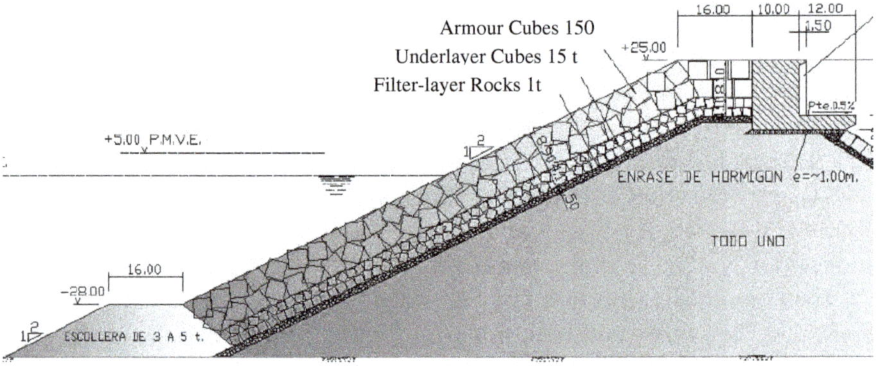

Figure 11.2. Concrete blocks for the 1st underlayer at a Coruna Port, Spain (International Breakwater Directory).

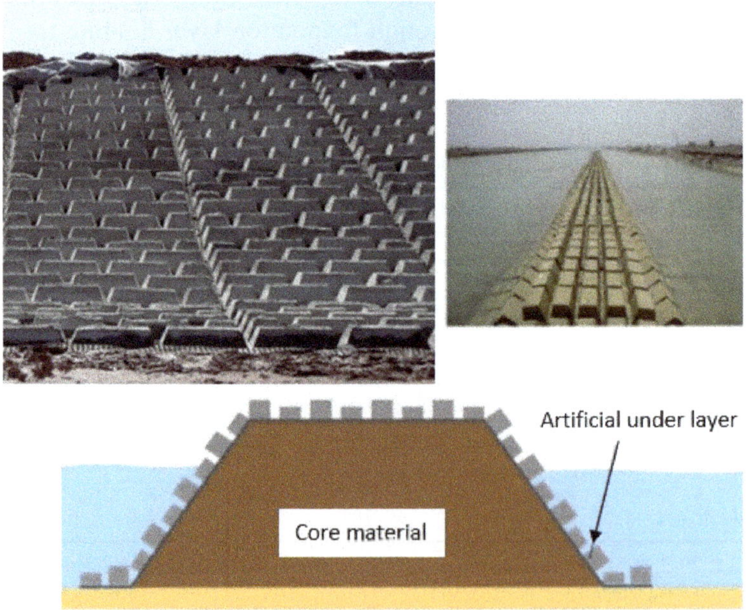

Figure 11.3. Artificial underlayer for CAU (CLAS Certification).

accumulated after-construction data. The underlayer shall consist of two layers of rock (minimum). Furthermore, and specially for the single layer CAUs, the following relationship (Hedges, 1983) needs to be satisfied:

$$D_{85\text{-underlayer}}/D_{\text{voids-armour}} > 2.0$$

Artificial underlayer could be used in places where large natural rocks are not available or in certain circumstances where the CAU is exceptionally large, which consequently requires large rock sizes. An example is shown in Figure 11.2 for concrete blocks used as an underlayer for massive 150 t concrete blocks armour layer.

Another type of an artificial underlayer is shown in Figure 11.3 designed to be used with single-layer CAUs. It is basically a concrete mattress which may or may not include a geotextile filter layer.

11.4. Underlayer Stability

During the construction stage, the underlayer shall function as the primary armour for a limited period of time, protecting the core material from external wave and current forces.

It is understandable that the construction may be required to be halted during storms, specially at rough sea season, therefore it could be difficult in some locations to find a long enough window of calm sea for the construction to progress without having a rough spell which could cause damage to the underlayer. Therefore, the sea state condition at the time of construction and the duration of underlayer exposure play an important part in the design of the underlayer. Thus, the design of the structure will need to satisfy two design conditions, the primary design event for the armour layer and the secondary design event for the underlayer.

The sequence of design normally starts from the outside layer, i.e. the armour layer, towards the inside. Thus, after deciding the type of armour and its size, the underlayer will follow and it is to be sized accordingly, this is as a first step. The second step is to check that the selected underlayer rock size is sufficient to satisfy the design requirements for the secondary design event, if not, then:

- For rock armour – use the largest recommended underlayer rock size with a slightly flatter slope, and this would normally resolve the issue.

- For CAU – use the largest recommended underlayer rock size with the flattest possible slope for the selected CAU, check if it will provide the required stability level. Otherwise, a possible alternative CAU could be investigated, which could possibly allow for larger underlayer rocks or a flatter slope.

Figure 4.13 is an example with 5.0m^3 Core-loc armour unit with 800–1700 kg underlayer (maximum recommended) and with 1:1.5 side slope (maximum permitted for Core-loc). Another good example is given by Gierlevsen (2019) where a Cubipod armour unit is used in an exposed location. The selection of this CAU is driven by two factors, the possibility of using larger underlayer rocks with a flatter slope.

11.5. Underlayer with Geotextile

The use of geotextile filters underneath the underlayer may reduce the permeability of the structure, and this means a considerable part of the wave energy will be reflected back instead of being dissipated within the voids of the rock material. Consequently, this situation will lower the stability of the armour layer. Thus, if geotextiles are used just underneath, or close to, the underlayer, more damage can be expected, unless this is accounted for in the sizing of the armour units.

The notional permeability, P (which is basically an index for the permeability of the layers underneath the armour layer) was empirically determined by Van der Meer (1988) as it constitutes an important input parameter in sizing the armour blocks in Van der Meer's stability formula. Three basic values for the P were determined for what are considered as standard situations, namely P = 0.1, 0.5 and 0.6 (see Figure 11.4). The fourth situation, i.e. P = 0.4, with a filter/underlayer and a quarry run rock core is determined by interpolation of the tested standard three situations.

Recent developments in coastal structures' construction, being driven by cost reduction, are seeing an increase in the use of dredged material/sand to form a considerable part of the core. This type of material requires to be covered by a geotextile to contain the fine material and preventing any washing out. Therefore, Kik *et al.* (2012) conducted a study to elaborate on case (b) in Figure 11.4 by limiting the depth of the quarry run core to include the sand covered by a geotextile sheet, i.e. wave reflective

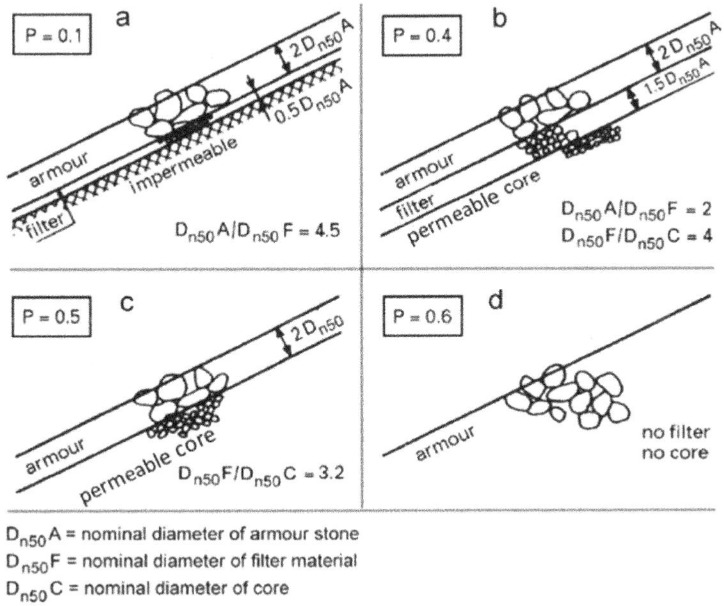

Figure 11.4. Notional permeability factor (Van der Meer, 1988).

surface. The finding from the study is shown in Figure 11.5, which indicates the need for a quarry run layer with a thickness approximately similar to that of the armour layer in order to achieve an almost similar

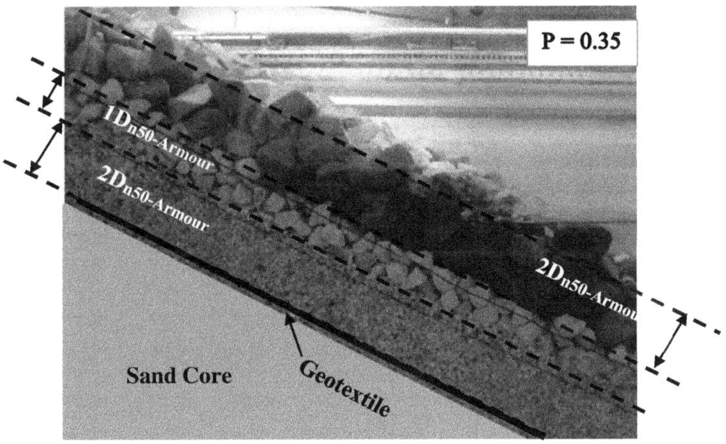

Figure 11.5. Rubble mound structure with sand core and geotextile (Kik *et al.*, 2012).

condition to that of case (b) with P value in the order of 0.35, i.e. reasonable overall porosity.

11.6. Underlayer Configuration

The underlayer on the seaside or the leeside faces of the structure could take different forms and shapes to achieve the design object of obtaining

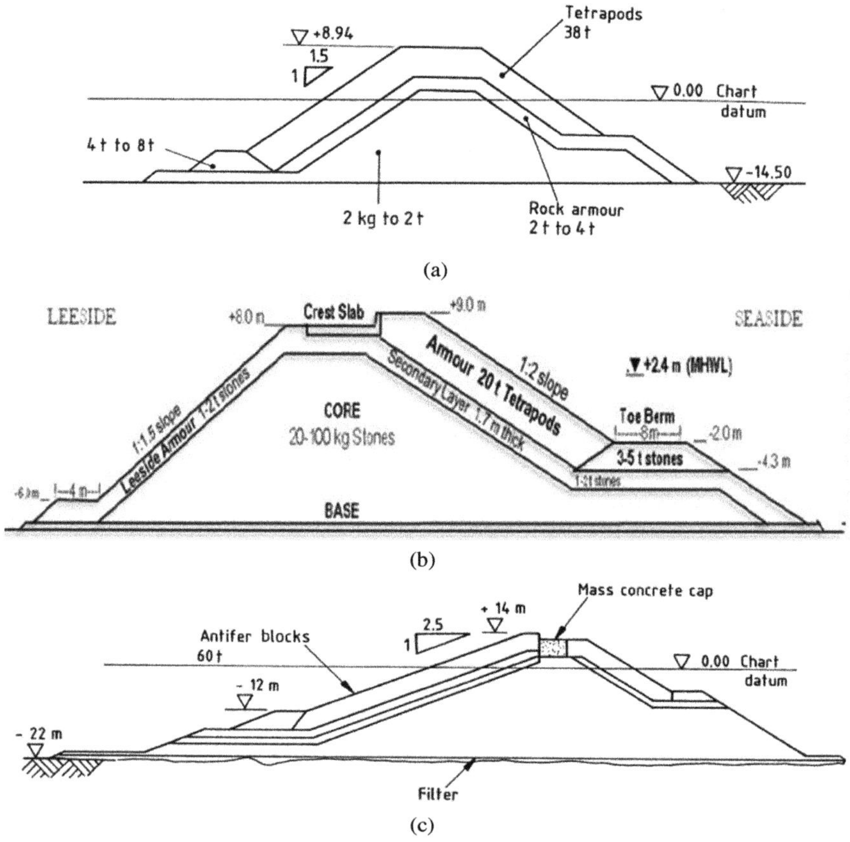

Figure 11.6. Examples of different underlayer arrangements in rubble mound breakwaters.

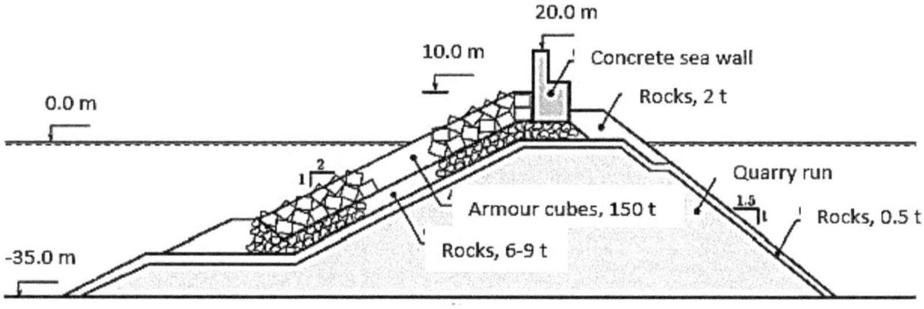

(d)

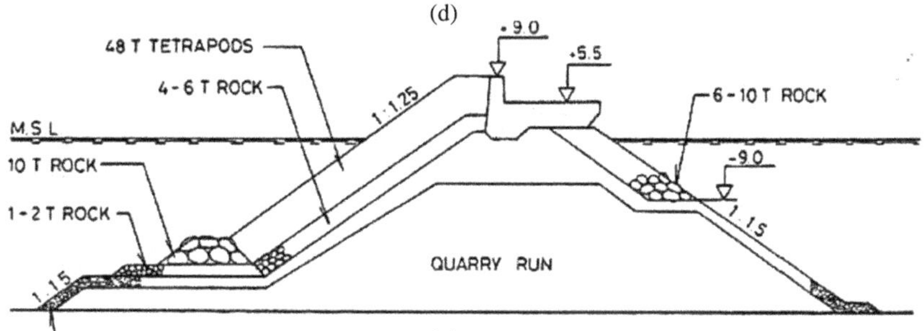

(e)

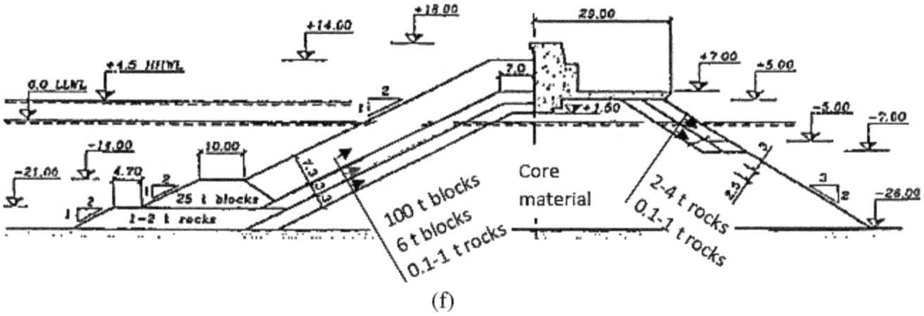

(f)

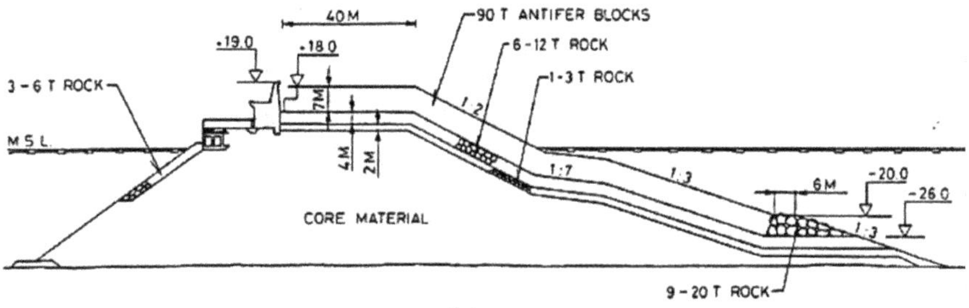

(g)

Figure 11.6. (*Continued*)

a robust, economical and preferably uncomplicated section. Different examples are shown in Figure 11.6 to demonstrate the importance of the underlayer in shaping and protecting the final section. Starting with a simple and basic single underlayer in (a and b) to double underlayers in (c, d, e, f and g).

Chapter 12

Core Material

12.1. Introduction

Core material normally constitutes the largest part of the material used in the construction of breakwaters and quite often the revetments as well. The selection and use of a certain type of material, or a combination of materials, to form the core needs to be evaluated very carefully as it will affect the integrity and performance of the structure.

The core should be structurally capable of supporting the internal and external loads. The material is also required to attain an acceptable compatibility level to avoid unacceptable post-construction settlement. The hydraulic properties of the core need to be carefully assessed as it has a large influence on the integrity and functional performance of the structure. With reference to Figure 11.4, the permeability of the core material is inversely related (for the same design conditions) to the armour block size (CAU/rock), run up and wave reflection, whereas the wave transmission through the structure will be directly related to the core permeability.

The core normally contains a certain percentage of fine material which should be prevented from being piped or washed out by hydraulic forces induced by waves, currents and water level fluctuation. Therefore, it is highly important to ensure that the core should only contain a very low percentage of fines or be protected by a filtering system (granular or/and geotextile).

Further to the above characteristics and effects of the core material, cost and construction methodologies could influence, or even dictate, the selection of the core material type.

12.2. Rock Core

Rock is the commonly used material to form the core, which will also give the basic configuration of the structure. Unlike the armour and the underlayer, there is no well-defined design requirement to be fulfilled for the core rock material. This type of material, widely known as "quarry run", generally has a very wide grading ($D_{85}/D_{15} > 2.5$). The used upper limit size is usually between 100 kg and 2,000 kg, with the common size in the range 500 kg to 1000 kg. The used lower limit size is usually between 0.1 kg and 10 kg, with the common size being1 kg. These limits are in practice considered as nominal limits; however, they still need to be written in the project rock specifications following the CIRIA class limits as presented in Section 9.3. Note that the NLL and NUL limits are to be designated as the class nominal limits, similar to that of the armour rocks, although some designers are, and incorrectly, using the ELL as a class limit.

Examples of quarry run class limits are given in Table 12.1, which could be used in specifying the rock gradation limits. For such wide graded material, it is impractical to keep the specification requirements limited to the four numbers of ELL, NLL, NUL and EUL. Therefore, it is recommended to provide an upper and lower accepted envelope for the

Table 12.1. Examples of rock core material class limits.

Rock class	Grading curve limits (kg)			
Core	ELL (<2%)*	NLL (<10%)	NUL (>70%)	EUL (>97%)
0.1–200	0.01	0.1	200	300
1–100	0.1	1	100	150
1–500	0.1	1	500	700
1–1000	0.1	1	1000	1200
10–1000	1	10	1000	1200

Note: *The recommended value for a wide graded material with NLL < 300 kg (as per the recommended limits of CIRIA, CUR, CETMEF, 2007).

material. Examples of such envelopes are shown in Figure 12.1. Note that the acceptable M_{50} range could be specified on these envelopes, as such it is shown as a broken line. A typical gradation curve is required to define the nominal M_{50}, M_{85} and M_{15} for the rock class to calculate the corresponding D_{50}, D_{85} and D_{15} which are required to verify the filter

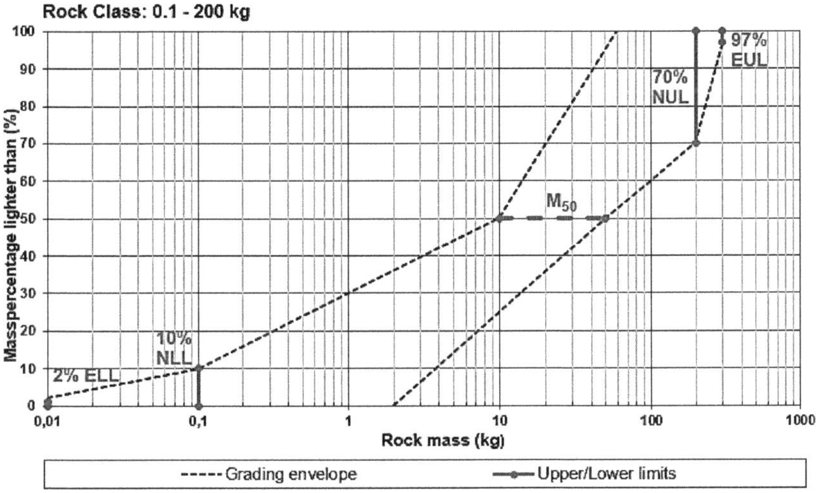

(a)

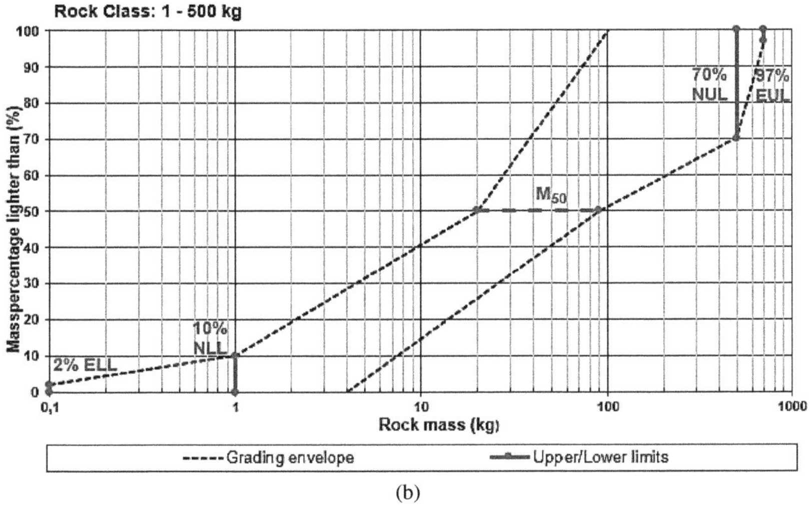

(b)

Figure 12.1. Typical mass distribution envelopes for well-graded rock core material (quarry run): (a) 0.1–200 kg quarry run; (b) 1–500 kg quarry run.

requirements. The designer is expected to select the quarry run class limits based on cost, quarry production, project rock classes, possibly filter requirements and constructability. However, other design-related issues may influence the selection of the core class limits, primarily the upper limit, as it affects the permeability of the rock material. Such issues could be the wave transmission through the structure (apart from the wave transmission due to wave overtopping) or tightness to sand drift through the structure if the structure is located in an active littoral drift location.

The amount of very small fragments of rock (<0.1 kg), fines and dust in rock core material is normally high, therefore it is recommended to have this fraction of the material not in excess of a 2% limit. This recommendation holds despite the fact that a significant part of the dust will be washed out during the material dumping, see Figure 12.5. The designer, however, would have preferred to have more fine rock fragments if the NLL limit is set at 0.1 kg, as such a requirement could probably be required to reduce the permeability of the core or to have a more compact material.

Contractors may request to include in the core other material, normally sourced at the project site. This material could be quarried at the site or nearby locations. If dredging is part of the project activities, then rock material could be part of the dredged material, this of course depends on the seabed formation.

Figure 12.2 shows two examples for structures, with quarry run (1–500 kg) being the core material in (a), and rock material of (1–25 kg) in (b) which became available as a side product from earth works at the project site.

Combining two types of quarry run is also a possibility if there is a cost saving. An example of such use is shown in Figure 12.3 where a low-quality quarry run, 0.1–500 kg (with high percentage of fines) is used in the central part of the core. The outer part of the core, with 5.0 m thickness, is constructed with a better quality 1–500 kg quarry run. Note that the 5.0 m layer thickness is on the conservative side as 3.0 m is normally sufficient, as in Figure 12.4(a), to yield a low hydraulic gradient which will not cause any washing out of fine material from the central core.

Figure 12.4 shows two examples using dredged caprock as the main core material encased by a 3.0 m layer of quarry run. The top level of the dredged material is to be kept at least 2 m below the DLWL. This is to have a reasonably thick layer of good quarry run under the crest and keep the dredged rock permanently under water (under the tidal fluctuation range, which could cause vertical migration of the fines), both of these

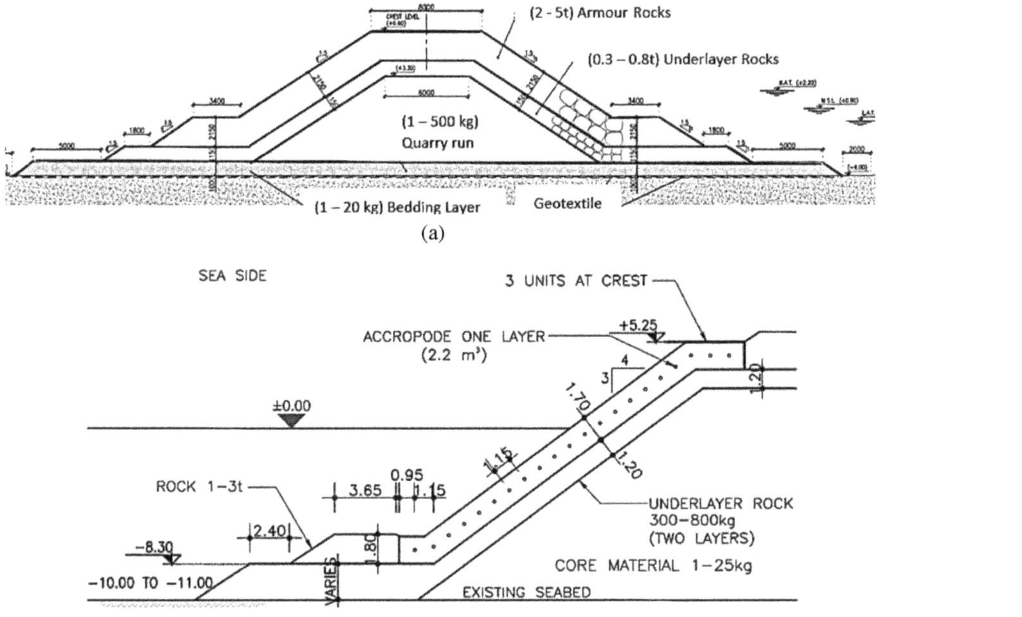

Figure 12.2. Example structures with quarry run core material: (a) Quarry run (1–500 kg) core; (b) Quarry run (1–25 kg) core.

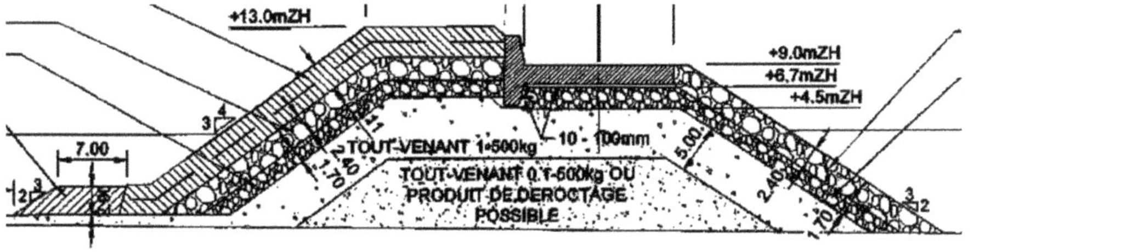

Figure 12.3. Example with two types of quarry run core material.

measures will reduce the risk of any relatively large settlement which might affect the level and integrity of the crest. The first example in Figure 12.4(a) doesn't isolate the dredged caprock with a geotextile filter layer, whereas in Figure 12.4(b) there is a geotextile fabric. To better illustrate the nature and composition of the dredged caprock material, few photos are given in Figure 12.6. The nature, composition, size and shape

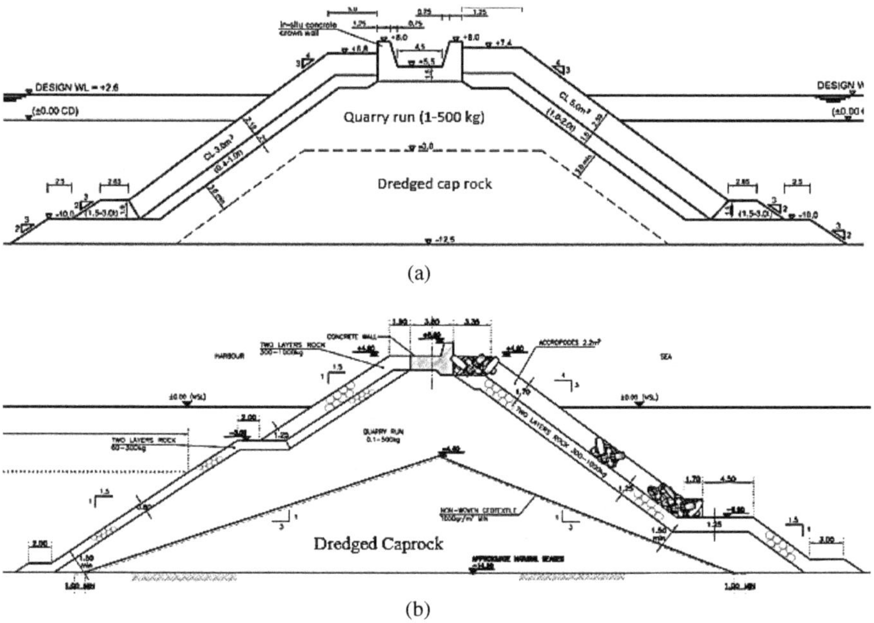

Figure 12.4. Core material with dredged caprock and quarry run: (a) Dredged caprock core encased in quarry run, no geotextile filter; (b) Dredged caprock core encased in quarry run, with geotextile filter.

of the dredged material depend on the thickness of the caprock layer, the hardness of the rock and the dredging type.

Mechanical dredging, i.e. such as by backhoe dredger (BHD) and grab dredger (GD), produces variable sizes with many large pieces, as shown in Figure 12.6(a, b). Hydraulic dredging on the other hand produces much smaller rock pieces. In relatively sheltered water, Cutter Suction Dredgers (CSD) are normally used, whereas in more open waters, i.e. offshore, the Trailing Suction Hopper Dredgers (TSHD) are to be used, however with a special draghead. A dredged material sample form TSHP is shown in Figure 12.6(c). The dredged material from CSD will be about the same with rock sizes ranging between 0.05 and 0.30 m, depending on the cutter head.

Based on the above, if the dredged rock comes from hydraulic dredging, then it would be composed entirely of smaller rocks, and this could be dumped directly in place with no precautions. However, if the dredged rock comes from mechanical dredging, then it could have some large slabs

(a)

(b)

Figure 12.5. Dust extraction during quarry run dumping: (a) Dust being washed out; (b) Dust being blown away by wind.

of rocks, these slabs could leave some large voids in the dumped material. Voids in the core absolutely cannot be tolerated as these will normally lead to large settlement which could endanger the integrity of the structure. Therefore, it is preferred to break the large slabs into smaller pieces, or if the slabs are not very large, then it could be accepted, but the dumping operation needs to be closely monitored with diving inspection to verify that no large voids are being formed.

Hydraulic dredging inevitably could produce dredged material with considerable amount of fine material, depending on the nature of the seabed. Therefore, the designer needs to carefully assess the composition of the dredged material before deciding on the need for a geotextile filter.

(a) (b) (c)

Figure 12.6. Dredged caprock: (a) Dredged caprock by Backhoe Dredger (BHD); (b) Dredged caprock by Grab Dredger (GD); (c) Dredged caprock by Trailing Suction Hopper Dredger (TSHD) (Neelissen *et al.*, 2010).

If the dredged material is mostly composed of rocks, then the fine material is expected to be contained within the voids between the rocks, thus no geotextile is required in this case, and the dredged material could be classified as "*dredged caprock*". There is no harm in allowing the fine material to be washed out as this will not affect the stability of the material, probably only minor tolerable settlement. On the other hand, if the dredged material is composed of fine material with fragments of rocks, then this material is better classified as "*dredged material*" and it needs to be protected by a geotextile filter. The example shown in Figure 12.4(b) should classify the core as "dredged material" if it really needs to be protected by a geotextile, otherwise, it could be classified as "dredged caprock" with no geotextile required.

One practical issue regarding the bulk density of the core rock material is worth mentioning as many engineers are likely to face it during the construction stage. Many construction contracts have the payment based on material volume as measured on the design drawings. Whereas the procurement of most of the rock material is carried out by weight as the quarries normally sell by the tonnage. In such contracts, the contractor will need to fix a conversion factor from tonnage to volume in order to facilitate the payment, and this factor is the bulk density (ρ_{bulk}) of the quarry run. Contractors are inclined to use either a low or high bulk densities to suit their payment terms.

The porosity (p) of the quarry run, and consequently the bulk density, will depend on many factors, mainly as follows:

- The density of the rock material (ρ_r).
- The upper/lower limits of the quarry run gradation, i.e. NUL/NLL.
- The amount of fines in the material, i.e. below the ELL limit.

The porosity (p) could be defined by Equation (12.1):

$$p = 1 - (\rho_{\text{bulk}}/\rho_r) \tag{12.1}$$

and,

$$\rho_{\text{bulk}} = (1 - p)\,\rho_r \tag{12.2}$$

The (p) value for the quarry run could be within 32–37%, or even outside this range in especial cases, however, more commonly it is found to be as 33–35%. Therefore, and especially in large projects, a measurement test is recommended to be carried out to determine the (p) value for the quarry run material. An example for field measurement is shown in Figure 12.7, where a sample is taken from the stockpile, and then all rocks smaller than 0.1 kg, as this is the ELL limit in this test, are excluded (note that this step is not always required if there is small percentage of fines, i.e. with the specifications). The clean sample is then transferred to a container with a volume, preferably, not less than 4.0 m^3. The container is to be weighed empty and full of rocks, after levelling the rocks' surface. Note that the volume of the rock material is actually slightly less than the container volume due to the container side effect. Therefore, a simple correction could be applied by pushing all the container surfaces, including the top surface level, inside by approximately 20 mm, i.e. reducing the dimensions of the container.

The bulk density of the quarry run rock sample could then be calculated as follows:

$$\rho_{\text{bulk}} = M_{\text{rock}}/V_{\text{container}} \quad \text{(with } V_{\text{container}} \text{ is the corrected volume)} \tag{12.3}$$

Making clean sample of quarry run.

Example rocks of 0.1 kg.

Filling the test container with the clean sample.

Leveling of the quarry run top plan in the test container.

Weighing of the full container with quarry run.

Measurement of the characteristic dimensions.

Figure 12.7. Bulk density field measurement of a quarry run rock sample.

12.3. Sand Core

Sand is not normally used on its own to compose the entire core of coastal structures, as demonstrated with different examples in Sections 12.4 and 12.5. This is especially true when the sand is placed as a hydraulic fill where it tends to assume a flat slope ranging between 1:5 and 1:10, normally taken as 1:7, see examples in Figures 12.9(a) and 12.10(d, e). It is however important to note that the tip of the sand wedge is not going to be as neat as shown on the drawings as the sand tends to flow many

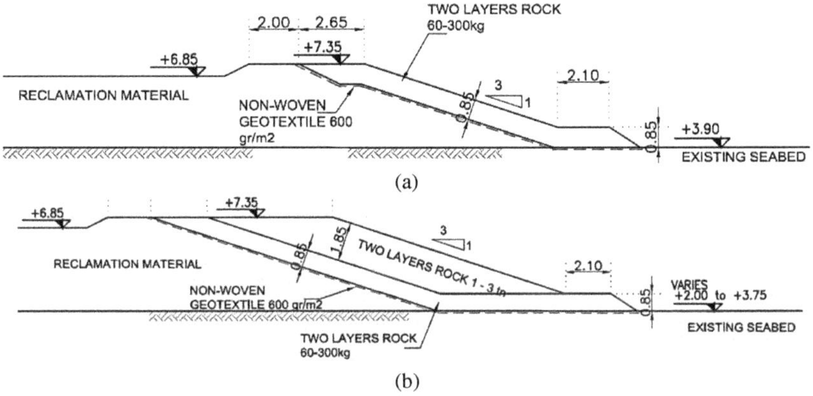

Figure 12.8. Revetment Sand core: (a) Sheltered section; (b) Exposed section.

metres beyond the theoretical tip line. Therefore, careful considerations should be taken in the design of the toe and the scour protection in this type of structures as it will be on newly placed, unconsolidated sand.

In projects where the sand could initially be placed away from the edge structure, it could be pushed at a later stage to take a steeper slope, up to 1:3, and as such it could form the entire core. This is however true in shallow and sheltered locations, as in Figure 12.8(a), or semi-sheltered, i.e. moderately exposed, locations as in Figure 12.8(b).

12.4. Composite Core in Breakwaters

Sand is often used in offshore breakwaters to form part of the core material replacing the more expensive rock material. This could take different shapes and configurations depending on the design, construction methodologies and local site factors which could influence the extent of sand use.

The main difference between rock and sand is the degree of self-compaction after placement, which constitutes an important factor in the overall integrity of the structure. The quarry run will sufficiently be compacted after placement/dumping, either wet (in water) or in the dry, whereas the sand is not, and this will make it prone to settlement and possibly liquefaction due to seismic loading. It is particularly challenging to design a geotechnically stable structure with untreated hydraulically placed sand core, as shown in Figure 12.9(a). Note that the side slope of

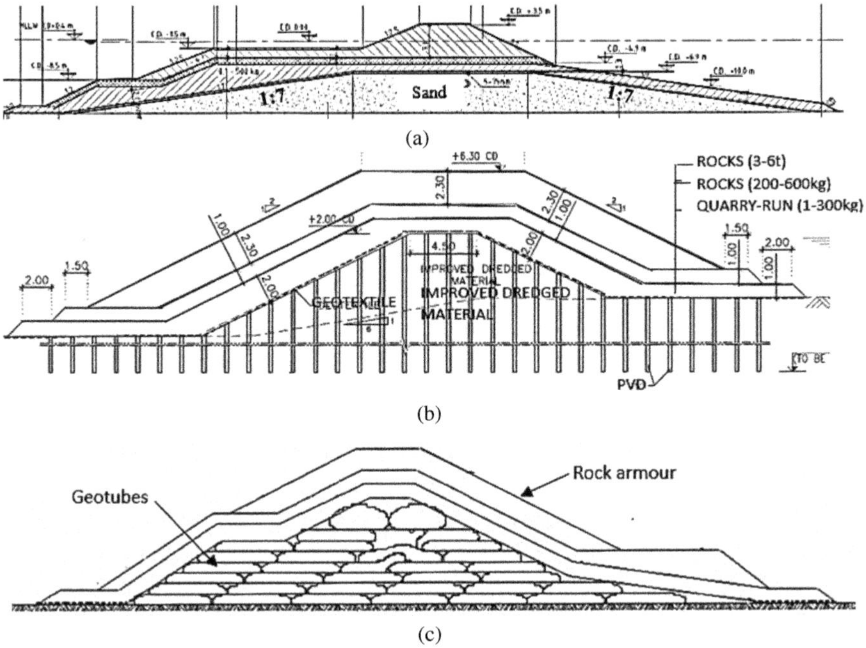

Figure 12.9. Examples of composite core in breakwaters.

the sand core is 1:7, which is the average for self-assuming stable slopes for hydraulically placed sand.

The sand core stability will be greatly improved if the sand is dumped and treated to increase its internal friction, as in the example shown in Figure 12.9(b).

In order to contain the sand material and prevent any loss through the rock fill, a geotextile, geometrically-closed granular filter or sometimes both are used.

In recent years, geotubes/geocontainers filled with dredged/sand material have been used to form the core, or part of it (Pilarczyk, 1996). By dividing the fill material into smaller isolated compartments, the need for soil improvement will be negated, and this will also allow to stack the geotubes to form steeper slopes as shown in Figure 12.9(c).

The geotextile, being a reflective layer, must have sufficient cover of rock material, see Section 11.5. However, the required cover thickness may differ depending on the location relative to the wave action zone.

Rock core material could also be replaced by rock material sourced by dredging a weak rock seabed (caprock). This material will contain rock fragments and possibly some sand, however, there are no gradation limits that can be specified for this type of material. Normally, no geotextile is used to wrap this material because the rock fragments often have sharp edges, which will tear the geotextile, and also the sand percentage is low. However, if the sand content in the material is high, then a heavy geotextile will be required.

12.5. Composite Core in Revetments

The core in revetments is more likely to be of a composite nature due basically to the fact that the revetment is an edge protection layer to an existing, or newly reclaimed, land. Figure 12.10 shows different examples of possible revetment core configurations ranging from simple to complex, depending on many factors as will briefly be explained in the following.

Figure 12.10(a) is an example of a simple revetment in a relatively shallow and sheltered location, i.e. small wave action, with the entire core made of fill material. The quarry run is partly used as an underlayer and partly part of the core in adjusting the flat slope of the fill material to be steeper.

Figure 12.10(b) shows the revetment section in a slightly more exposed location, i.e. heavier armour with underlayer, with the core being of fill material. No quarry run core material is used in this example as the underlayer is placed directly on the geotextile with a varying thickness to adjust the flat slope to a slightly steeper one for the armour layer.

Where the fill material is not of pure sand material and it can assume steeper side slopes, especially in sheltered locations, as in Figure 12.10(c), the quarry run core will be a uniform layer rather than a wedge. The minimum thickness of this layer is as recommended in Section 11.5.

Figures 12.10(d) (18 m crest to toe) and 12.9(e) (25 m crest to toe) show two examples with the reclamation fill in deep waters and in exposed locations. The sand fill in such cases (being hydraulic fill) will assume a flat side slope in the order of 1:7, accordingly the revetment section will be more complex than in the previous examples. Furthermore, maintaining a low crest level, as a design criterion for this project, dictated having a long berm to minimise the wave overtopping, which also helped in enhancing the stability of the sand wedge (which cannot be improved

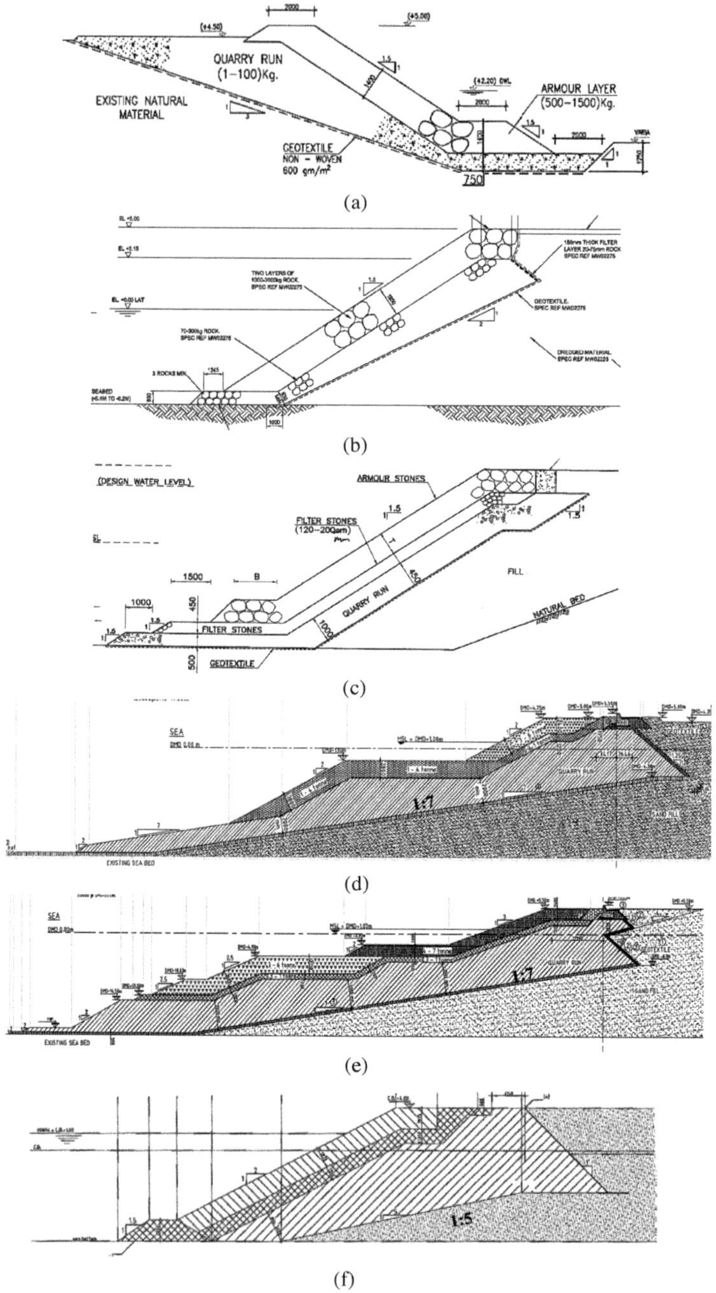

Figure 12.10. Examples of different composite core configurations in revetments.

by vibro-compaction). Where no such long berm is required, as in Figure 12.10(f), the sand wedge needs to be limited in depth under the revetment.

In all cases where fill material forms part of the core, there comes a problem to be addressed, and that is the stability of the sand wedge which supports the revetment. All reclamation sand needs some form of soil improvement to safeguard against settlement, insufficient shear resistance strength and the risk of liquefaction. Therefore, a meticulous geotechnical assessment of the stability of the revetment should be performed to establish the overall section configuration to facilitate soil improvement where required. Input from construction team regarding construction methodology, sequence of construction, equipment and capabilities all should be considered in the design process.

12.6. Bunds

The sand fill or the reclamation sand might be required to be contained to avoid having a long sand wedge underneath the rock revetment. Such a requirement could be due to design and stability issues, construction methodology and/or time, cost or to fulfil some other design criteria. Rock bunds are to be constructed first, in full or in sequential steps as illustrated in Figure 12.11, to contain the fill material.

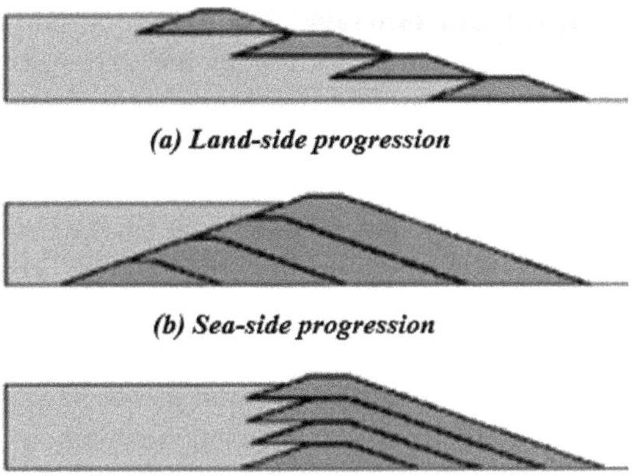

Figure 12.11. Types of sequential bund construction.

The seaside slope of the bund(s) will take the same slope of the armour layer, whereas the reclamation side will be made as steep as possible to optimise the rock material. In practice, slopes steeper than 1:1.2 or 1:1.15 could hardly be possible, however construction teams will frequently push for slopes as steep as 1:1 to be specified on the design drawings, with a note saying, "or flatter".

The simplest and more traditional form of bunds is the periphery single bund, as shown in Figure 12.12(a). This type of bund is usually constructed by land-based operation, which is cheaper than the marine born equipment, and this will partly compensate for the cost of the increased quantities of rock material.

The rock quantity in the single bund could be reduced by splitting it into two bunds, as shown in Figure 12.12(b), and this will require double the construction stages than the single bund. The principle of multiple-bunds could be used with as many bunds as practically possible, as shown in Figure 12.12(c), and only after detailed discussions with the construction team. However, the stability and settlement issues of the small sand wedges under the rock bunds need to be carefully examined. Geotubes, specially in relatively shallow areas, have been used to replace the rock material in containing the reclamation fill (Fowler, 2002), as shown in Figure 12.12(d).

Causeways are another type of coastal structures where bunds are widely used. This is because the causeway is sufficiently wide to justify using sand fill material contained between two bunds, or one bund if the other side is within sheltered waters. Different causeway arrangements are shown in Figure 12.13. The rock bund could be used on both sides, as in (a), with armour protection when both sides are exposed to wave action. If the top level of the causeway is much higher than the water level, then the bunds could be made low to support the main armour whereas the upper slopes require only light rock protection. One rock bund is used in (c) towards the most exposed side, whereas the smaller armour rocks and the underlayer are placed directly on the fill material. This is possible if the fill material could be profiled to such a steep slope, though it is generally not recommended to place rock armour on sand fill. One bund is also used in (d), however this is a truncated bund as the sand fill is placed in two parts. The triangular part is dumped first as a hydraulic fill, which means the sand is expected to assume an average slope of 1:7

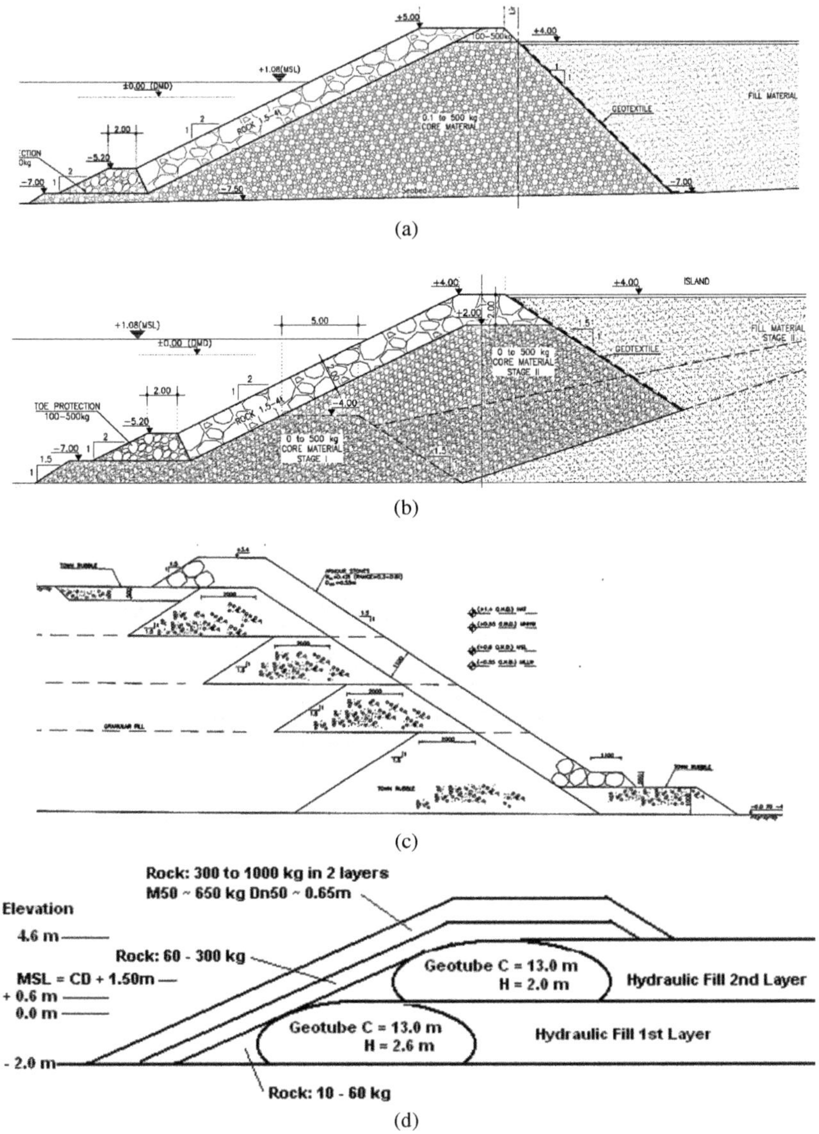

Figure 12.12. Examples of bunds in revetments: (a) Full rock bund; (b) Truncated rock bund; (c) Stepped small rock bunds; (d) Geotube bund (Fowler *et al.*, 2002).

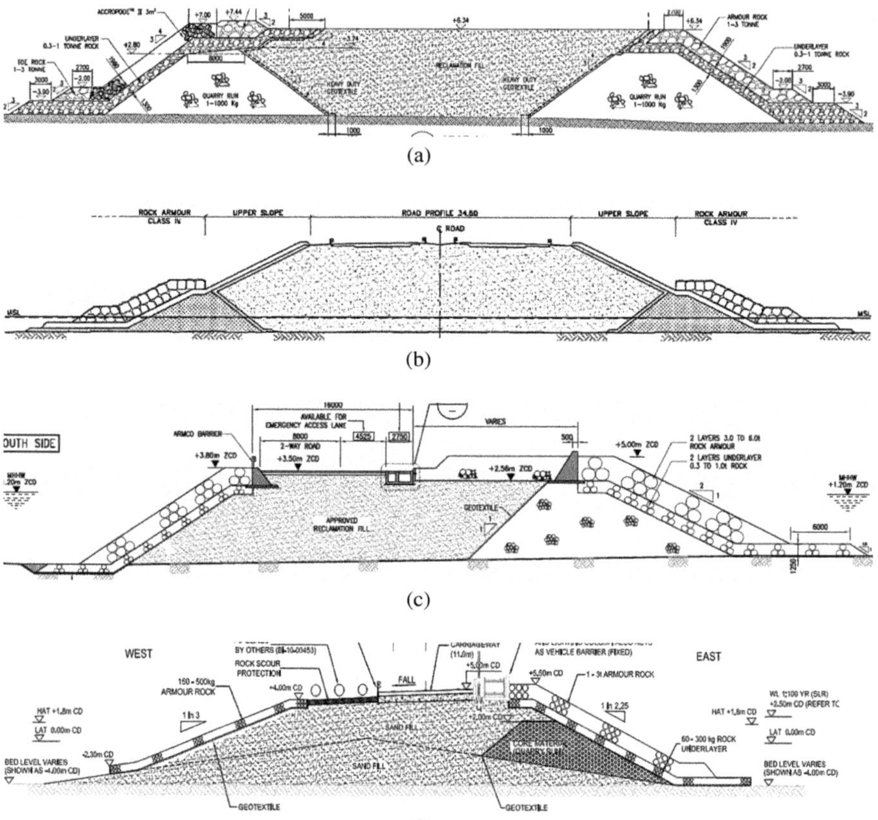

Figure 12.13. Examples of Rock bunds used in Causeways: (a) Two full bunds; (b) Two low bunds; (c) One full bund; (d) One truncated bund (Close *et al.,* 2009).

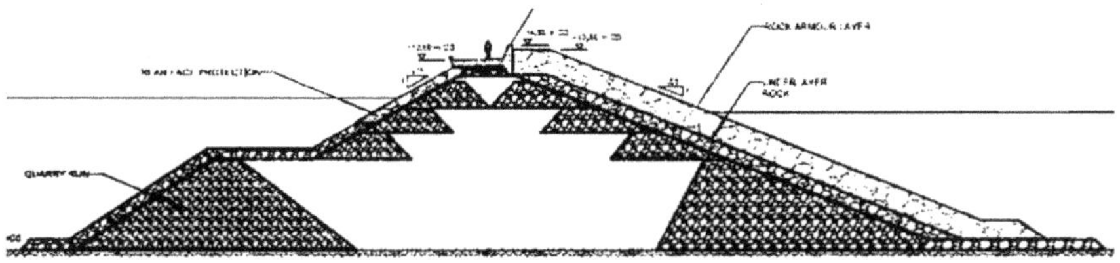

(a)

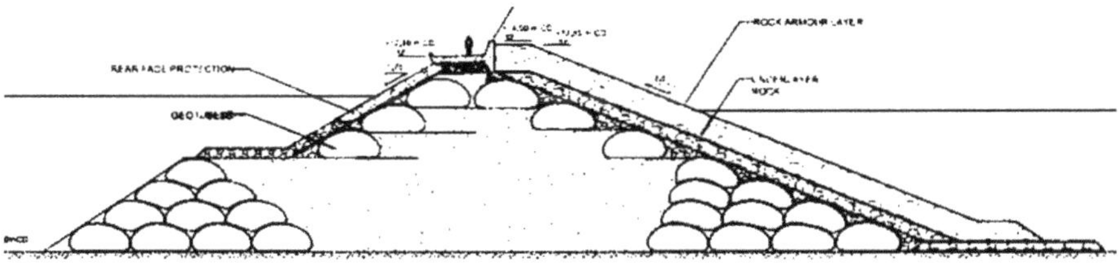

(b)

Figure 12.14. Examples of Bunds in Breakwaters: (a) Combined rock bunds; (b) Combined geobag bunds.

(which could range between 5 to 10). The upper part of the sand fill will be placed after having the truncated rock bund in place. The rear slope is profiled to 1:3 and protected by light rock armour.

Bunds are also utilised in breakwaters if the core could be made of fill material. Two examples are shown in Figure 12.14 with rock material and geotubes forming the bunds.

Chapter 13

Granular and Geotextile Filters

13.1. Introduction

Generally, two different types of filters are used in coastal and marine structures, namely granular filters and geotextiles. These two types of filters could of course be used each on its own or together by either forming a composite or a combined layer. The selection of the filter type will be dictated by design criteria, type of used construction material, construction methodology, configuration and location of the section, with possibly other particular requirements specific to the project.

Filter Layers are defined as layers that protect the underlaying base material from erosion by waves and currents without excessive build-up of pore pressure in the underlying material. This means two performance criteria are to be met in designing the filter layer: permeability and retention.

13.2. Granular Filter

The design criteria of the granular filters, which are composed of gravel or crushed stones, is well presented in *CEM* (2011), vol. VI, Ch. 5, which is basically intended to satisfy the requirement for:

- Retention
- Permeability
- Internal stability

Since the filter design criteria were first developed by Terzaghi for unidirectional flow, it is recommended to adopt a conservative approach for the cyclic loading caused by wave action (BS6349-7, 1991). That is because the reversal flow within the filter layer will cause some disturbance of the finer material and possible migration through the overlaying material. The filter material can finally reach a stable condition after settlement of the layer.

Filters are therefore expected to prevent the erosion of the fines but allow for the transport of water. This type of filter is termed as "geometrically closed" filter, see Figure 13.1, which is in practice more difficult to produce by quarries. Furthermore, depending on the particle grading distribution of the fill and the rock material, more than one filter layer could be required to meet the geometrically closed filter requirements, see Figure 13.2.

In light to moderate hydraulic loading situations, a less tight filter criteria, i.e. "geometrically opened", could be tolerated and used as in

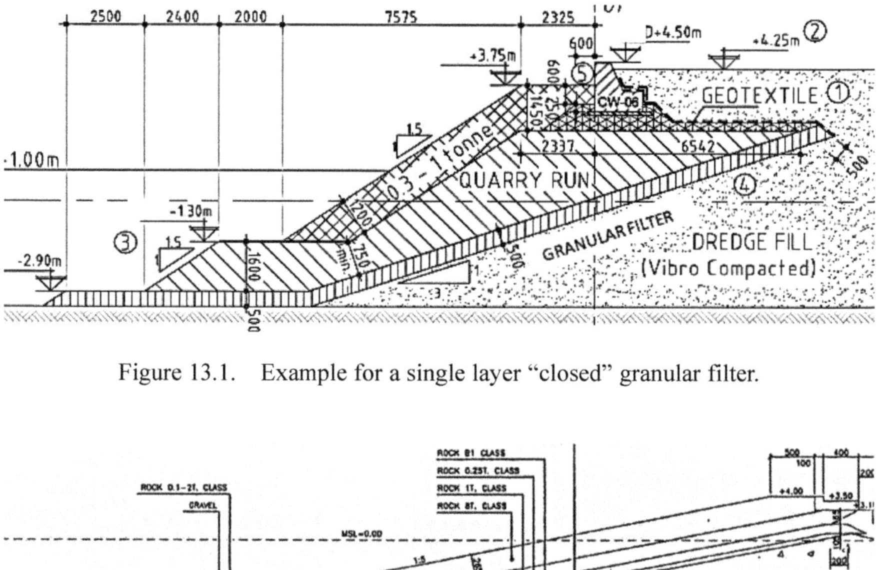

Figure 13.1. Example for a single layer "closed" granular filter.

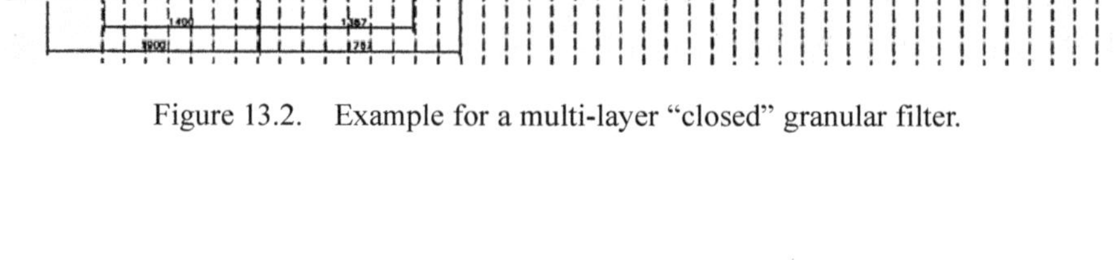

Figure 13.2. Example for a multi-layer "closed" granular filter.

Figure 13.3. A geotextile has been used with the granular filter on the side slope of the quarry run bund where the filter is required to be a "closed filter", whereas only the granular open filter is used at the base of the quarry run bund. A second example for a "geometrically open" filter is shown in Figure 13.4 where the upper part of the rear slope of the quarry run bund is made tight by using granular filter and a geotextile, whereas the open granular filter has been used on its own at the lower part of the slope. The upper part of the rear slope of the quarry run, being exposed to a relatively large hydraulic gradient, a granular filter is used as a bedding layer for a geotextile filter to ensure the tightness of the filter. However, at the lower part of the slope only an open granular filter is used as it is a zone well below the tidal fluctuation levels with low hydraulic gradient.

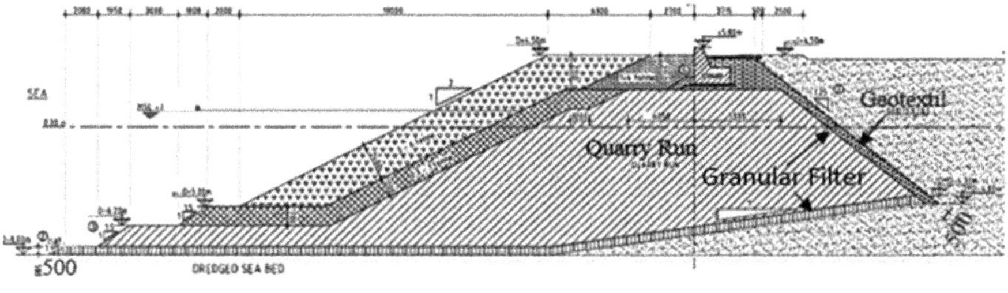

Figure 13.3. Revetment cross-section with a deep "open" granular filter.

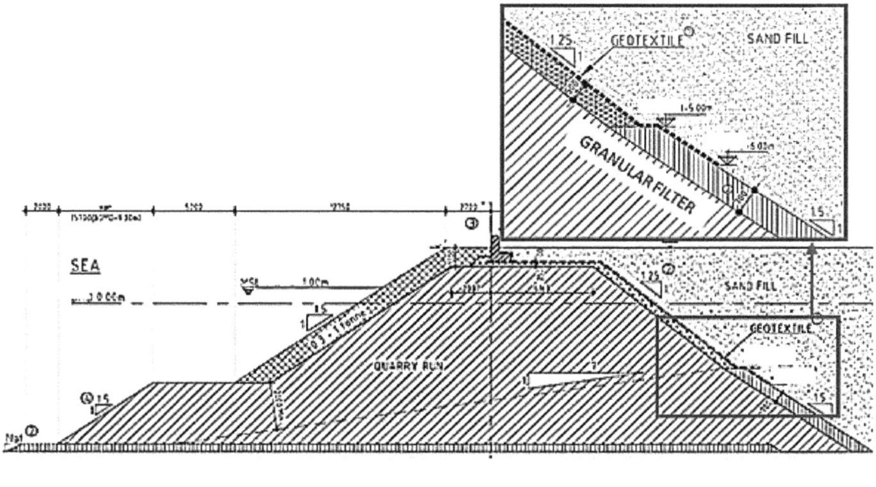

Figure 13.4. Revetment cross-section with a partially "open" granular filter.

The open filter comes with a slightly larger particle size, which cannot stop the fine particles of the base material from passing through, even with an increased layer thickness as in Figure 13.4. Note that a slope of 1:7 is shown in the section, starting at the top level of the open filter, to indicate "approximately" the expected natural slope of fine material in case of seepage, which needs to be entirely within the structure.

The deciding factor in choosing between the two types of filters is the hydraulic loading at the interface, expressed by the hydraulic gradient (i), which must be too low to initiate erosion of the fine particles from the base material. For more on the filter type issue, reference is to be made to Section 5.4.3.6 of CIRIA-CUR-CETMEF (2007).

The minimum theoretically required layer thickness for the granular filter, i.e. to ensure the layer function as a filter, is $2D_{85}$. However, practically the minimum layer thickness is recommended to be (CEM, 2011):

- 200 mm for controlled above-water construction.
- 300 mm for shallow depth underwater sections.
- 500 mm for deep depth underwater sections (to account for difficulties in placement and thickness control).

13.3. Geotextile

Geotextile systems (geotextiles – woven/non-woven, geocomposites (geotextile + geonet/geogrid), geomattresses, geobags, geotubes, geocurtains and screens) became an integrated part of almost all designs in coastal, shoreline and marine structures (Pilarczyk, 2000). However, the use of geotextiles primarily as a filter or a separation layer will be discussed in this and the following sections.

Geotextiles are made primarily from Polypropylene (PP) or Polyester (PET). The fabrics are as either Woven (mono/multifilament or slit-film mono/multifilament) or Non-woven (continuous filament needle-punched/head-bonded). Woven geotextiles are generally preferred for applications where high strength properties with low elongation are needed, where filtration requirement comes second. However, for soil with high silt content, the non-woven geotextiles may experience clogging or blinding, then using a woven monofilament geotextile could be considered as a better option. In situations where the filtration is to be ensured, then Non-woven needle punched geotextiles are the preferred type. This is especially important where the geotextile is in contact with sand to silty-sand materials, hence this type is normally specified by most of the designers for coastal structures.

Many factors must be considered in designing the geotextile, including: the section configuration, type of rock, fill and/or seabed material, construction methodology, water levels, construction equipment loading, soil improvement loading, etc. (Koener, 2005). The outcome from the design shall be a comprehensive specification for the specific use of the geotextile, which should cover the following relevant properties:

- Physical properties (specific weight, mass/unit area, thickness).
- Mechanical properties (tensile strength, burst strength, tear strength, puncture strength, etc.).
- Hydraulic properties (porosity, apparent opening size, permittivity, transmissivity, etc.).
- Endurance properties (creep response, stress relaxation, abrasion, hydraulic conductivity ratio, etc.).
- Degradation and weathering properties (sunlight exposure, chemical and biological pollution, etc.).

This point is actually very important to consider not only in the technical specification but also during the detailed design stage. The location of the structure must be well studied for possible presence of chemicals which might affect the geotextile fabric, for example ports and industrial sites. In case of any such risk then, it is better either to select a geotextile material which is more resistant to such chemicals, or to use granular filters. The other issue is the weathering of the geotextile, which will primarily depend on sunlight exposure with some other contributing factors (CUR 243). The allowable exposure time during construction will largely depend on the construction site geographical location. If the geotextile has retained >80% of its strength, after being tested according to EN 12224, the maximum exposure duration (uncovered during installation) is one month (in Europe). However, and as mentioned before, other factors could contribute in accelerating the geotextile weathering process, like the temperature, humidity, possible pollutants in the fill material, etc. It is therefore highly recommended to specify a maximum of two weeks in mild weather locations (like Europe and North America), and not more than one week, or even 3–4 days, for locations with high UV radiation levels (like Africa, Middle East, India, China, Australia and South America).

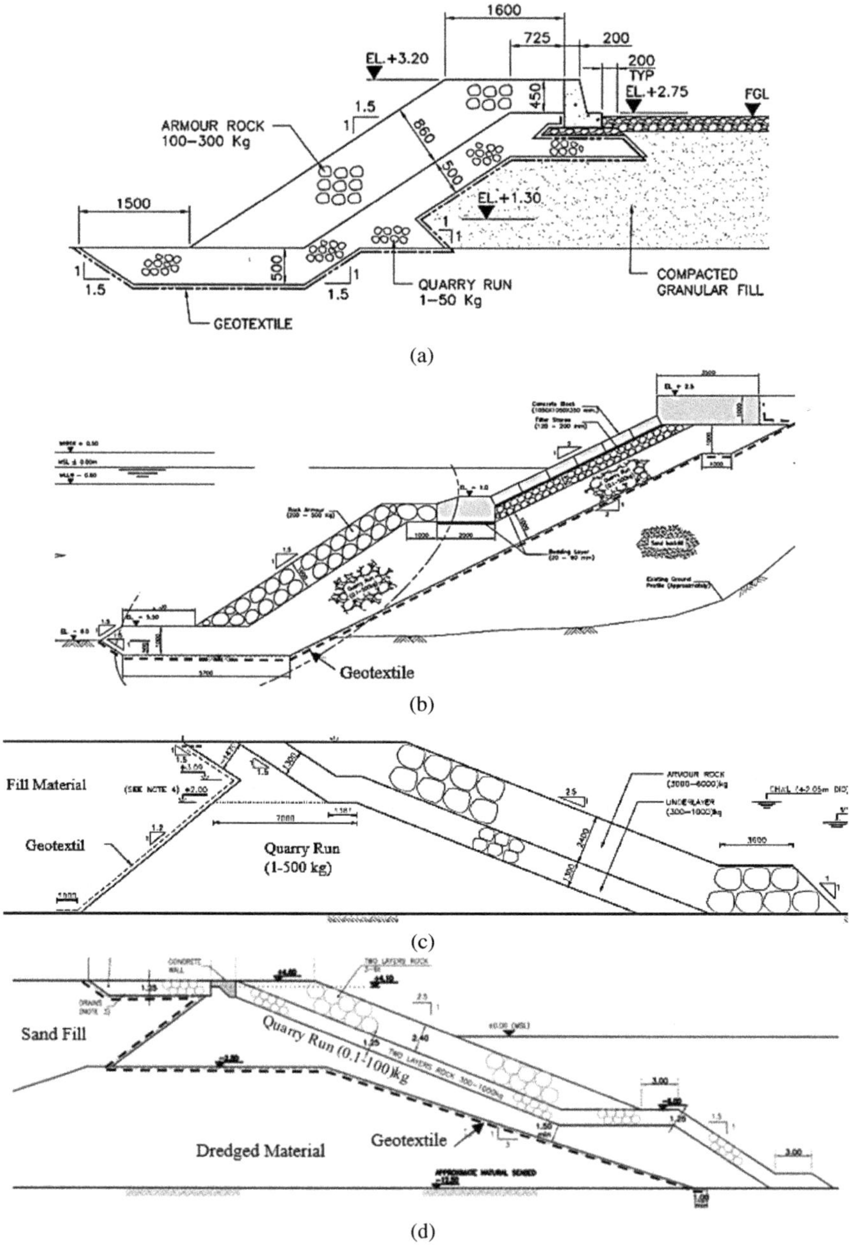

Figure 13.5. Examples of geotextile use in revetments of different configurations: ((a, b) cover type, (c) bund type, (d) bund-cover type).

Note that most geotextile manufacturers are recommending a maximum exposure time of two weeks.

A technical specification template, as given in Section 3.16.5 of CIRIA, CUR, CETMEF (2007) could be used as basis for the project specification.

Geotextiles are supplied in rolls, and these will be placed in stripes. Adjacent stripes may be joined by sewing or overlapping, the latter is the preferred option for most contractors. The minimum overlap should be 0.5 m or 1.0 m for above- and under-water installations, respectively. It is important to highlight the importance of having a clean surface of the lower geotextile sheet at the overlap area, especially from any rock fragments, before placing the upper sheet. The geotextile shall be securely fixed during the placement operation, preferably by sandbags or other non-damaging means until covered by fill/rock material. Few illustrative examples are given in Figure 13.5, where it is shown how the geotextile encloses and separates the rock material and crest wall, if there is one, from the fill material and the seabed (in case it is a sandy bed). Examples (a and b) are both with the rock material on top of the fill, i.e. cover-type revetments (see Figure 4.5). Example (c) is a typical example for a bund-type revetment with the fill material on top of the rear slope of the rock bund. In this case it is preferred to have a trimmed surface of the rock slope before placing the geotextile to remove any protruding sharp rock edges. Example (d) is for a mixed type of revetment, i.e. bund-cover type. In this example the rock material is on top of a dredged rock (mixed with fine material) at the lower part of the section and the sand fill is on top of the rock material in the upper part of the section.

13.4. Composite Filter Layer

The geotextile is assumed to remain more permeable than the fill material, with the pore sizes small enough to prevent the migration of the larger particles of the fill material. However, it must be noted that geotextiles cannot always replace the granular filter completely. A granular filter layer can often be needed to reduce the internal hydraulic gradient at the interface (Pilarczyk, 2000(b)), so that the geotextile performs as required.

The combined use of both filters, i.e. granular and geotextile, may be needed due to reasons other than the one mentioned above. In Figure 13.6(a), it can be seen that if the trimming of the rock bund cannot

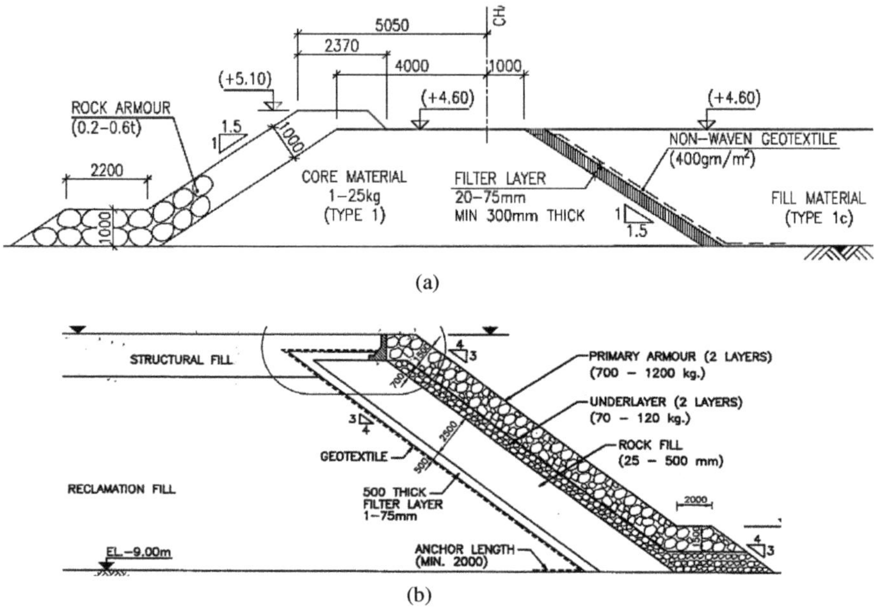

Figure 13.6. Composite filter layer (granular filter + geotextile): (a) Fill material at top of filter layer; (b) Rock fill at top of filter layer.

be achieved, or the rock material has many pieces with sharp edges, then a blinding layer of gravel is required to separate the geotextile and the rock bund.

Another example is shown in Figure 13.6(b), where the geotextile is placed on the fill material but needs a protection layer, which is a granular filter/bedding layer. This protection layer is to withstand the impact of the larger stones when they are placed/dumped on the filter layer during construction. The designer of this example was probably too cautious as the rock sizes are relatively small and one underlayer would have been sufficient instead of the three used.

13.5. Combined Filter

In more complicated sections than just a straight slope, the designer may choose only a granular filter in one part, only a geotextile in another part, or a combination of the two, i.e. composite filter. This is well illustrated in Figure 13.7. The selected option in these examples, which is also the recommended practice, is a composite filter where the fill material is on

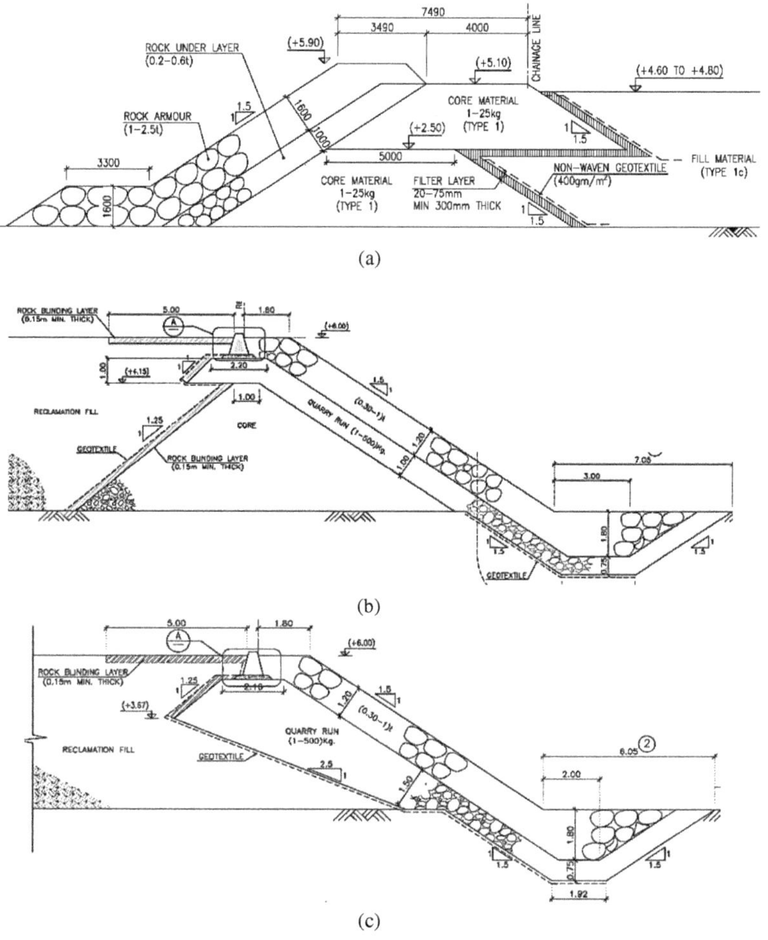

Figure 13.7. Examples of combined filter arrangements.

top of the interface. Whereas either granular filter (as in Figure 13.7(a)) or a geotextile (as in Figure 13.7(b, c)) is used where the rock material is on top of the interface. There is no preference between these two alternatives, i.e. geotextile or granular filter, but cost and construction methodology will be the deciding factor in selecting the filter type.

13.6. Vibrocompaction Near Geotextiles

In reclamation projects, there is almost always a need for some sort of compaction work to stabilise the fill material against the risk of

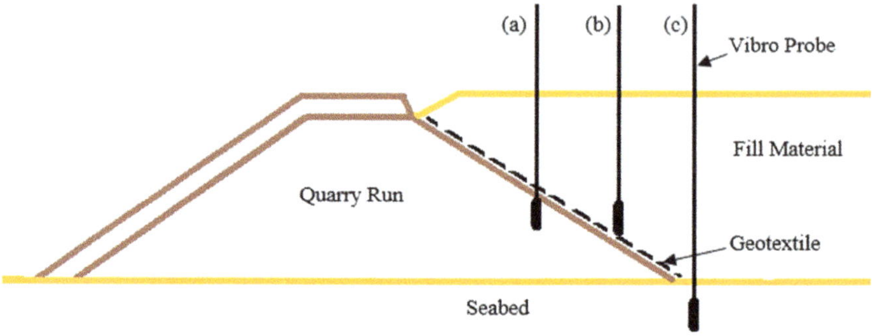

Figure 13.8. Geotextile possible damage scenarios due to vibro compaction.

liquefaction. Usually, the reclamation depth in most projects is large enough to warrant the use of Vibrocompaction.

Vibro and surface compaction near rock structures will be performed basically in the same way as for the regular inland areas. The presence and geometry of the rock structure introduce some risks that need to be accounted for in the compaction process, as schematically illustrated in Figure 13.8, and summarised as follows:

(a) A direct contact between the vibro unit head and the rock structure will frequently lead to rupture of the geotextile fabric.
(b) Excessive displacement of the rock material and/or other components of the revetment due to compaction/densification of the rock material (rearrangement of the quarry run/rocks) by vibrations due to the close proximity of the vibro probe.
(c) Excessive (vertical) displacement of the rock material and/or other components of the revetment due to compaction/densification of the natural seabed under the level of the revetment toe. Note that this risk may be very small or does not exist if compaction to the seabed is not required.

The vibration waves produced by a vibrator probe are primarily shear waves that propagate horizontally. Therefore, the vibro probe lateral distance from the geotextile is the relevant critical distance and not the vertical distance.

In order to define a safe lateral distance from the vibro probe, the question arises, what degree of influence from the vibro probe would

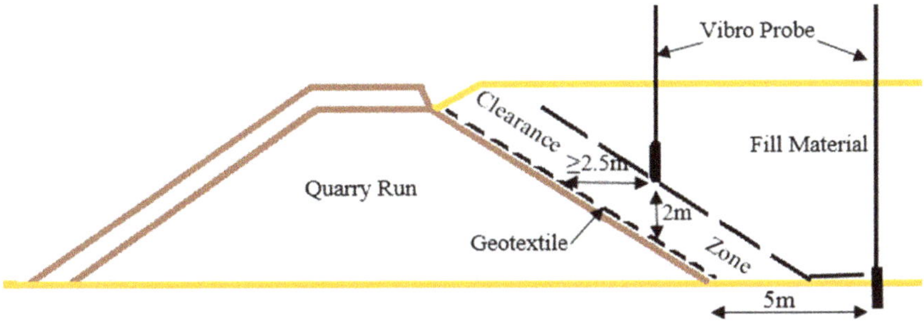

Figure 13.9. Vibro probe clearance zone from geotextile.

harm the geotextile? A very conservative assumption would be that the minimum lateral distance from the tip of the vibro probe to the geotextile is equal to the distance in which no influence from the vibro probe is measurable. This distance is approximately 5.0 m for a typical dredged sand while using a relatively large probe (high energy).

By considering that the geotextile has some resistance against vibrations due to its large elasticity, practical clearance zone for the vibro probe could be outlined as shown in Figure 13.9. However, if the seabed requires some improvement, then a 5.0 m will be applied so as not to induce any additional settlement to the structure.

13.7. Bedding Layer

A granular filter/bedding layer, a geotextile or both are used in many structures to protect the seabed against scouring and to enhance the stability of the structure, as illustrated in the examples given in this section.

A granular filter/bedding layer, as shown in Figure 13.10, is recommended to be used under the main structure if the seabed is composed of

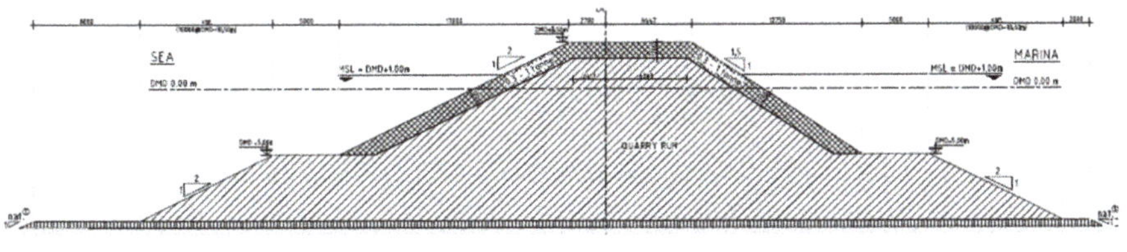

Figure 13.10. Granular filter/bedding layer.

a thick layer of cohesionless soil, especially sand (CEM, 2011). The filter/bedding layer is to prevent differential wave pressures, currents and groundwater flow, which could induce an unstable foundation condition due to washout of the fine particles. As large pressure gradients normally exist at the seaward part of the structure, it is therefore important to ensure placing a geometrically closed filter layer/layers under toe/berm of the seaward side of the structure. Bedding layers are practically not required where the depth is greater than three times the design wave height as the bottom current velocities are expected to be below the incipient motion level for the bed material. However, placing a filter/bedding layer will also help to distribute the structure's weight over the seabed to provide a more uniform settlement.

A non-woven type of a geotextile is often placed under the toe in a sandy seabed instead of the granular filter to prevent washing out of bed particles, which could undermine the toe and destabilise the structure. The geotextile is preferably extended slightly inside the main structure, as shown in Figure 13.11.

A woven type of geotextile could also be used not to serve as a filter layer but as reinforcement layer beneath the full width of the section, as shown in Figure 13.12. When the seabed soil is a low draining type,

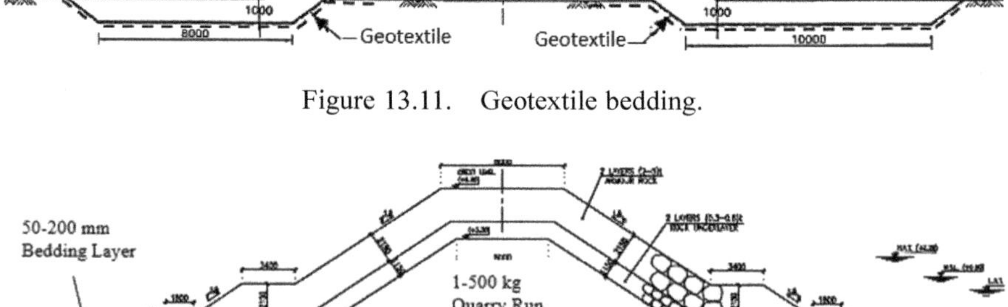

Figure 13.11. Geotextile bedding.

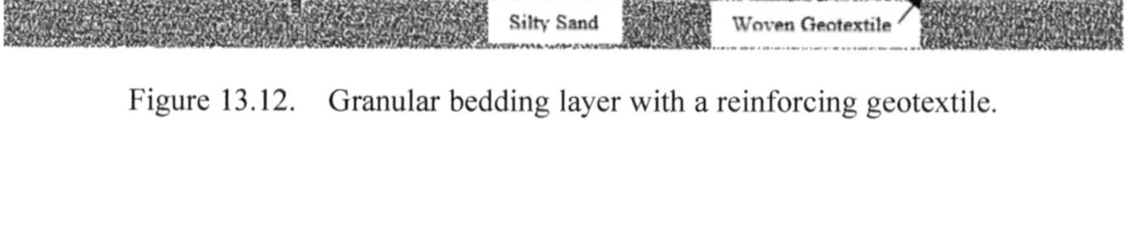

Figure 13.12. Granular bedding layer with a reinforcing geotextile.

i.e. with high silt/clay content, then the use of such a high tensile strength geotextile layer is highly recommended to reduce and control the settlement, and to stabilise the whole structure.

Quite often the bedding layer is constructed as an open type of granular filter, that being easier to produce, easier to construct and consequently less costly. The effectiveness of such a bedding layer will depend on the hydraulic loading, the relative grain diameter sizes between the bedding/filter layer and the seabed base material, i.e. (D_f/D_b), the thickness and the grading characteristics of the bedding layer.

The granular filter/bedding layer will also serve as an apron, or a scour protection layer, extending few metres in front of the toe, see Section 14.6.

13.8. Selection of the Filter Type

In some situations, the designer might prefer to use granular filters over geotextiles, or vice versa (PIANC, 1992). However, such a selection is better made after discussions with the construction team as there might be some consequences overlooked by the designer.

13.8.1. *Advantages and disadvantages of the granular filters*

There are some general advantages and disadvantages of using granular filters, which might influence the designer's selection in favour of this type of filter over geotextiles, as follows:

13.8.1.1. *Advantages*

- The filter materials (gravel or crushed stones) are usually very durable.
- High resistance, or not affected by exposure to chemical or biological agents or to ultra-violet light.
- Granular filter provides a good contact interface between the filter and adjacent layers, which is very important for long sloping structures.
- The granular filter is a porous medium, which helps damping the wave energy and enhancing outer layers' stability.
- The loose nature of the filter material allows the layer to better withstand impacts when larger stones are placed/dropped/dumped on the filter layer during construction, or due to settlement afterwards.

- Granular filter layers are relatively easy to repair, and in many cases may be self-healing.
- Granular filter is the preferred option when a filter layer is required to cover dredged rock material, which normally comes with many sharp edges, see Figure 13.13.
- If piles are to be driven after the placement of the filter layer, as shown in Figure 13.14, then it is recommended to use a granular rather than a geotextile filter as the geotextile will be ruptured and severely damaged by the piles.

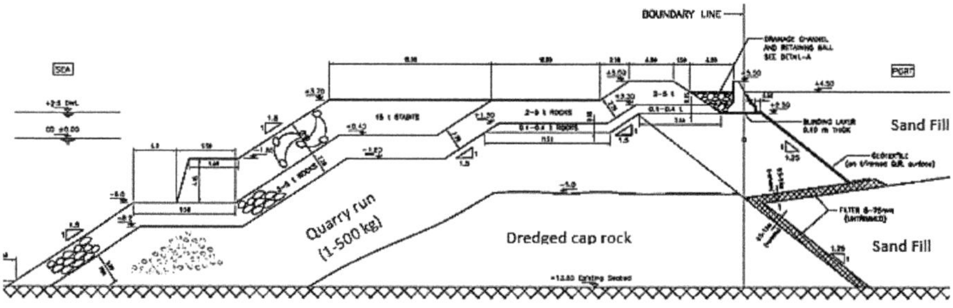

Figure 13.13. Granular filter to dredged cap rock.

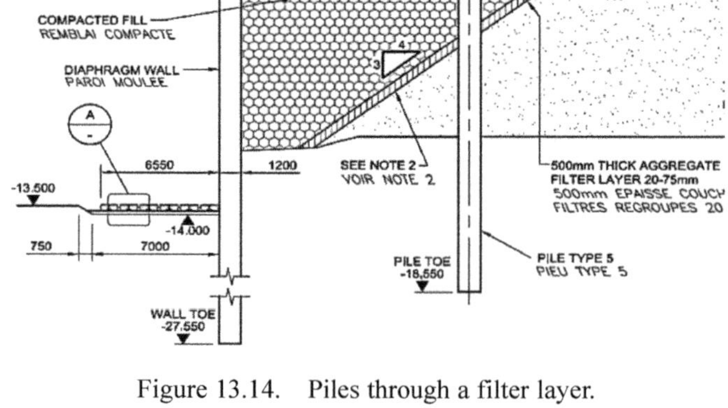

Figure 13.14. Piles through a filter layer.

13.8.1.2. *Disadvantages*

- Difficult to place underwater with the required design layer thickness. Therefore, it is a normal practice to specify a thicker underwater layer to ensure meeting the minimum requirement.
- The aggregates may segregate during handling and placement; hence, the resulting grading may deviate from that assumed in the design with the filter tightness possibly becoming less effective.
- Depending on the particle distribution curves of the core and cover material, it is difficult sometimes to satisfy the close/tight filter requirement by one filter type. Either more than one filter layer is to be used or a combination of one layer with a geotextile.
- Quarries tend to sprinkle water on the crushed rock to wash out the dust, however, this process might lead to coagulation of the dust particles forming clumps, as shown in Figure 13.15. These clumps could pass undetected and will be placed as part of the filter material, though they will gradually disintegrate under water, and this will create voids and gaps in the filter layer.

Figure 13.15. Coagulated clay clumps.

13.8.2. *Advantages and disadvantages of the geotextiles*

The selection of geotextile fabric as a filter may be influenced by the following.

13.8.2.1. *Advantages*

- Faster and easier installation than the granular filter, particularly underwater.

- Although the granular material is likely to be cheaper than the geotextile fabric (depending on the construction site location), the cost of transportation and placing operation will escalate the price dramatically.
- Geotextiles are more consistent with well-defined properties than the aggregate filter as they can be made to meet strict specifications.
- Effective filters for fine materials, specially when subjected to variable flow conditions.
- As the thickness of the granular filter layer is relatively thin, it makes it difficult to place within the allowable tolerance especially in deep sections, which needs to be verified by an echo sounding profile survey. Whereas no such elaborate work or survey is required for the geotextile, apart of course from a detailed final diver survey.

13.8.2.2. *Disadvantages*

- If a geotextile fabric is used close to the armouring layer, then it will reflect a significant part of the wave energy, which means larger armour block will be required.
- Geotextile fabric is not to be used where the material in contact has too many sharp edges, see Figure 13.13.
- Geotextile fabric cannot be used where piles are to be driven through the filter layer, see Figure 13.14.
- Damaged or punctured geotextile fabric is difficult to discover after being covered by the subsequent layer(s). However, even if the damage is discovered, it will be difficult to repair.
- Limited exposure time to sunlight, hence construction should follow a tight schedule to avoid leaving the geotextile uncovered longer than what could be acceptable.

13.9. Inferior Filter Design

In attempting to cut cost or to simplify the construction, designers are sometimes producing what could be considered “inferior” designs for the filter layer, as explained in the following examples.

As a general advice, geotextiles are not to be placed directly under armour layers, as shown in Figure 13.16. A normal revetment is shown in (a) with rock armour layer, although being relatively of small rocks in this

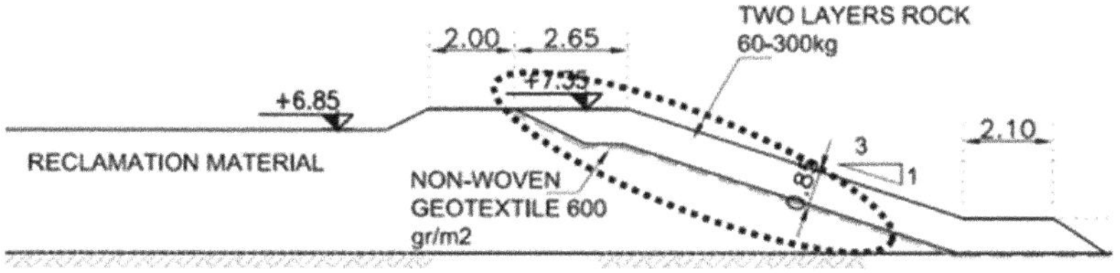

(a)

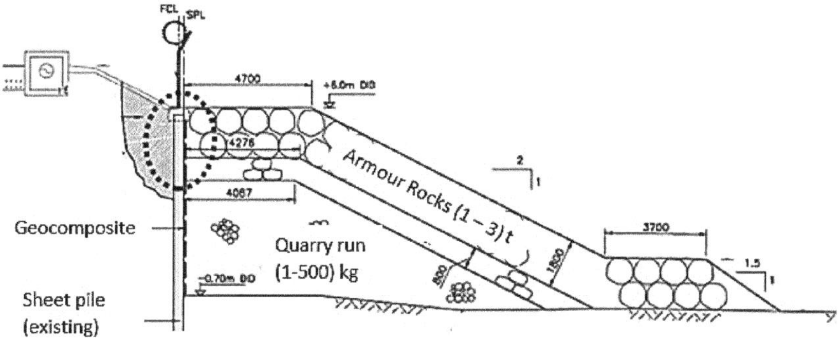

(b)

Figure 13.16. Geotextile under main armour.

example, it has been placed directly on the geotextile. Open voids between the rocks will let the UV light through, besides it will expose the geotextile to a variety of dirt (possibly contaminated) and floating objects. All these factors will adversely affect the geotextile. Apart from the exposure to the outside elements, the larger rocks could have some sharp edges which could puncture and damage the geotextile.

In Figure 13.16(b), the geotextile covers a vertical edge with the top part separating the larger rocks, both underlayer and armour, from the vertical face. This is expected to lead to a rapid deterioration of the upper part of the geotextile and it will not perform as it should.

Another example where a geotextile is not recommended to be used is shown in Figure 13.17, with a geotextile being placed under a precast concrete wall. This will lead to losing any contact between the concrete wall base and the bedding rock, which means there will be low frictional force resisting the lateral movement.

In Figure 13.18, the example shows a design where the granular filter layer is actually in contact with the underlayer. Although being at the seabed level and not at the wave action zone, i.e. not affecting the stability

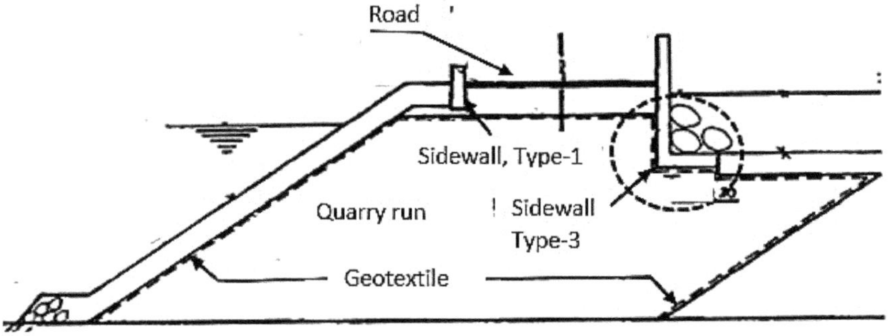

Figure 13.17. Geotextile under concrete wall.

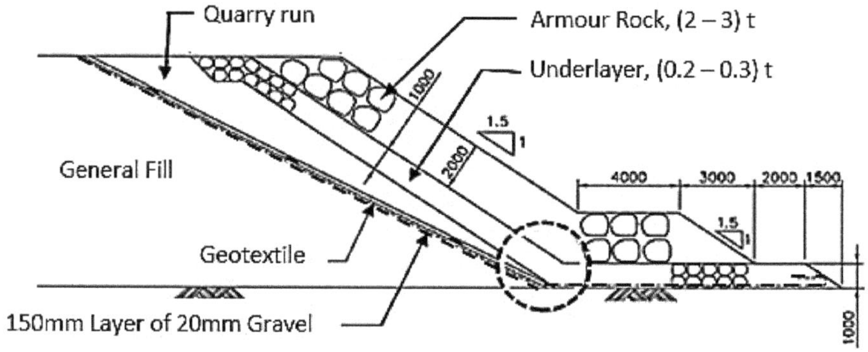

Figure 13.18. No sufficient clearance.

of the armour rock, it is however recommended that a minimum distance of 1.0 m is to be maintained. This distance will help keep a good quarry run cover layer over the filter and it will provide some construction tolerance because of the uncertainty of the exact slope of the fill material.

Chapter 14

Toe and Toe-Berm

14.1. Introduction

The toe-berm-apron/scour protection are three components which secure the foundation for the armour layer in rubble mound structures, and simultaneously form a transition from the sloping face to the horizontal toe/berm. The design of each of these three components should ensure sufficient protection against scouring and undermining of the structure, as well as toe stability to secure the armour face against possible sliding. The general required design input data will include wave characteristics, water depth, seabed material, armour block type, rock material, optimised configuration and the preferred construction methodology (which depends on available machinery and construction cost and time).

The term "toe" is used to refer to the toe of the armour layer if it is on, or near, the seabed. However, if it is relatively elevated above the seabed, then it is referred to as "toe-berm". The term "berm" in this context has no relation or connection with "berm breakwaters", see (Van der Meer and Sigurdarson, 2017), and as illustrated in Figure 4.1(3), or with a composite slope configuration consisting of a wide berm to reduce the overtopping, see (EurOtop, 2018), and as illustrated in Figure 4.7(d).

The basic configuration of the toe depends on many factors, as illustrated in Figure 14.1. The figure classifies the basic toe configurations based on the type of the seabed material (sandy or rocky) and on the depth of the toe (deep or shallow).

The toe must be stable and able to resist the acting forces from the waves (breaking waves and/or wave down-rush), currents, and the horizontal weight component force from the armour layer. To account for the wave forces, some guidelines are given in BS 6349-7 (1991), which are schematically presented in Figure 14.2. These guidelines are based on the depth of the toe bund relative to double the design significant wave

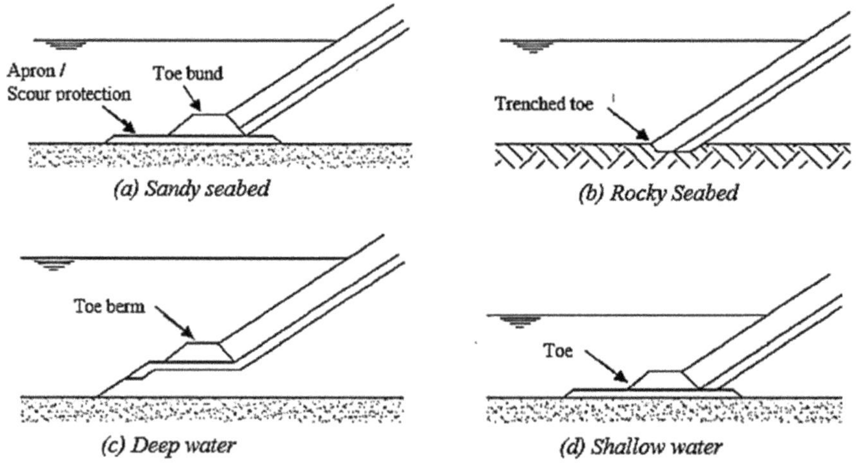

Figure 14.1. Generic toe classification.

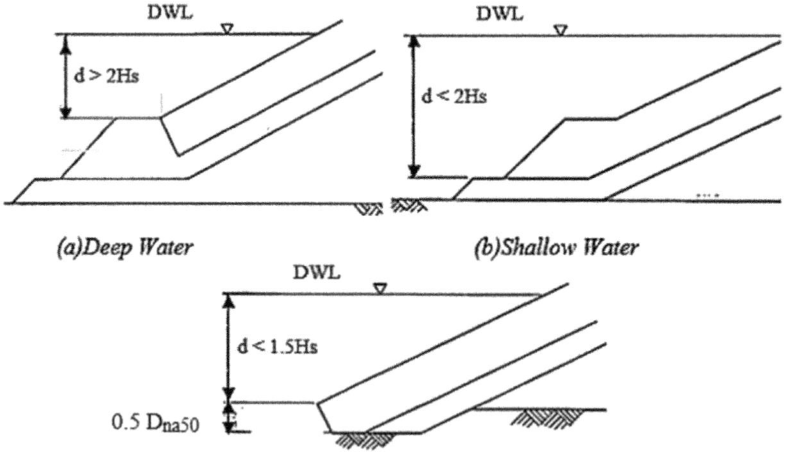

Figure 14.2. Toe bund and trenched toe guidelines for depth requirements (Based on BS 6349-7).

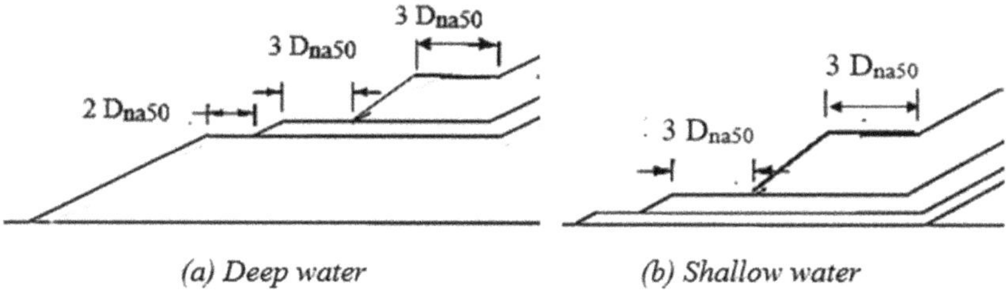

Figure 14.3. Recommended toe/toe-berm minimum dimensions.

height. It will help in the preliminary stages of the design to size up the rocks for the toe bund either by using the underlayer rocks or the armour rocks. For shallow rocky seabeds, it is recommended to trench the toe to a minimum depth equal to 0.5 D_{na50}.

The toe bund is recommended to be 2–3 layers thick and 3–5 stones wide at the top layer. However, it is commonly designed with satisfactory performance, using double layer thickness, 2 D_{na50}, and three stones wide (top layer), 3 D_{na50}, see Figure 14.3. If the toe is of a different rock class than the armour rocks, then Dna50 is to be substituted by D_{n50} of the toe rock class.

14.2. Toe for Rock Armour Layers

The toe for a rock armour layer can take different forms and shapes depending primarily on the wave characteristics, the depth of the seabed and its nature. Example designs of toes placed on the seabed, i.e. no berm, shall be presented in this section in two categories, rocky seabed and sandy seabed.

Figure 14.4(a) shows a basic example of a breakwater structure on bedrock. A rock bund has been used on the seaside where there is little to moderate wave action, with no breaking waves, and the base of the armour face needs to be protected and secured against possible sliding. Whereas on the port side, with very limited wave action, there will be no need for extra support from a toe bund.

Figure 14.4(b) shows a rock armour face with the toe bund made of the underlayer stones. This option could be used where the toe is deeper than the case shown in Figure 14.4(a). The smaller stones do have another advantage when used as a toe or as an extra layer between the larger

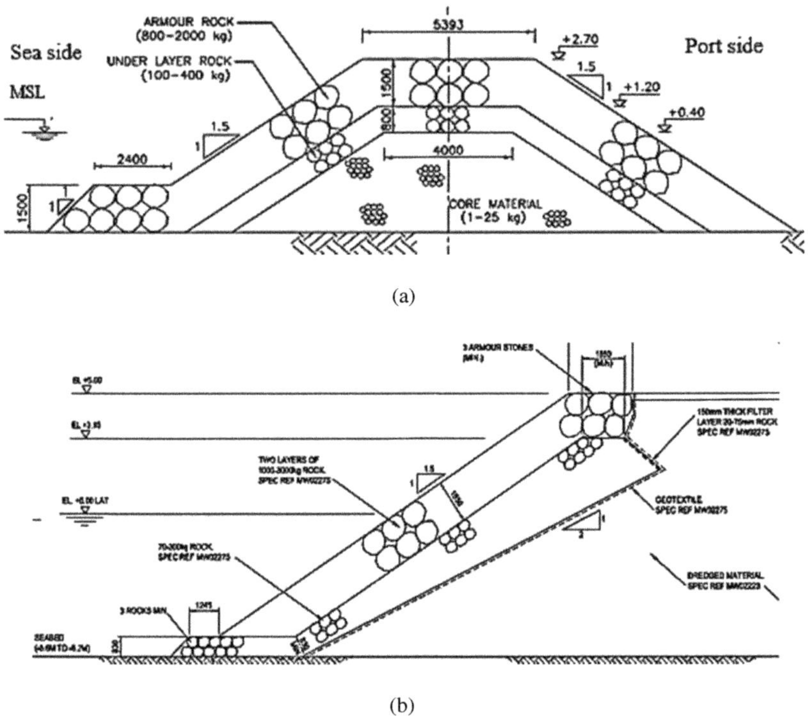

Figure 14.4. Toe on bedrock: (a) Toe bund (seaside) and no toe (port side); (b) Toe bund with underlayer rocks.

armour stones and the seabed, the advantage is to increase the surface area of contact between the stones and the seabed. This will greatly reduce the risk of sliding, specially on a slightly slippery surface. For this reason, a layer of quarry run could also be used as an added safety factor.

It is important to note that sometimes, especially near the shoreline, the bedrock could be covered by a layer of sand. No special precaution is required if this sand layer thickness is approximately less than 0.40 m. If a thicker layer is encountered, then either it should be removed to have the toe resting on the rock layer, or the seabed is to be treated as a sand bed and proper measures are to be considered to protect the toe against erosion and scour.

When the toe is exposed to breaking waves, or is placed on a relatively steep foreshore, or even if the seabed is slippery, which can cause toe slippage, then it is necessary to have the toe anchored and placed in a

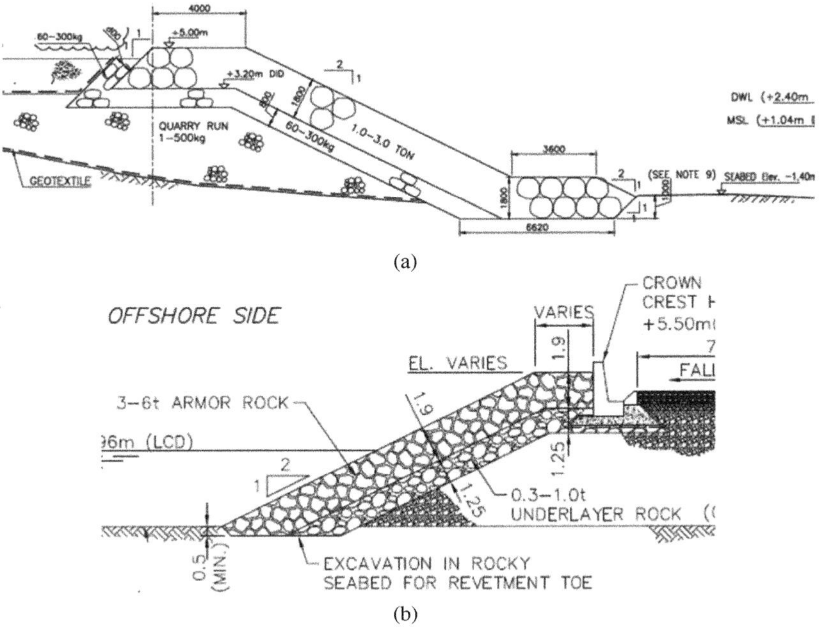

Figure 14.5. Trenched toe in bedrock.

trench. Figure 14.5 shows two possible arrangements for a trenched toe. More possible designs for this type of toe are given in Figure 14.6. The trench type shown in Figures 14.5(b) and 14.6(i) are used when the rock layer of the seabed is dense and strong enough to excavate a trench with a near vertical edge with a fore-slope >60°, to a minimum depth of $0.5D_{na50}$. When it is not possible to sustain a near vertical fore-slope due to a weaker rock layer at the seabed or subsurface layers of loose material (as the surface seabed rock layer may be thin), then alternative trenches could be used as shown in Figures 14.6(ii–iv). Note that breaking a thin caprock layer to construct the toe is not always advisable, as it is preferred to keep the layer intact. Therefore, other options are worth investigating to secure the toe before resorting to the trench.

The second category of toes are those placed on sandy seabeds. Figure 14.7 shows three examples of toes placed on the seabed. Figure 14.7(a) shows a typical toe bund with a rock class smaller than the armour rocks and larger than the underlayer, placed on a thin layer (0.50 m) of quarry run which acts as a filter layer on the sandbed.

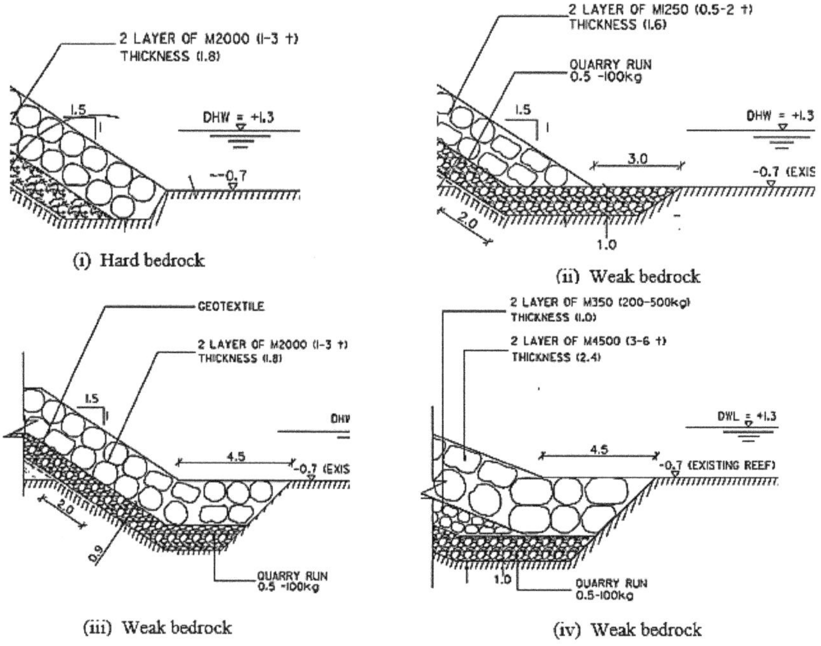

Figure 14.6. Possible types of trenched toe in bedrock.

This example presents a basic revetment toe design where little wave action is expected. The sand layer is most likely not thick and little sand erosion and scour are expected to take place.

Figure 14.7(b) is a typical toe design with three classes of rock layers. The toe bund is constructed from the armour rock (3 rocks wide), the underlayer and the quarry run extend 1.5 m and 1.0 m, respectively. This will ensure that 3.0 m of the seabed is covered by rock material in front of the rock bund. A geotextile is used to prevent any loss of the seabed sand, which could undermine the toe structure.

Figure 14.7(c) is for a revetment with no underlayer using a relatively thick layer of quarry run as the apron/scour protection layer with a 0.5 m granular filter to prevent any sand loss from underneath the structure.

One important distinction between the toe being placed on a rock seabed or a sandy seabed is the need for an apron/scour protection for the latter. More will be discussed about this in Section 14.7.

Where the toe is very exposed to wave action or/and the possibility of breaking waves or even high-water currents, it is recommended to have

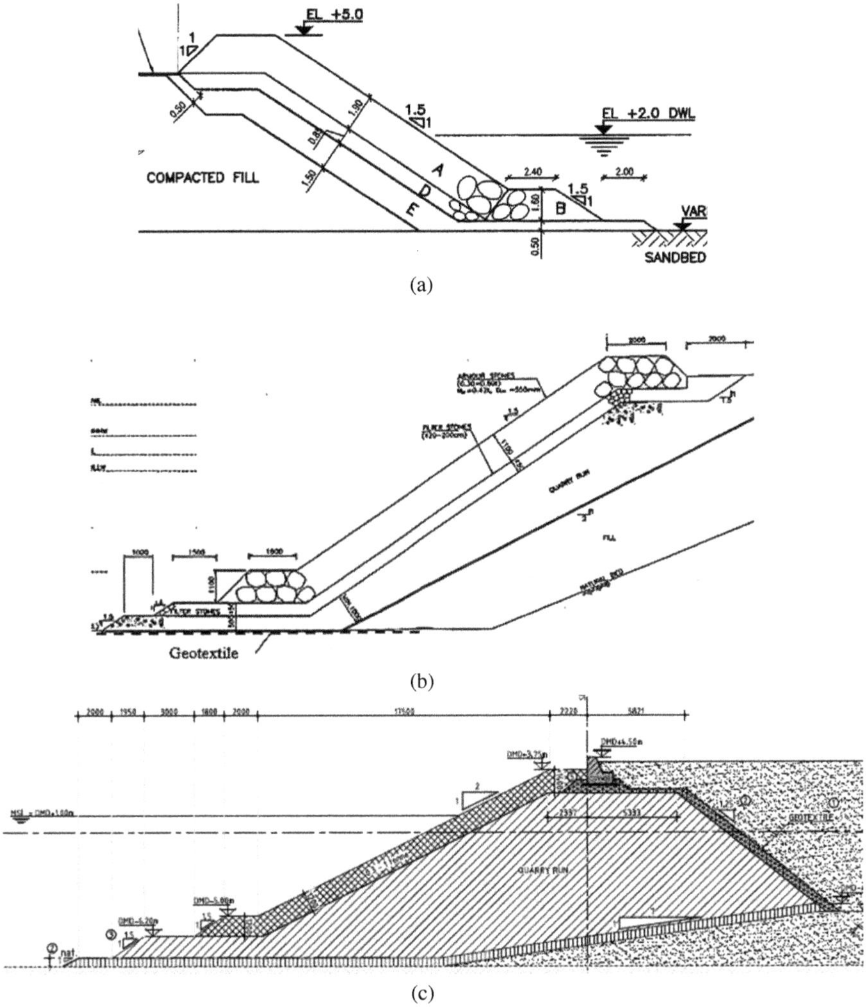

Figure 14.7. Toe on sandy seabed.

the toe trenched in the sand bed. Figure 14.8 shows four different possible designs for such a trench arrangement, ranging from the relatively less exposed in (a) to the most exposed in (d). Note the necessity to use a geotextile filter lining the trenches in all the different designs. This type of toe could also be considered as a type of seabed soil replacement, however, on a limited scale as being only under the toe.

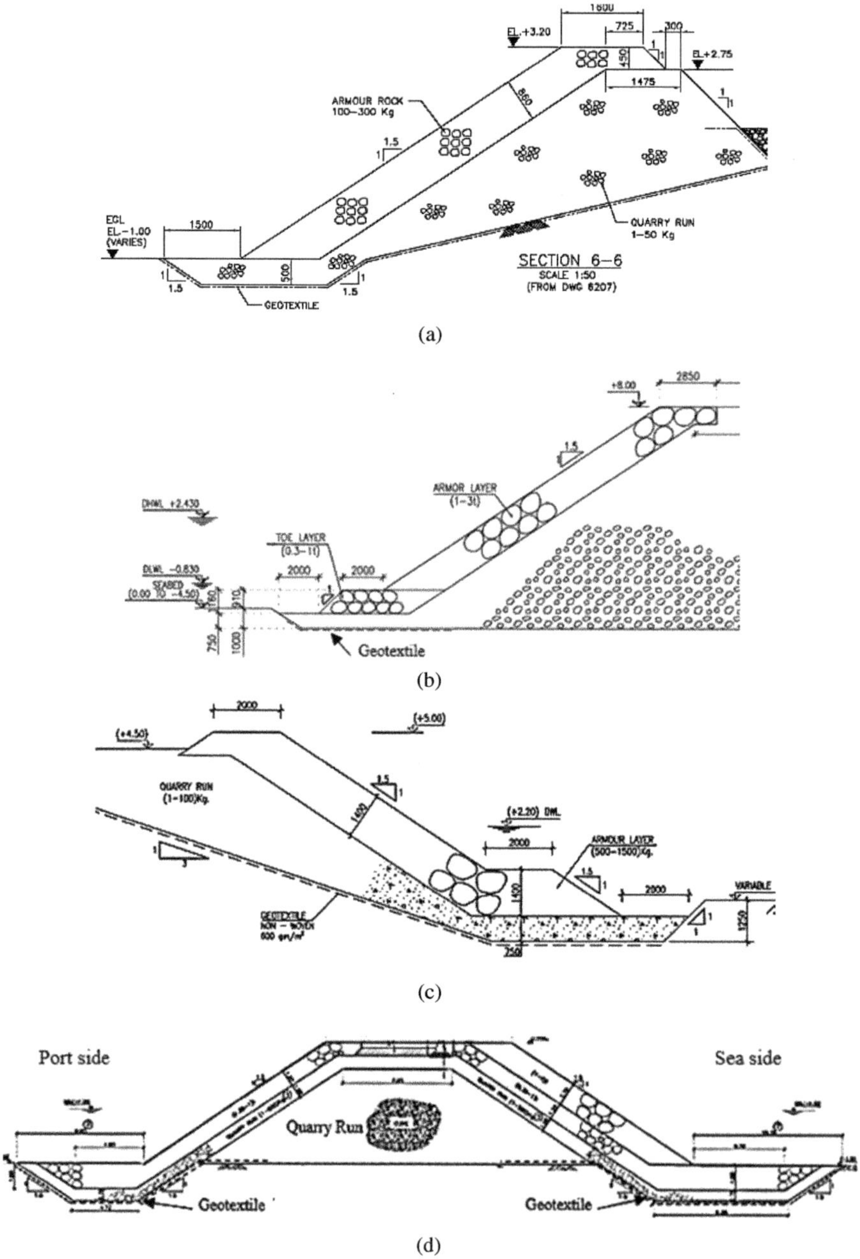

Figure 14.8. Trenched toe in sandy seabed.

14.3. Toe to CAU (Concrete Armour Unit) Layers

The toe arrangement for the CAU shall not be presented in this section in categories based on the nature of the seabed as in the previous section but on the type of the CAU.

Accropode and Core-loc armour units are very similar in their toe arrangements and accordingly they will be presented together. It is important to highlight the required stability features for the toe unit, or the first row, before discussing the various possible configurations for the toe. It is obvious that the stability of the armour face relies on the toe being stable. In this case it means that the toe unit, or the first bottom row, is heavy enough not to be displaced from its position, and the unit should be restrained against rocking, otherwise rocking will cause the units in the above rows to settle and open up and this will comprise the integrity of the armour face.

An early toe arrangement is shown in Figure 14.9, which was commonly used when these units were first introduced, and still some designers prefer to use it. It is basically similar to the rock armour toe bund, as it is to be constructed before placing the armour units. However, it is more complicated to construct due to the V-shape between the underlayer and the toe bund, which must be perfectly formed to hold the first row of the amour units. The actual toe will be the rock bund, and the unit in the first row will rest on the underlayer and the toe bund, with its limbs embedded in between the rocks, and this will prevent the unit from rocking.

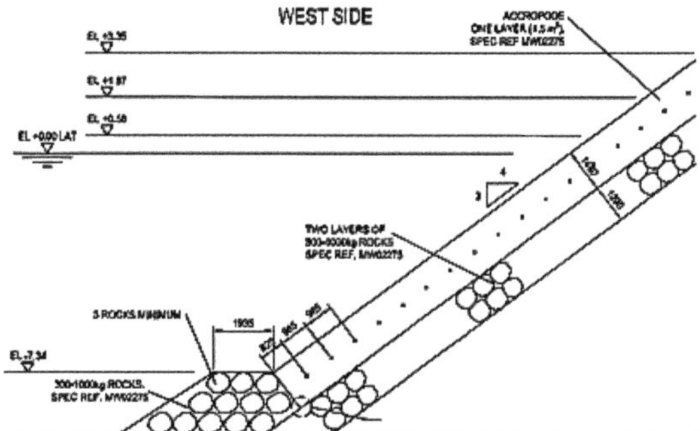

Figure 14.9. Rock toe to the CAU (Accropode/Core-loc).

An alternative early toe configuration was also in use, and that is by using the units in the first row, being placed on the flat rock layer, as the toe, as shown in Figure 14.10. After several years of use, the potential instability of the first row was associated with increased armour settlement on the face. It was then recommended to buttress the toe unit of the CAU with a rock bund, as shown in Figure 14.11. The rock will minimise the adverse effects of construction inaccuracies in placing the toe unit by stabilising and restraining the toe unit from any movement. When the interlocking between the toe unit and the first row (the first bottom unit on the slope) is efficient, the toe unit will most likely remain stable. The inefficient interlocking could not have been detected or not been clear during the construction inspection, but it can take place after the structure has been exposed to few storms. Therefore, adding the rock toe bund reinforcement is certainly a good practice for the long-term toe stability assurance.

The rock bund shall be of two layers in thickness and at least three rocks wide at the top. The construction sequence of the rock bund

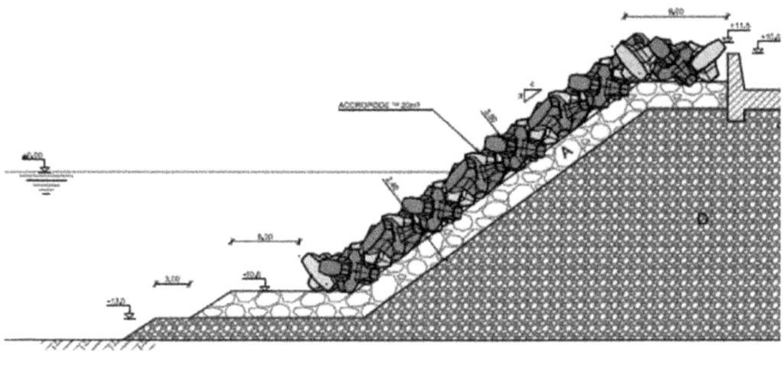

Figure 14.10. One-unit CAU (Accropode/Core-loc) toe.

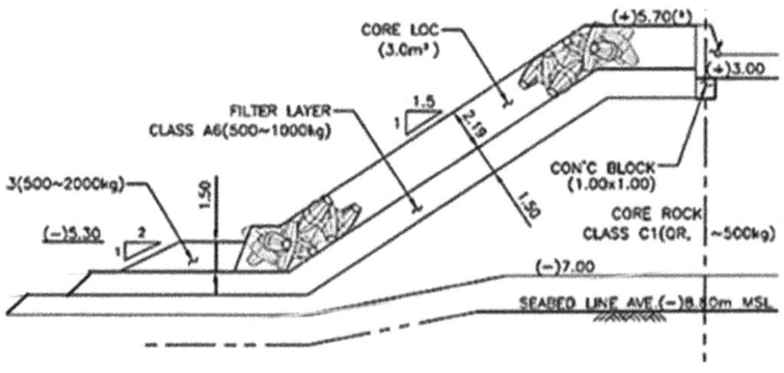

Figure 14.11. CAU (Accropode/Core-loc) toe unit reinforced by a rock bund.

reinforcement is as indicated in Figure 14.12, which is to be followed to ensure good support from the rocks in securing the CAU toe unit in place and preventing it from any rocking movement. Step (a) indicates that the placement of the CAUs will start from the toe up, and only for few rows (4–5 rows). Then step (b) follows in placing the rock bund against the CAU toe unit, and the rocks must be properly wedged between the units. This will be followed by step (c), which is to continue placing all the remaining rows and this is to tighten any gaps left between the CAU toe units and the rock bund.

In locations where the toe is exposed to high wave loads, and large enough rocks could not be supplied, the toe could be made of two toe units. The second unit will act as a reinforcement to the first toe unit, as shown in Figure 14.13, and also in Figure 14.37(a), which shows a 2-D model testing in 2003 with a successful two toe units on the seaside.

It is worth highlighting that although the toe arrangement shown in Figure 14.12 is being used by many designers, it is however not the preferred design to follow. It is always preferred to have the CAU toe unit and the first row placed on a solid rock layer rather than on a thin layer of

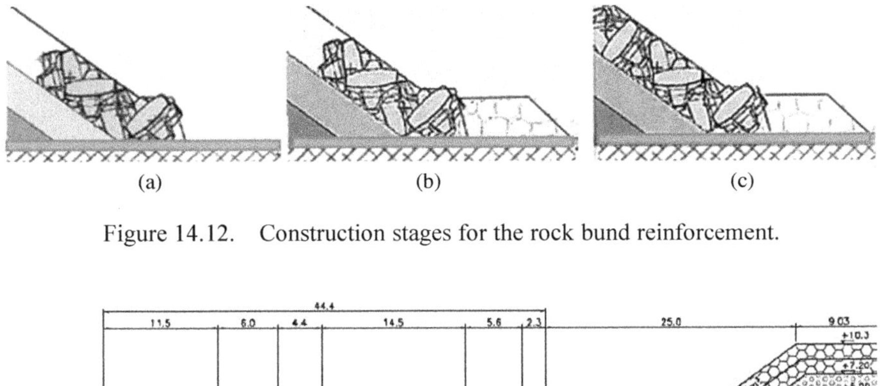

(a) (b) (c)

Figure 14.12. Construction stages for the rock bund reinforcement.

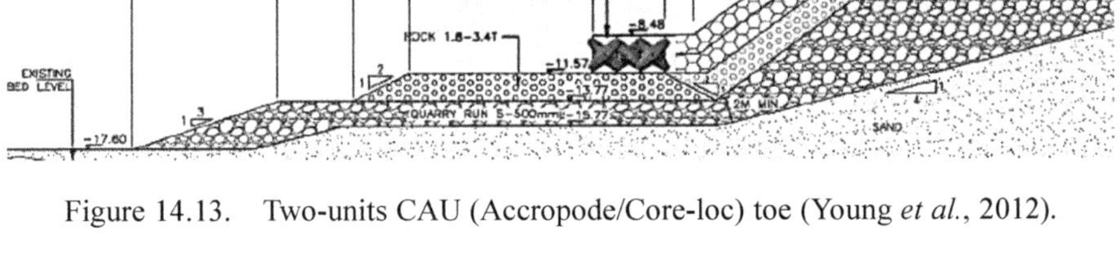

Figure 14.13. Two-units CAU (Accropode/Core-loc) toe (Young *et al.*, 2012).

smaller rocks. A design example is shown in Figure 14.14 to clarify this point, as in this example the thin layer of the small rocks is laid on a sandy seabed. The larger rocks are preferred because it will not allow the toe unit either to sink into the layer or to move (rocking movement), whereas with a layer of smaller rocks the toe unit is vulnerable to both, i.e. sinking and rocking.

Recently, The Concrete Layer Innovation, CLI (www.concretelayer.com) announced the introduction of two toe units for the Accropodes, the Accroberm I and its updated version Accroberm II, both are schematically shown in Figure 14.15.

For structures founded on bedrock, it is preferable to have the toe trenched in the seabed. Three examples are presented in Figure 14.16, in (a) the bedrock is strong enough to allow a clean and well-defined trench with a minimum depth of 0.6H (where H is the unit height) and a cut angle of 45°–60°. As the first row unit (as there is no toe unit in this case)

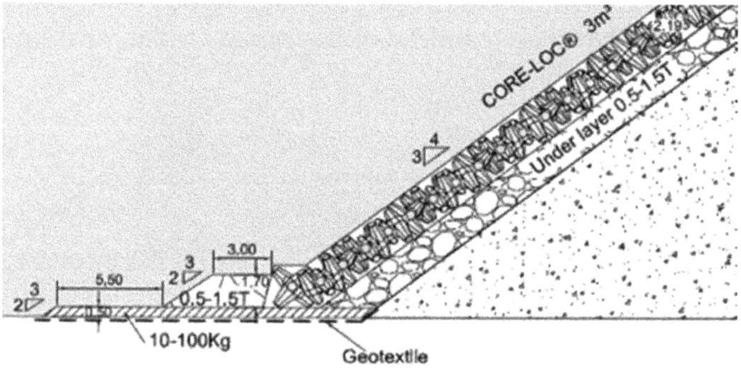

Figure 14.14. Toe unit placed on a thin layer of small rocks.

(a) (b)

Figure 14.15. Accroberm toe unit for Accropodes (www.conceretelayer.com): (a) Accroberm I™; (b) Accroberm II™.

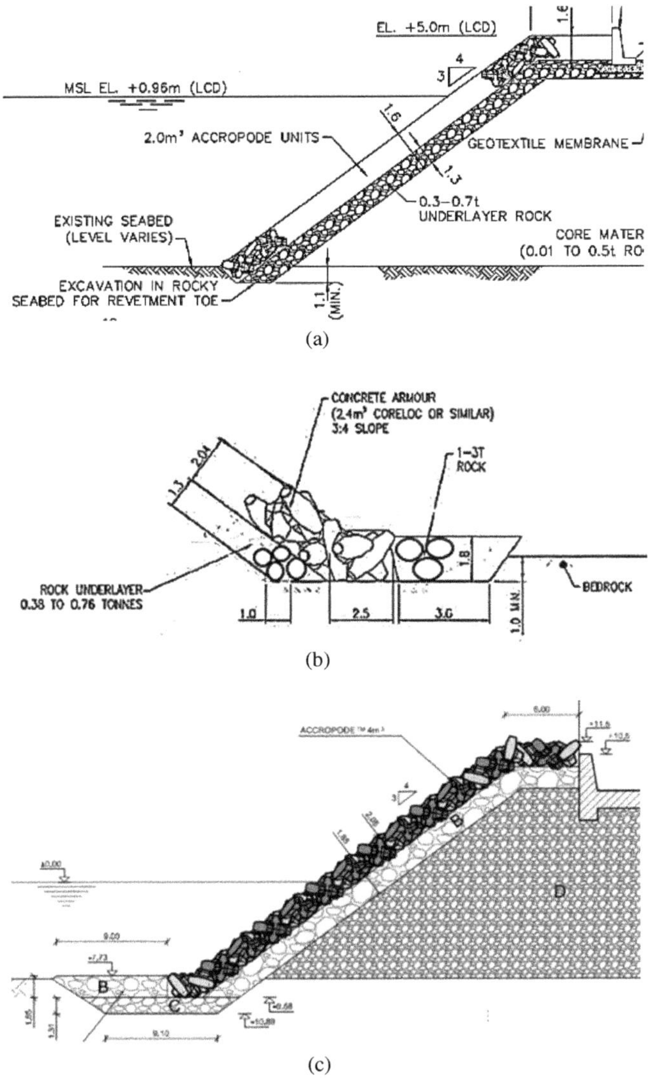

Figure 14.16. Trenched CAU (Accropode/Core-loc) toe: (a) Trenched CAU in hard bedrock; (b) Trenched CAU in weak bedrock; (c) Trenched CAU in weak bedrock with underlayer.

doesn't have its limbs embedded in rocks, i.e. not restrained, it could suffer from some rocking movement. This could be prevented by filling the trench with cement grouting.

In Figure 14.16(b), where the bedrock is weak, no precise dimensions could be expected from the trenching operation. Thus, the trench is to be filled with rocks large enough to withstand the hydraulic loading. No need for grouting in this case, as the toe unit will be better restrained from any movement. However, the toe unit and the first row are not fully restrained because they are resting on smooth bedrock, hence there could be some rocking movement. Therefore, in this type of trenched toe it is recommended to adopt the arrangement shown in Figure 14.16(c) where the limbs of the toe units are embedded between the underlayer rocks and they will be well-restrained.

All the examples with the toe above the seabed, i.e. not trenched, are basically for structures on a sandy seabed, however, with minor adaptation it could also be used for a relatively flat bedrock seabed instead of having a trench. For example, the toe arrangement shown in Figure 14.11 could still be used, however without the quarry run layer.

It is to be emphasised that the comment made in Section 14.2 regarding the 0.40 m approximate maximum thickness of the sand layer that may be covering the bedrock is valid for all toe types.

The Xbloc toe was made of rock bund after the introduction of the unit, as shown in Figure 14.17. However, this type of toe is not recommended anymore after the development of the Xbloc toe unit, called Xbase, as shown in Figure 14.18. Note that the use of this toe unit, i.e. the Xbase, still requires a rock bund reinforcement. The Xbloc design guidelines (Xbloc, 2018) does recommend the toe arrangements as shown in Figure 14.19.

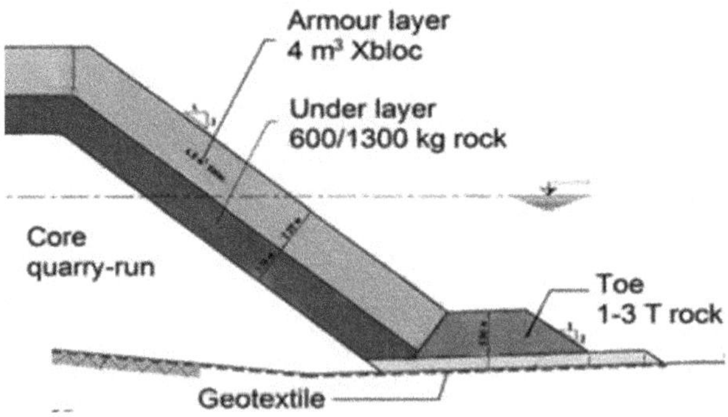

Figure 14.17. Xbloc with rock toe bund.

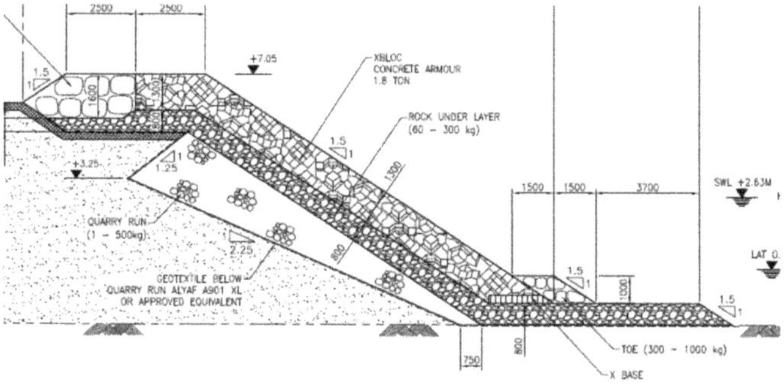

(a)

(b)

Figure 14.18. Xbloc toe unit (Xbase): (a) Xbase toe unit with a supplementary rock bund; (b) Illustrative photo from a model for the Xbase toe unit.

There is no need to keep the toe level constant for the interlocking single armour units (Accropode, Core-loc and Xbloc), if the seabed level varies along the breakwater/revetment alignment, as in Figure 4.15. Here the armour layer side slope length will also vary, i.e. the armour face gets longer as the seabed gets deeper. The placement grid can well absorb such a change in depth by gradually increasing the number of rows. It is recommended that the toe follow the gradual change in the seabed level with no sudden drops. The maximum gradient for the toe that could be accommodated is 1V:10H. For steeper gradients, the seabed should either be levelled by rock material fill or be dredged.

The Cubipod unit could be used as a single- or double-layer armour, in both cases the toe shall be made of a rock bund, as shown in Figure 14.20.

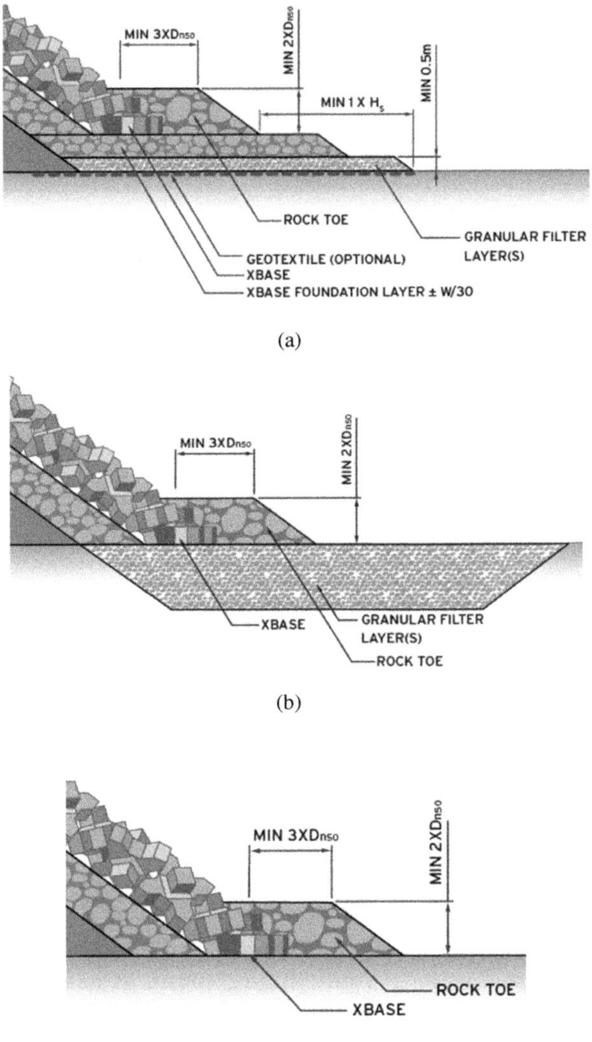

Figure 14.19. Typical toe to Xbloc toe unit (Xbloc, 2018): (a) Toe on Sandy seabed; (b) Toe on shallow sandy seabed; (c) Toe on rocky seabed.

The Tetrapod, being one of earliest most popular units, has different toe arrangements developed to meet different project specific requirements. Figure 14.21 shows some of the widely used toe arrangements.

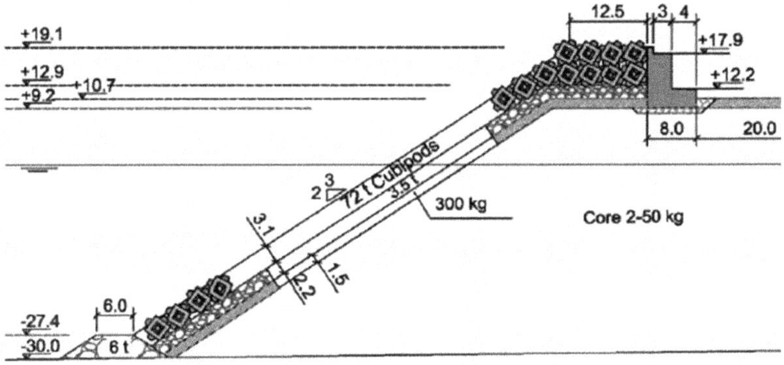

(a)

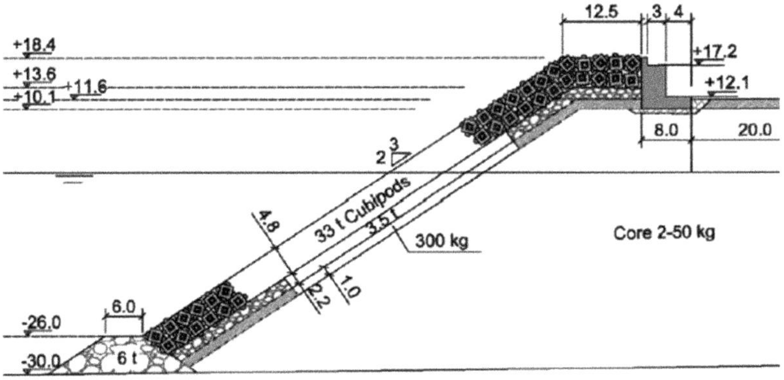

(b)

Figure 14.20. Typical toe configurations for the Cubipod unit: (a) Rock bund toe to a single layer Cubipod (Cubipod, 2016); (b) Rock bund toe to a double layer Cubipod (Cubipod, 2016).

The Dolos unit, similar to the Tetrapod unit, is one of the early developed units. This unit could have various toe arrangements, two examples are shown in Figure 14.22, where in (a) the armour layer rests on the first row with no toe and in (b) the armour layer rests on a rock bund toe.

The Stabit unit has been used in projects over the years and different toe arrangements have been used successfully with this type of CAU. Three different toes are shown in Figure 14.23, with no toe in (a), Stabit toe units in (b) and rock bund in (c).

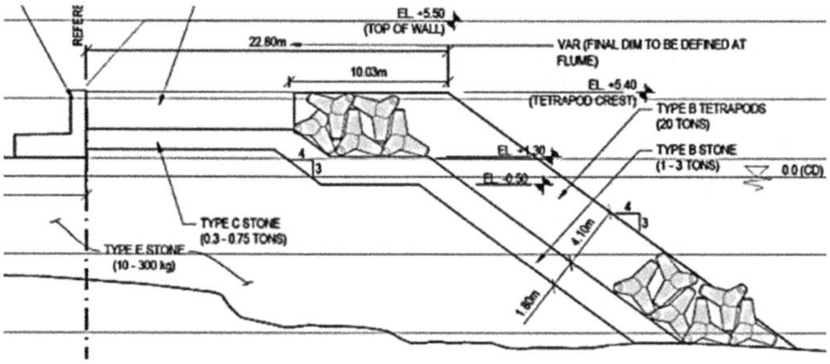

(a)

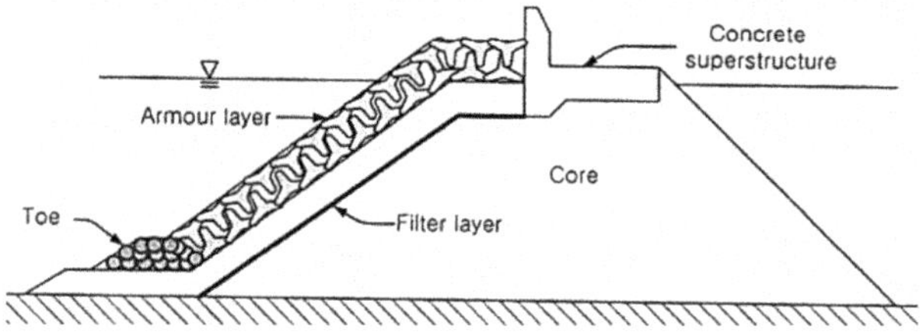

(b)

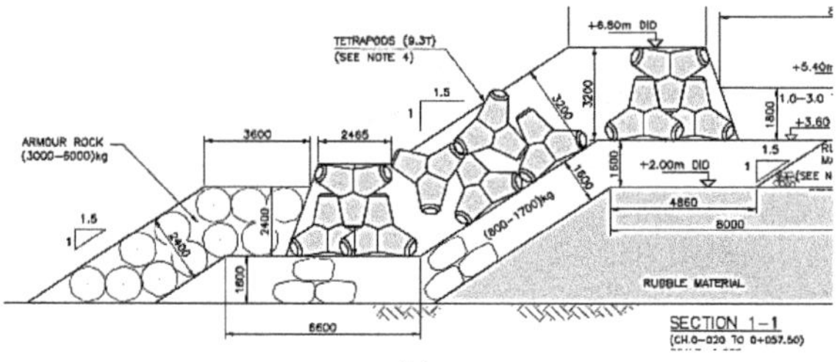

(c)

Figure 14.21. Selected toe configurations for the Tetrapod unit: (a) No toe units in front of the armour layer (Wesson *et al.*, 2018); (b) Rock bund toe; (c) Tetrapod toe units with rock bund reinforcement.

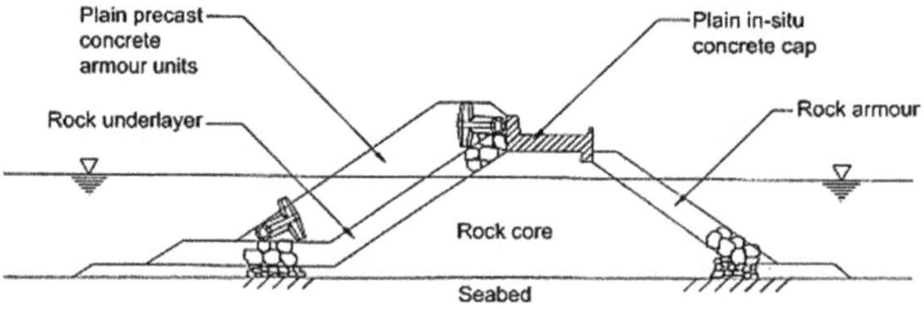

(a)

Figure 14.22. Selected toe configuration for the Dolos unit: (a) No toe units in front of the armour layer; (b) Rock bund toe.

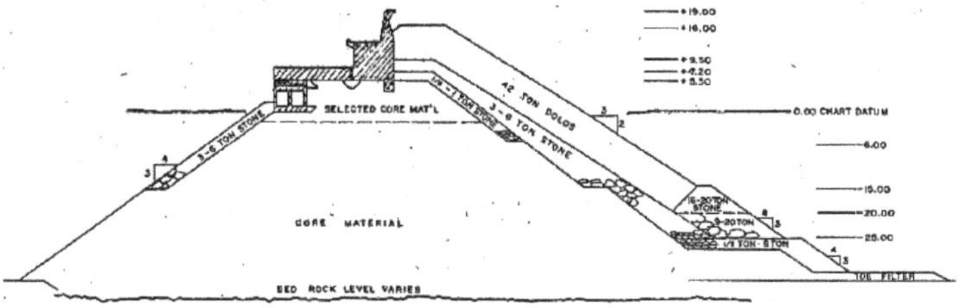

(b)

Figure 14.22. (*Continued*)

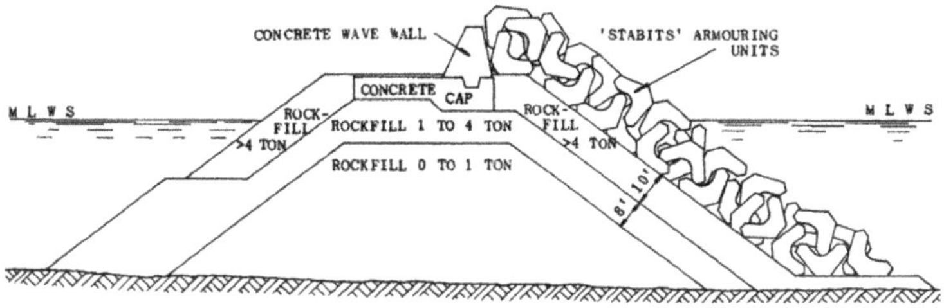

(a)

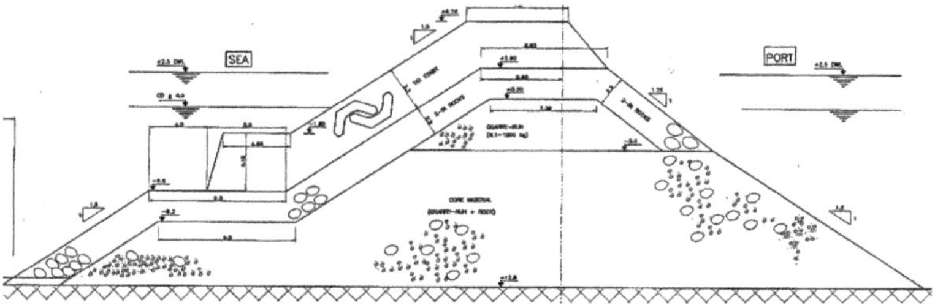

(b)

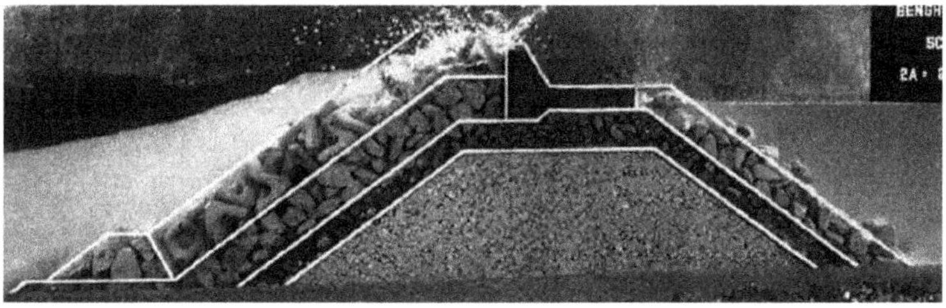

(c)

Figure 14.23. Selected toe configurations for the Stabit unit: (a) No toe units in front of the armour layer (Singh, 1968); (b) Stabit toe units; (c) Rock bund toe.

The Cube is the first unit to be used and still the preferred one when exceptionally large units are required. In many cases elongated cubes, i.e. rectangular blocks, are used to obtain better performance and higher roughness. As such, a wide range of toe configurations have been developed to suit the location, exposure and other project specific characteristics. Figure 14.24 shows some of the used toe arrangements. Starting with no toe in (a), a small two blocks toe in (b), a two-layer thick toe (on the seaside) that is three blocks wide in (c), a rock bund toe in (d), blocks and rock bund reinforcement in (e) and blocks of 25t used to form the toe bund for the 100t blocks primary armour.

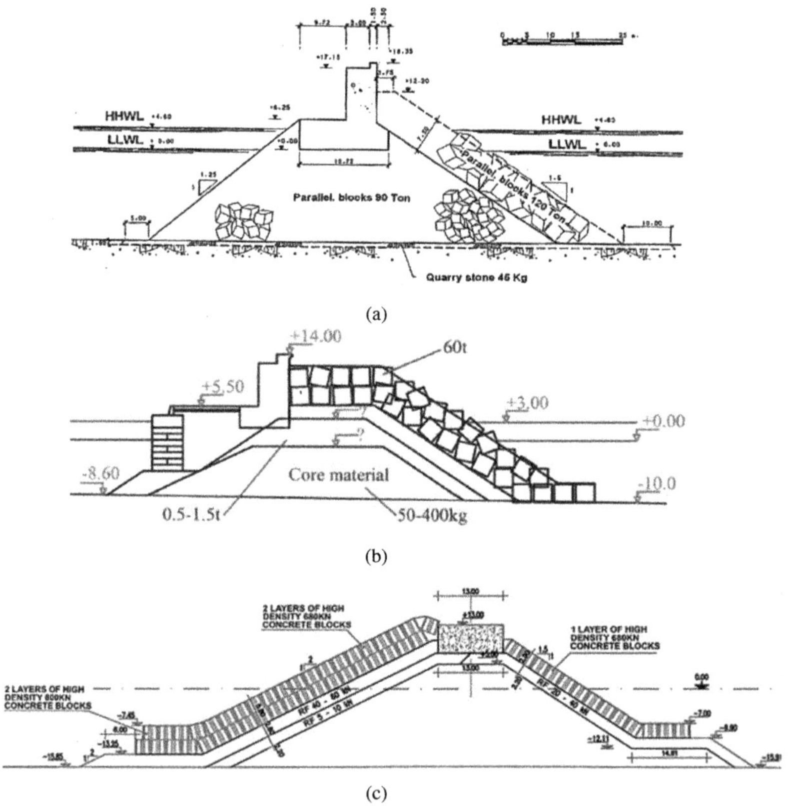

Figure 14.24. Selected toe configurations for the Cube unit: (a) No toe units in front of the armour layer (Martin *et al.*, 1999); (b) Toe is made of two cubes in a single layer; (c) Toe is made of multiple cubes in double (front) and single (rear) layer; (d) Rock bund toe; (e) Cube toe units with rock bund reinforcement; (f) Cube blocks bund toe (Martin *et al.*, 1999).

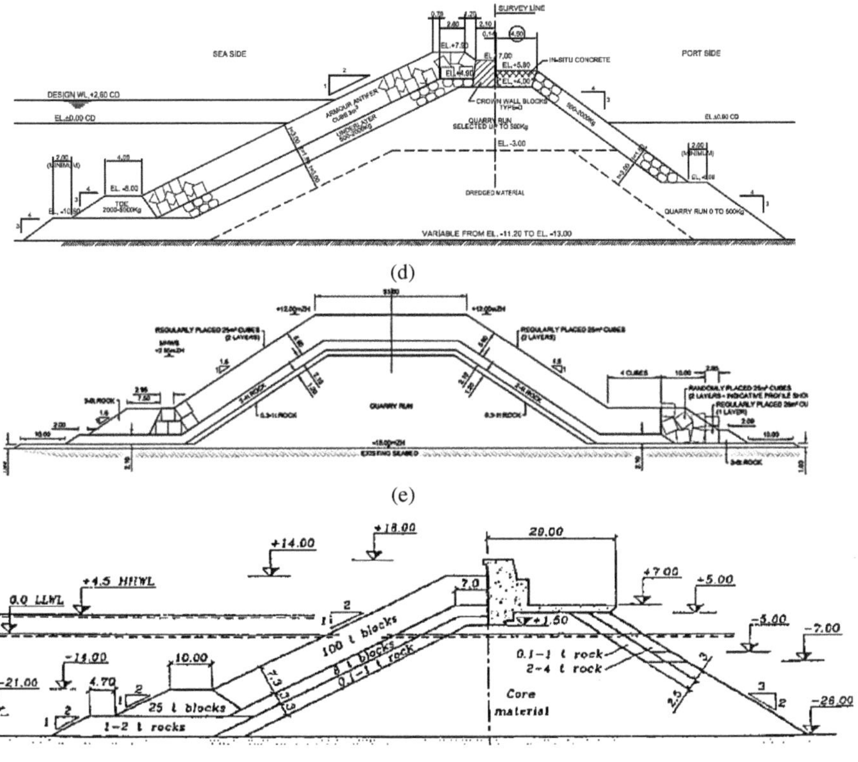

Figure 14.24. (*Continued*)

14.4. Toe Blocks and Other Toe Support Structures

It was very clear from the previous sections that the basic form of the toe is a rock bund, or the toe will be reinforced by a rock bund. There are however situations where the rock bund either cannot accurately provide the required angle and surface, or it is not heavy enough to act as a toe or as a reinforcement for the toe. A concrete toe block shall therefore be used as a substitute to fulfil the required function of the toe.

Hollow cube armour units are placed like tiles, therefore the requirement for a robust and well-aligned toe is highly important. As shown in Figure 14.25, such an arrangement definitely requires an accurate

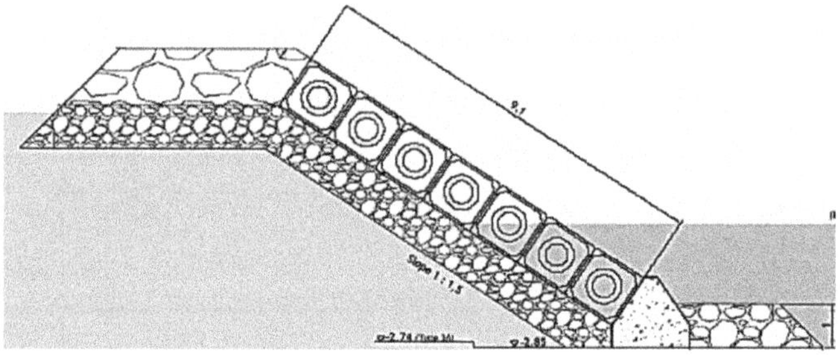

Figure 14.25. A toe block for hollow cube revetment (Allsop, 2006).

construction. A rock reinforcement is most likely required, as shown in the figure, to secure the toe block against possible sliding.

An example of a concrete blockwork revetment is shown in Figure 14.26, a general cross-section is given in (a) and the details of the used toe block, with a photo, are shown in (b). The details of the toe block include those for the side key, which is necessary in this toe block to keep the blocks in-line, and also the serrated base of the block, to ensure a mechanical grip in the base material and not just relying on the frictional force. The toe block, as in the previous example, is supported by rock reinforcement.

Toe blocks are likely to be required for CAU armour layers especially if the toe is relatively shallow and exposed to high wave energy. An example is shown in Figure 14.27 for a revetment with Stabit armour. The revetment was designed at first by using 10 t Stabit units, being heavy enough, to form the toe. However, after running the model test for the initial design as shown in (a), no damage was recorded to the armour on the revetment slope, but the toe units were severely damaged under the design storm with (Hs = 4.0 m). The only solution was to replace the toe units with a heavier toe block, as shown in (b and c). It is not possible to accurately calculate the required weight of the toe block, but it needs to be dimensioned to be heavy enough to support the CAU and within the construction team's lifting capacity on site. The selected shape and size are as shown in (c) with a unit weight of 52 t, (35 t/m), which proved to be stable in the model test.

The advantage of such a toe block is not only the weight but also the shape, which offers less resistance to the waves downrush than the armour units at the toe. Note that a rock reinforcement is used to support and protect the block against possible undermining at its base.

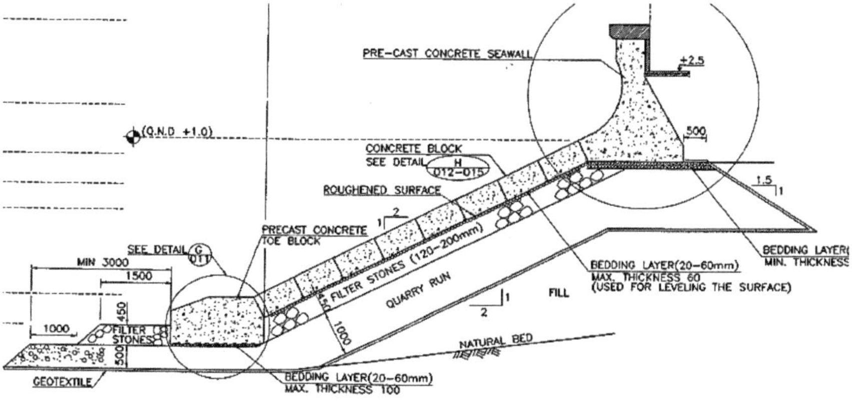

(a)

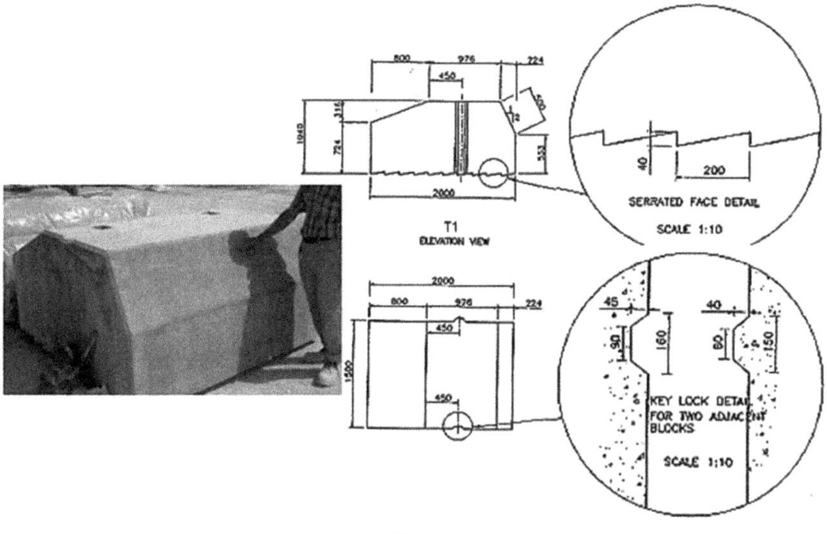

(b)

Figure 14.26. Toe block for concrete-blockwork revetment: (a) Cross-section of concrete-blockwork revetment; (b) Toe block details and photo.

Another example of a toe block to a CAU armour layer is shown in Figure 14.28, by using a 47 t antifer block as the toe block, which is supported by another smaller type of a concrete block instead of using a rock reinforcement.

Concrete toe block is not normally used for structures with rock armour, however (Van Gent, 2014) has reported the modelling and use of an interlocking concrete toe block, as shown in Figure 14.29.

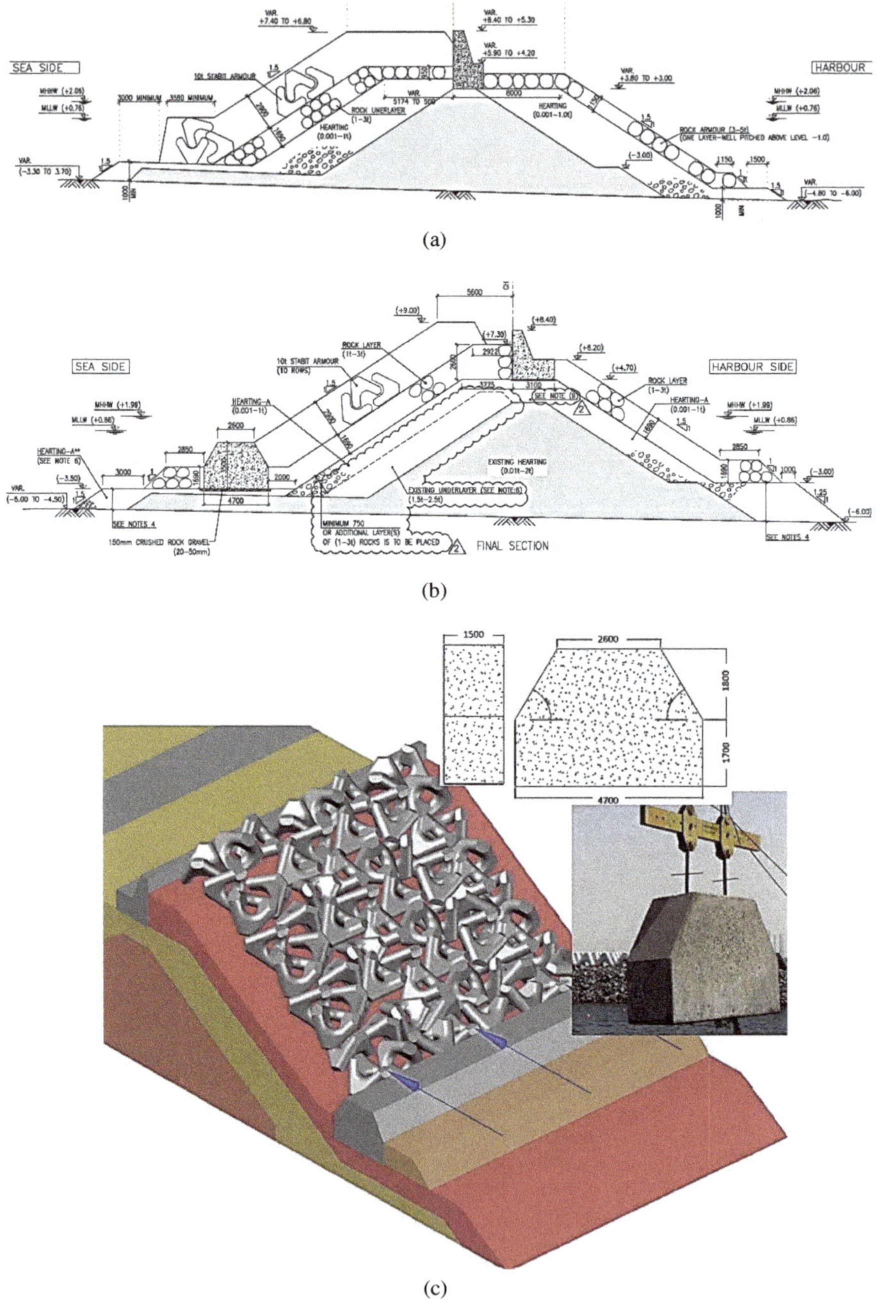

Figure 14.27. A toe block for Stabit unit armour: (a) Initial design — unstable toe units; (b) Final design — stable toe block; (c) 3D sketch, block details and a photo.

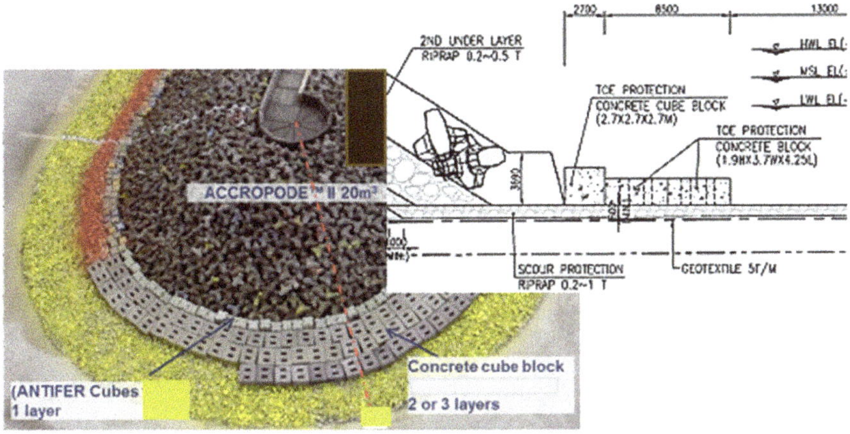

Figure 14.28. Toe block to Accropode armour (Coutos, 2016).

Figure 14.29. Interlocking toe block for rock armour (Van Gent, 2014).

It is always preferred to place the concrete toe blocks on a bedding granular layer with a corrugated, serrated or even roughened base (as these base forms will produce some grip and a higher coefficient of friction than the standard 0.6 for a flat base). Placing a bedding layer may prove to be rather difficult in certain locations, therefore the toe block may be resting on larger stones or on the bedrock.

There are however situations where the stability of the toe block cannot be ensured and it cannot confidently be secured in place, especially in highly exposed shallow locations, especially if coupled with a rocky seabed. A good example is shown in Figure 14.30, where the breakwater is located on a reef with a tidal range of 9.0 m and a design storm of (Hs = 7.0 m), with the toe exposed and dry at low water. The toe block

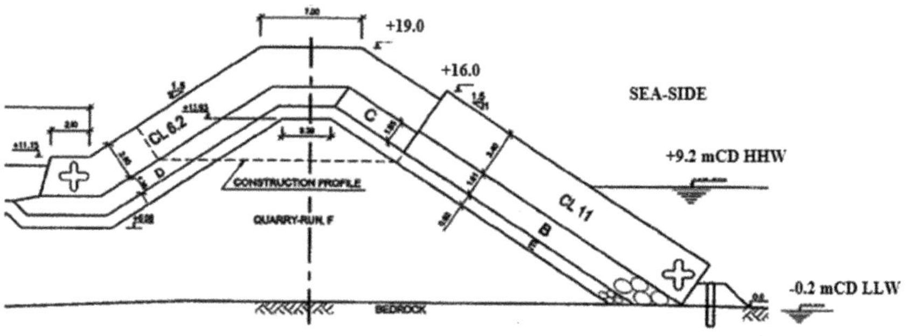

(a)

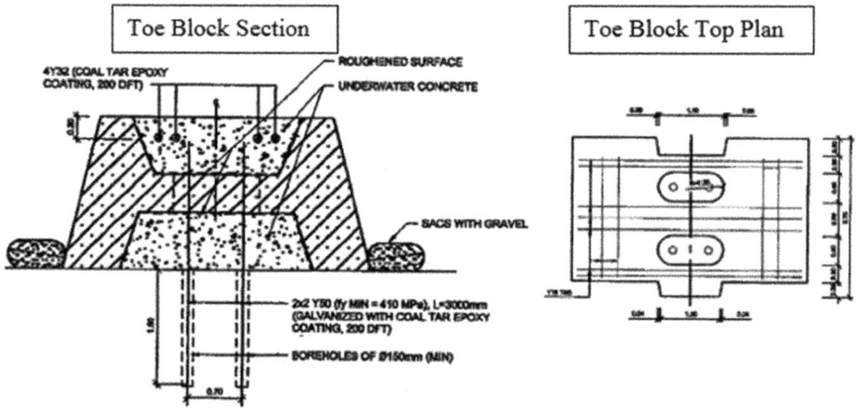

(b)

Figure 14.30. Toe block pinned to the bedrock (Mohammad *et al.*, 2002): Breakwater cross-section; (b) Toe block details.

dimensions are: 2 m high, 3.0 m wide at top and 4.0 m wide at bottom. The toe elements are keyed to each other, pinned to the bedrock by two 1.5 m long steel bars, with the top and bottom voids in the block filled by cast in-situ concrete. The bottom row of the Core-loc is placed by the canon method, see (Turk *et al.*, 1997), and not the random method, to minimise possible unit rocking which evidently leads to settlement. However, if the toe units are to be placed randomly, then they need to be restrained by grouting.

In 2002, the construction of Scarborough revetment started to cover and strengthen an old seawall, as shown in Figure 14.31. The revetment has a relatively short side slope length as the seabed is not deep. The used armour for the revetment is 6.0 m^3 Accropode. As the seabed is shallow and rocky, the designer selected short concrete piles to support and secure

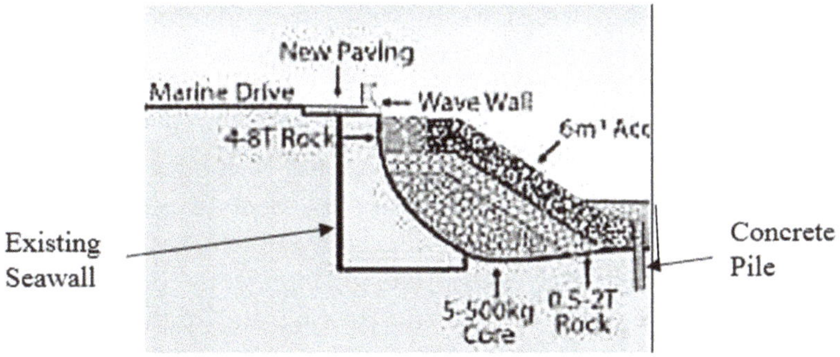

(a)

(b)

(c)

Figure 14.31. Scarborough revetment-design sketch and photos: (a) Design sketch; (b) Placement of the first row (toe units) against the piles; (c) Final constructed revetment.

the toe unit, see Figure 14.31(b). In 2004, many Accropode units were reported to have cracks in them, with few broken ones.

It is apparent that the toe units were not restrained against rocking, and as a result of the continuous rocking of the units, including those in the revetment face, the units kept knocking into each other, which eventually caused a number of them to break, see Sections 10.2.2 and 10.4. Investigation studies have revealed that the problem for the structure is that the Accropode units were too small, apparently designed for the high K_D factor of 16. This led to the rocking and reorientation of the Accropodes in the armour layer.

Piles are also used to support a beam toe to blockwork revetments as shown in Figure 14.32. Individual piles may be used to support a tie beam

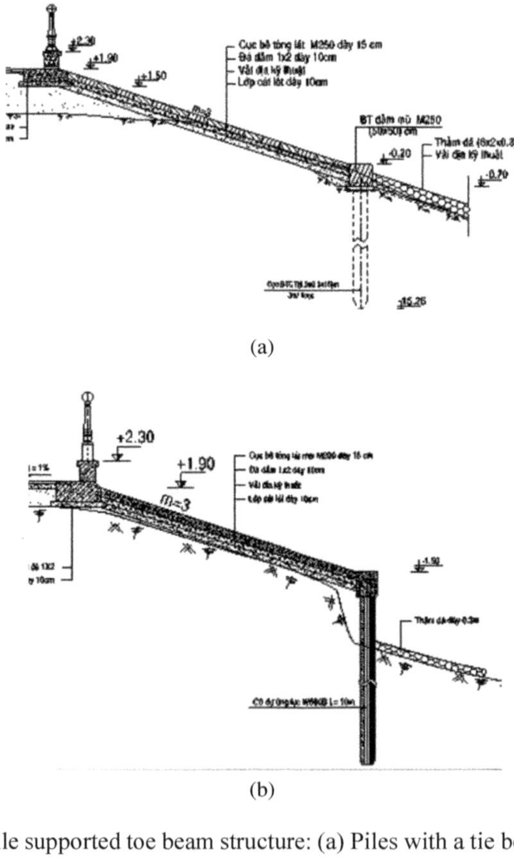

(a)

(b)

Figure 14.32. Pile supported toe beam structure: (a) Piles with a tie beam; (b) Sheet piles with a tie beam.

toe, as in (a), or a sheet pile to act as a supporting pile and as cut-off wall if there is a difference in level, as in (b).

14.5. Toe-berm

A toe-berm, or an elevated toe, is used in structures due to many different reasons. Selected examples are given in the following figures to illustrate few of the different types and configurations of toe-berms. Note that revetments and the seaside face of breakwaters shall be considered in this section, whereas toe-berm on the rear, or port side, of the breakwaters shall be discussed in Chapter 17.

A thick wide berm may be used because of a geotechnical requirement to stabilise a weak sandy seabed. In Figure 14.33, two examples are shown for comparison purposes, in (a) the seabed is firm and can withstand the weight of the revetment rock bund, whereas in (b) the seabed is relatively weak and the bund weight needs to be distributed over a wider

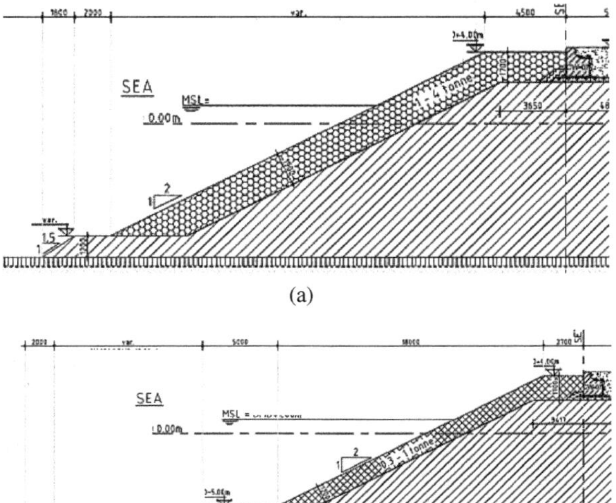

(a)

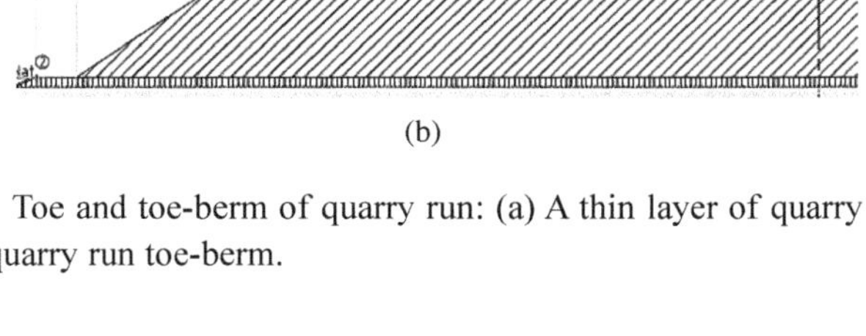

(b)

Figure 14.33. Toe and toe-berm of quarry run: (a) A thin layer of quarry run toe; (b) A thick layer of quarry run toe-berm.

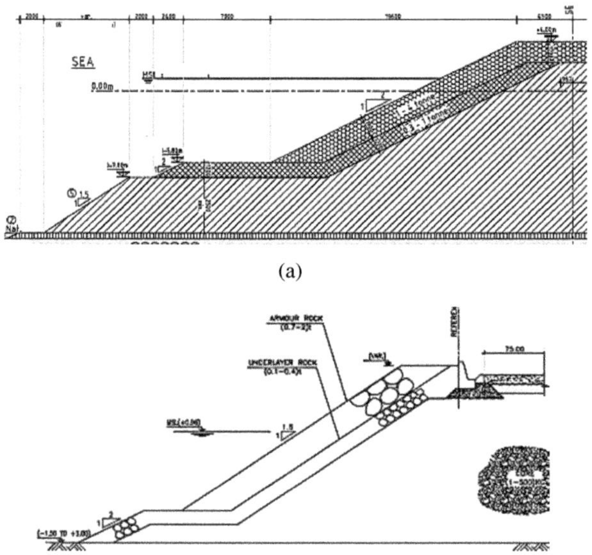

(a)

(b)

Figure 14.34. Toe-berms with partial and total protection: (a) Quarry run berm with partial top protection; (b) Quarry run berm with total (top and slope) protection.

area than the rock bund base to counteract any risk of a slip circle failure.

Two more examples for the toe-berm are shown in Figure 14.34, where in (a) the top surface of the quarry run berm is protected by the underlayer rocks. In (b), however, the underlayer has been extended to protect the quarry run top surface as well as the slope.

Figure 14.35 shows very wide toe-berms on both sides of a breakwater trunk constructed on a soft seabed. The wide toe-berms are intended to stabilise the breakwater against possible slip circle failure. However, this breakwater did suffer from continuous sinking of rock material and localised collapses. This could be attributed, among other possible causes, to the missing geotextile, which should have been used under the full width of the breakwater, see Section 13.7.

The stability of the rock on the toe-berm obviously needs to be checked using the appropriate design method. Therefore, the degree of exposure to hydraulic loading shall determine the appropriate size of rocks to be used and to what extent to ensure the stability of the toe-berm.

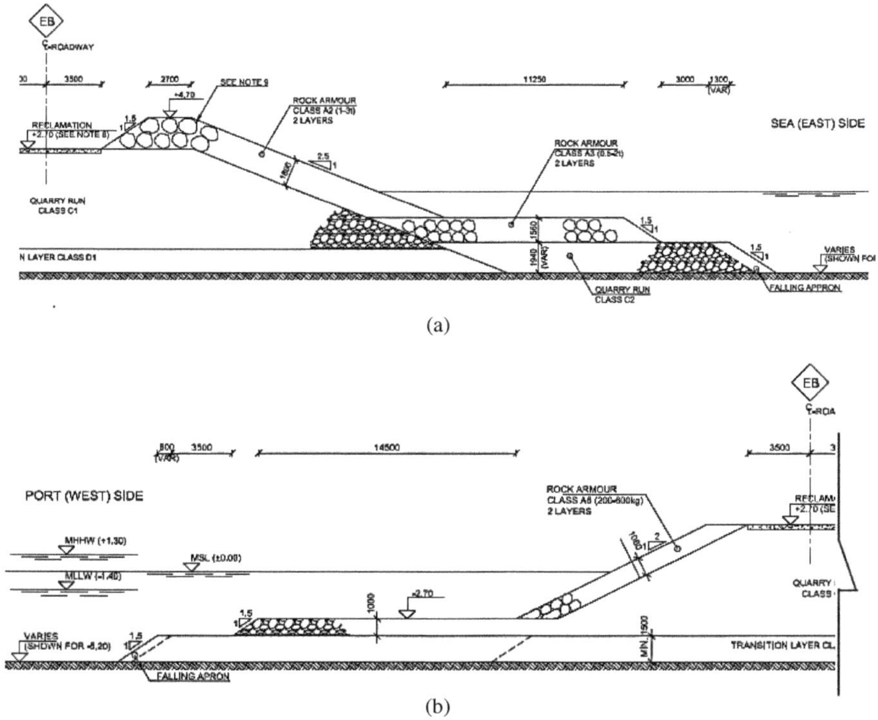

Figure 14.35. Wide toe-berm on soft seabed: (a) Seaside part of the breakwater (cross-section in trunk); (b) Port-side part of the breakwater (cross-section in trunk).

A toe-berm could be introduced to split a long slope face to form a stepped slope, which will reduce the hydraulic loading on the upper and lower slopes. The toe-berm is located below the design water level, the effectiveness of the toe-berm is related to its width and depth, see CIRIA, CUR, CETMEF (2007). An example for this type of toe-berm is shown in Figure 14.36, in which the designer managed to reduce the size of the tetrapod unit to 8.0 t only for a design wave height, Hs, of 5.5 m, and 2.0–3.0 t rock armour for the berm.

Other than the seabed geotechnical requirement, a toe-berm may be required when using interlocking single layer armour units. This type of armour unit has a maximum number of rows which cannot be exceeded, see Section 10.4, and if the revetment face is too long, then the slope is to be broken up into two slopes by introducing a toe-berm. Figure 14.37

shows three examples, in (a) the toe to the armour layer consists of two CAUs, and the berm consists of three layers of different rock sizes, a fine and coarse quarry run and a layer of 2–5 t rocks. In (b), a rock bund is used to reinforce the CAU toe units and extends to protect the slope of

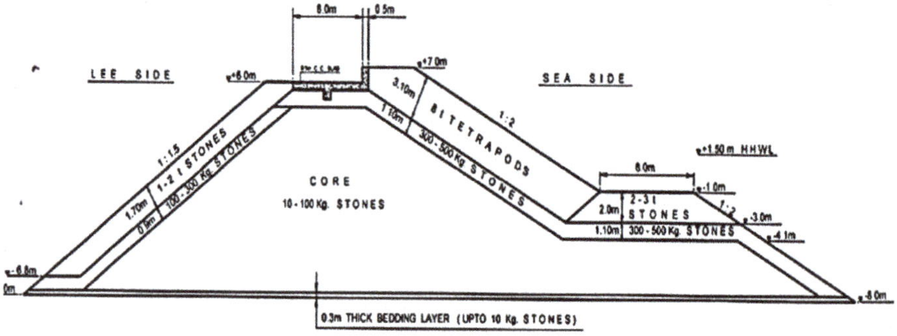

Figure 14.36. A toe-berm to reduce hydraulic loading on the armour layer (Shinde *et al.*, 2017).

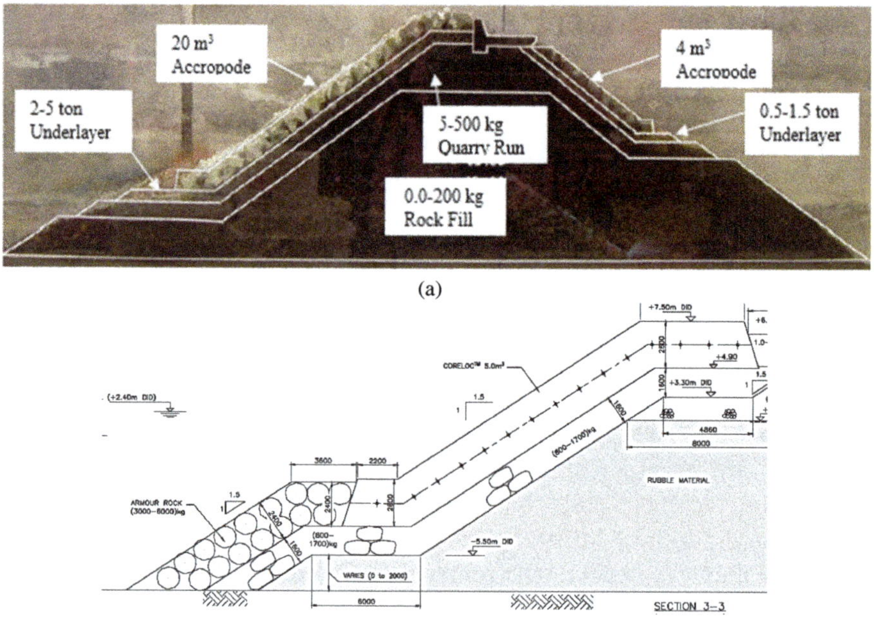

(a)

(b)

Figure 14.37. Toe-berm due to limited number of rows for Interlocking Single-Layer Armour Units.

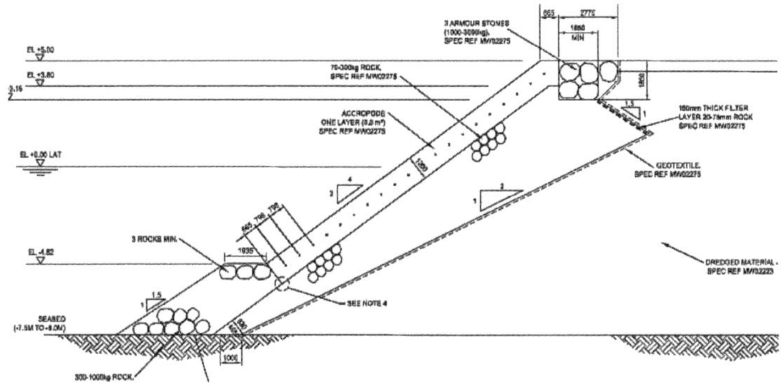

(c)

Figure 14.37. (*Continued*)

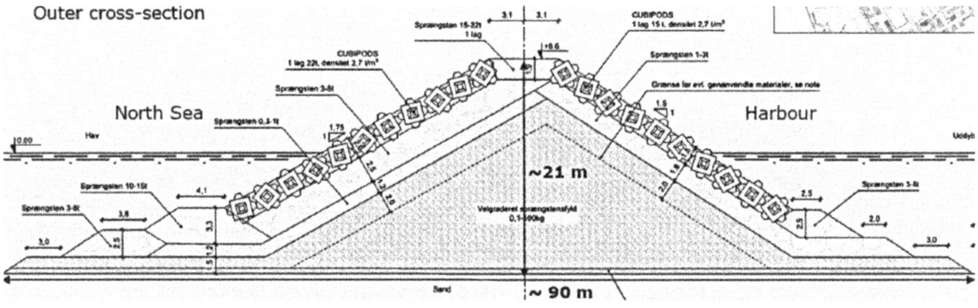

Figure 14.38. Toe-berm (compound toe) for construction and material availability reasons (Gierlevsen, 2019).

the berm. In (c), a different arrangement is used to form a rock bund toe, which is part of a berm.

A toe-berm, or a compound type of toe, is shown in Figure 14.38 on the seaside of the breakwater. The upper rock bund toe is made of 10–15 t rocks and the lower rock bund toe is of 3–6 t rocks. The designer has used this type of toe-berm arrangement not because of limitations to the armour number of rows (as the Cubipod, or the cubes in general, have no such limitations) but probably due to construction issues, material procurement limitations for the heavier rocks and instability of the 3–6 t rocks in the main toe.

A different reason for using a toe-berm is cost saving by optimising the use of the concrete armour units, which are generally more expensive than rock material, see the example in Figure 14.39. In long slopes, the starting point in armour optimisation is to identify the most exposed zone

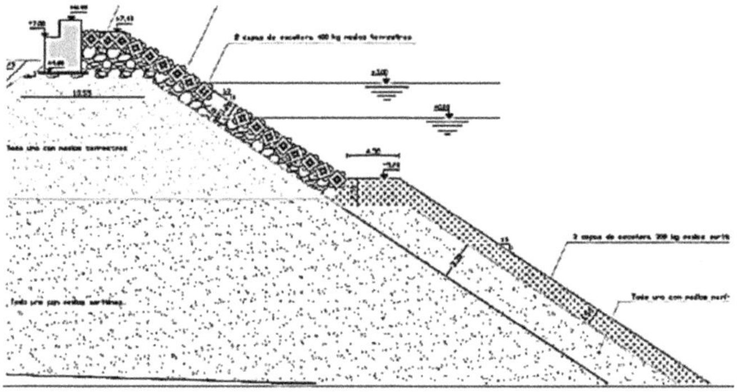

Figure 14.39. Toe-berm for cost saving (Cubipod, 2016).

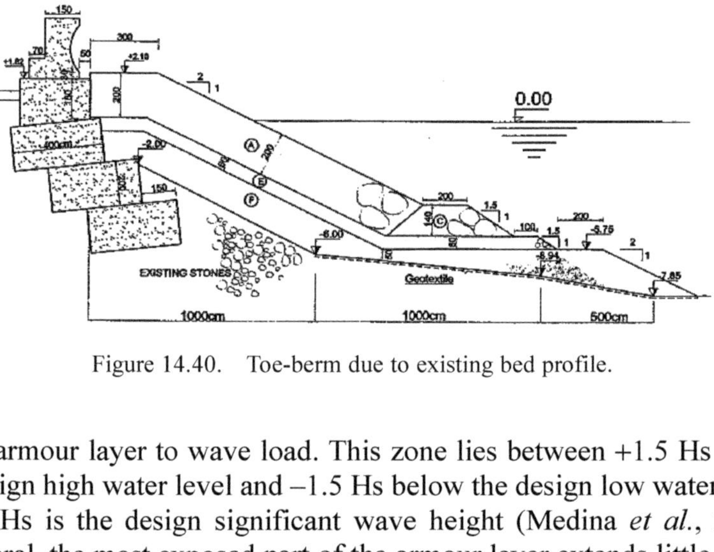

Figure 14.40. Toe-berm due to existing bed profile.

of the armour layer to wave load. This zone lies between +1.5 Hs above the design high water level and –1.5 Hs below the design low water level, where Hs is the design significant wave height (Medina *et al.*, 2016). In general, the most exposed part of the armour layer extends little below SWL and the largest forces on the armour occur during the wave run-down. In practice, however, the armour could be optimised outside the zone of ±Hs from the design high water level. The optimised design shall utilise a smaller armour unit or even a different type of armour, and it can only be approved after being verified by physical modelling.

A different reason, which could necessitate the use of a toe-berm is the construction of a revetment on an existing old seawall on a sloping beach front where the bed is not flat. An example for this situation is shown in Figure 14.40.

14.6. Apron/Scour Protection

An apron, or a scour protection layer, in front of the toe is an important element of the toe arrangement in all rubble mound structures founded on sandy/erodible seabeds. It is to provide protection against loss of support from beneath, and in front of the toe. If a scour hole develops, it will cause the toe to slump down, and in turn this could destabilise the whole face of the armour layer. An illustrative sketch is shown in Figure 14.41 to mark the difference in location between the bedding layer (see Section 13.7) and the apron, and also to indicate the possible location of a scour hole that might develop if the seabed is not protected. Note that if the bedding layer extends beyond the tip of the toe (if judged to be suitable), then it also serves as an apron.

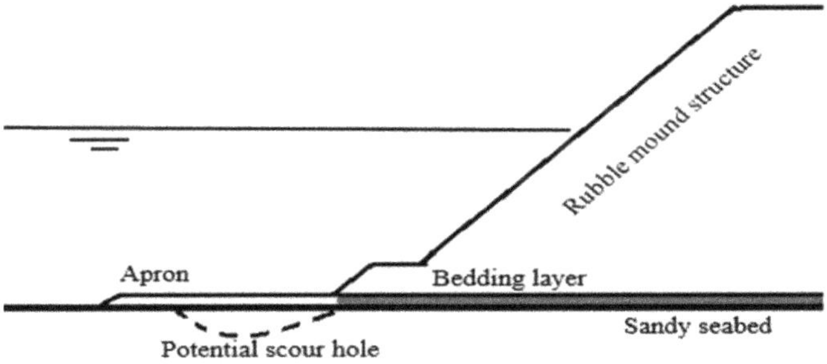

Figure 14.41. Illustrative sketch for a potential scour hole at the tip of the toe.

The risk of scour formation depends on the structure wave reflection coefficient, C_r, the wave properties, the relative water depth (h/L), local currents and seabed material characteristics.

The initial dimensioning of the apron in length and thickness, together with the proposed grading of the rock material and the possible need for a geotextile, are definitely to be fully evaluated in the detailed design. As a general rule, scour potential due to waves is greatest where the water depth at the toe is less than twice the design wave significant height, Hs. This is schematically illustrated in Figure 14.42 for a simple and basic toe configuration. In (a), a deep seabed is considered with no toe bund (as it is not required, however not advised with sandy seabeds), the recommended minimum apron width in this case is 3.0 m. Whereas in (b) a shallow seabed with a toe bund is considered, the recommended minimum apron width will be 1.5 m. The thickness of the apron layer is generally set to be in the range 0.5–1.0 m.

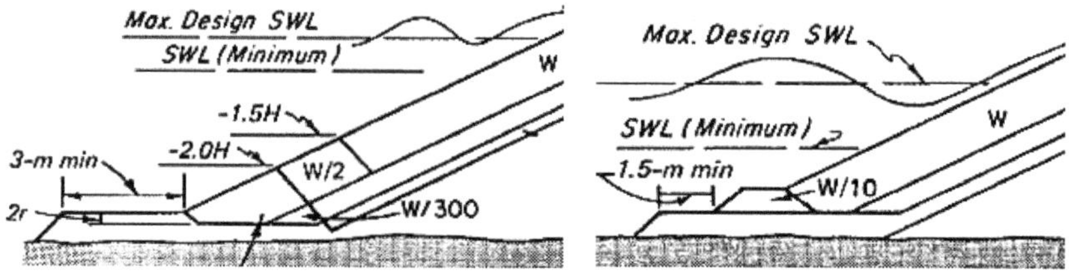

(a) (b)

Figure 14.42. Scour protection recommended length (SPM, 1984): (a) Deep toe; (b) Shallow toe.

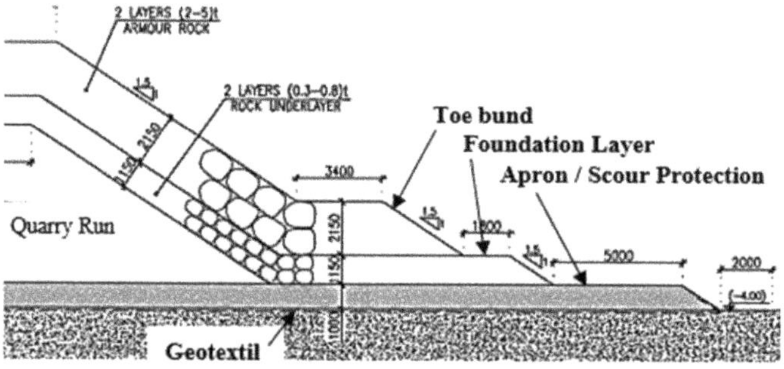

Figure 14.43. Basic configuration for the toe and supporting layers.

Various factors are to be considered that could contribute positively or negatively towards the exposure of the seabed and the susceptibility to possible scour. Significant aspects are to be carefully assessed, which include:

- ***Toe configuration*** — this is an important aspect, especially in shallow toes, as it affects how much of the breaking waves or the wave downrush energy reaches the seabed level, i.e. the apron. Figure 14.43 shows an example of a very common arrangement for the toe where there is a toe bund (by using the primary armour stones or a smaller rock class), a foundation layer (which is an extension for the underlayer), an apron (scour protection) layer and a geotextile filter. The addition of the foundation layer, or could be layers, will help in separating the intense turbulent zone and the apron farther apart.
- ***Currents and turbulence*** — currents are to be evaluated very carefully at the project location site as they could generally be attributed to seasonal changes, storms, tides, water density changes. These currents could intensify locally due to local shoreline or seabed formations or

due to certain features in the structure layout plan or sections. This would include breakwater roundheads, curvatures, corners and highly reflective structures like block walls or caissons. Sections of straight stretches of the structure are obviously the first to be evaluated, then sections in localised high currents and turbulence are to be assessed next. In these special zones, the apron will normally need to be elongated at least by two times, if not more, of that of the straight stretch. The foundation layer(s) will sometimes require to be elongated as well.
- ***Breaking waves*** — the wave characteristics and the structure configuration should jointly be checked for possible wave breaking conditions. If such conditions exist, proper enhancement to the apron should be considered, see Baart *et al.* (2010).
- ***Armour layer slope*** — if the structure face slope is steeper than 1:3, then additional precautions should be considered for the apron due to the reflected wave energy and wave downrush.
- ***Armour roughness*** — this depends on the type of the primary armour; more roughness means less wave downrush.
- ***Seabed material*** — two aspects are to be considered when evaluating the seabed sand material, the sand grain average size and the condition of the seabed, which could be cemented, partially cemented, loose drifting sand or newly placed as a soil replacement or as part of the structure's core material.
- ***Apron rock material*** — the selection of a material grading in relation to hydraulic stability, filter performance and the corresponding layer thickness.

The selection of the rock gradation of the apron depends on two factors, the hydraulic loading and the selected type of scour protection layer, i.e. open or closed filter type. If the hydraulic loading is mainly due to wave action, then it is more likely that larger stones than filter stone size will be needed. Whereas if the hydraulic loading is due to currents, then filter size stones will normally be sufficient, however with a wider apron layer as normally is the case with scour due to currents. If the selected apron stone gradation can be a closed filter layer for the seabed material, then there will be no need for a geotextile, otherwise a geotextile shall be used to prevent loss of seabed material from under the apron layer. The latter case, i.e. an apron with a geotextile, is frequently used as the quarry run core material is easier to extend to form the apron rather than using a closed filter, i.e. additional material type. Note that larger stones (heavier than 100 kg) are to be removed from quarry run material used for the apron.

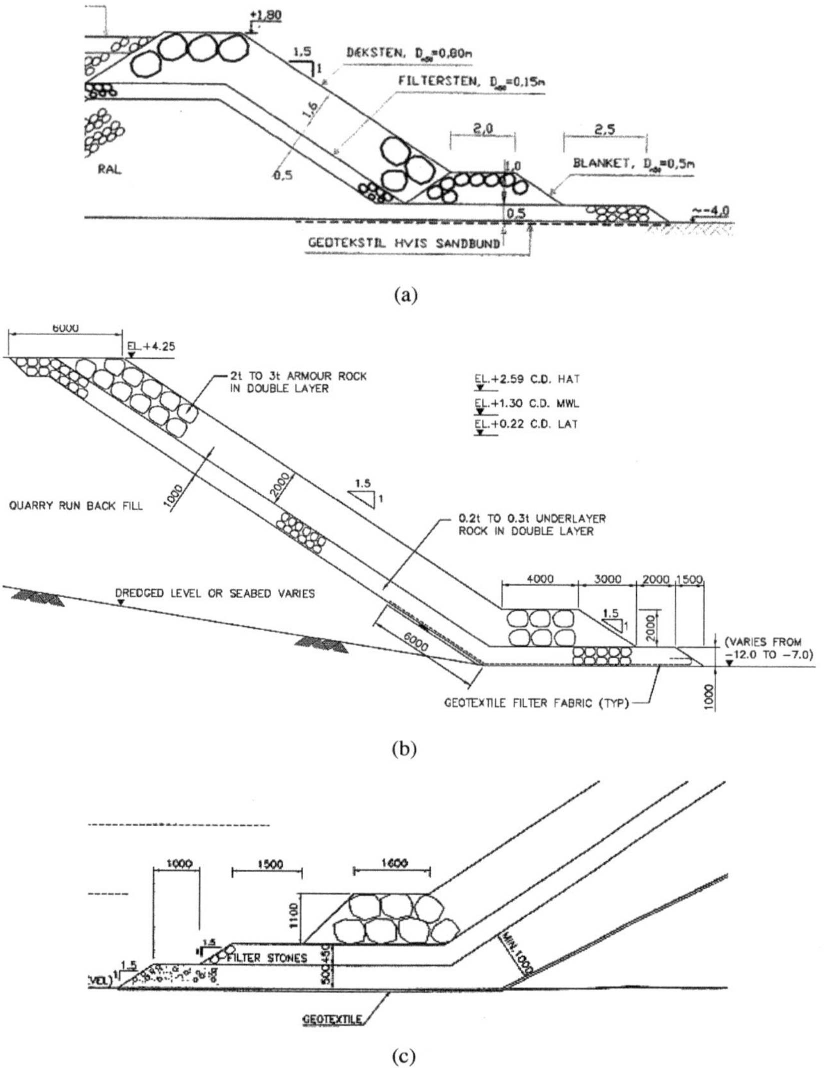

Figure 14.44. Examples of an apron with a geotextile: (a) Quarry run apron with geotextile; (b) Underlayer stones used for an apron with a geotextile; (c) A foundation layer and an apron with a geotextile.

Illustrative examples are given in Figure 14.44 for an open type apron with a geotextile, in (a) and (c) the apron is made of quarry run. However, in (b) the underlayer stones are used, and this is not a recommended practice to place larger stones on a sandy seabed, albeit on a geotextile. Aprons

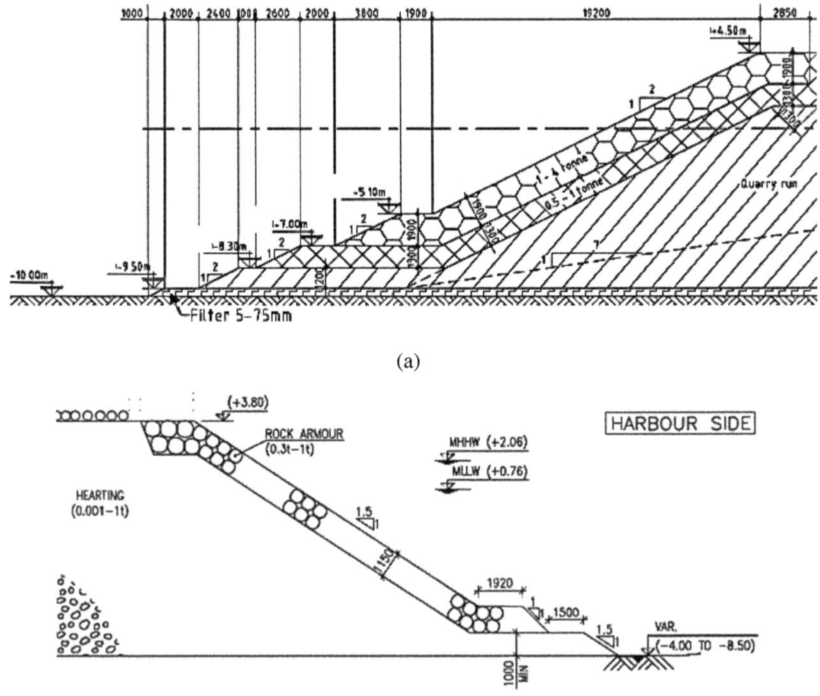

Figure 14.45. Examples of aprons without a geotextile: (a) An apron of filter stones with two foundation layers; (b) A relatively thick apron of quarry run inside the harbour basin (no waves or currents).

without a geotextile are given in Figure 14.45, in (a) for a closed filter apron and in (b) for an open filter apron, however, it was made of a thick layer and located in a calm harbour basin, both will effectively render any possibility for a scour to be obsolete.

14.7. Falling Apron

A flexible and dynamic type of apron was initially developed to protect river training works and estuaries' banks against scour caused by high currents instead of a well-defined static apron, i.e. a scour protection layer, as presented in the previous section. This is because the river-bed level is highly influenced by the amount of flow and currents. Hence, a static scour protection could only be constructed if the bed is to be dredged down to the expected scour hole depth to place the scour protection, then reinstating the dredged material to cover the scour protection layer. Such an operation is very difficult, or sometimes practically cannot be achieved, as the dredged trench may be filled very quickly while dredging is on going and consequently it will be very expensive.

The falling apron is a self-deployed scour protection following the development of the side slope of the scour hole. A sufficient store of rock material on the seabed must be ensured, either in the form of blanket or a bund, see Figure 14.46. It will fall in place, or be launched, progressively as the scour hole gets deeper until it reaches its maximum stabilised depth. Note that the required rock material for the falling apron could also be either partially or completely trenched in the seabed (Pilarczyk, 2003). The scour protection layer will eventually protect the scour hole side slope as it develops by gradually assuming a slope of approximately 1:2 (CIRIA, CUR, CETMEF, 2007), preventing it from progressing any further to undermine the structure. However, the natural development of the rocks takes place in one layer. This is, in most cases, not sufficient to stabilise the sand underneath as it may be sucked out through the rock layer.

A falling apron does not completely protect the sand underneath the apron from being washed away, but it does retard the scouring process significantly. However, different rock gradings do provide different degrees of protection (Verhagen *et al.*, 2003). The resulting layer thickness could range between $1D_{n50}$ (as observed in laboratory tests) to $5D_{n50}$ (field measurements). If relatively large stones are used, then the layer would have large voids which cannot retain the sand. It is therefore recommended to use wide graded rock material, which can produce a layer nearly fulfilling filter rules. Thus, quarry run would be a suitable material to be used as a falling apron, though it is preferred to have the upper grading limit not heavier than 200 kg. However, the size of rocks should still be selected to ensure they are stable for the waves and currents.

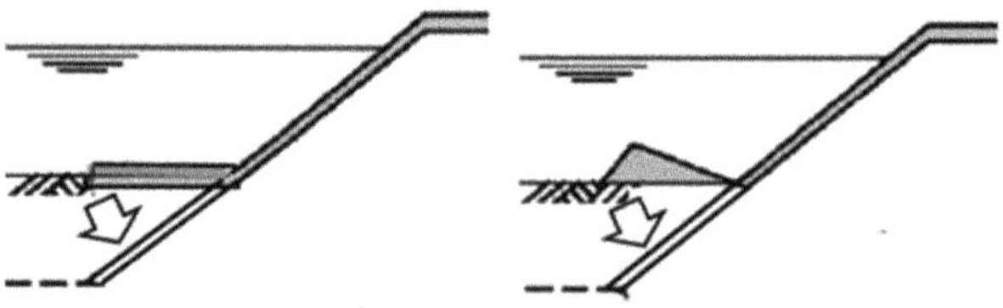

(a) *Blanket* (b) *Bund*

Figure 14.46. Rock material for the falling apron.

A static apron is widely used in coastal structures, however, a dynamic and flexible type of scour protection, i.e. falling apron, has been adopted and used in some coastal structures where the structure type and seabed condition allow for such a scour protection.

Figure 14.47 illustrates the basic dimensions for a falling apron type of scour protection to ensure the existence of sufficient rock material required to provide protection to the developing scour hole. The figure is based on recommended dimensions as given by Kirby *et al.* (2015) with slightly more elongated apron length following the recommendation of CIRIA, CUR, CETMEF (2007).

An example is shown in Figure 14.48 to illustrate the use of a falling apron type of scour protection with partially entrenching the tip of the

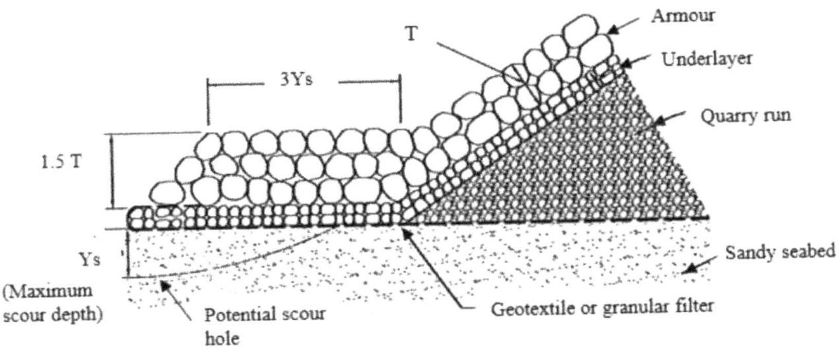

Figure 14.47. Falling apron basic dimensions and configuration.

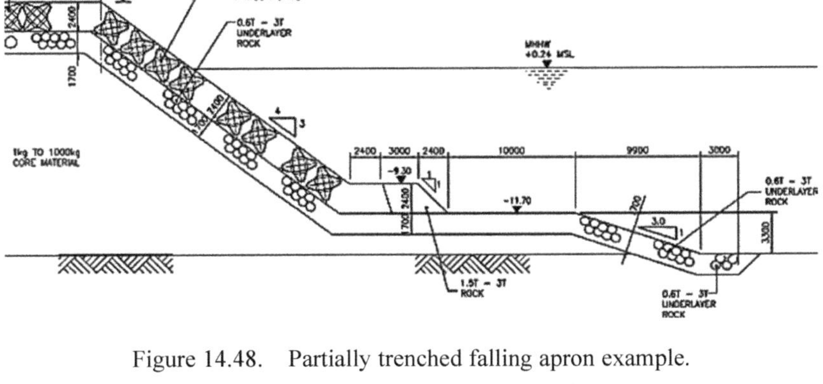

Figure 14.48. Partially trenched falling apron example.

apron layer, with no geotextile used. It is clear from this example that the apron was not made of quarry run only but needed the protection of larger stones (0.8–3.0 t) because of the wave downrush and the possibility of breaking waves. The trenched part of the falling apron was made of the larger stones without the quarry run because these will sink faster into the sand, due to waves and currents, prompting the launch of the apron into the scour hole.

It is, however, important to note that at least 1.0 m of the apron must remain stable, i.e. flat, and not be part of the rocks falling into the scour hole. This is to ensure the stability of the toe bund and consequently the armour layer face.

A smaller version of the previous example is shown in Figure 14.49 where a relatively shallow toe is to be constructed on a sandy shoreline. The expected scour hole in this example is in the order of 1.0 m, hence the placed material is relatively short horizontally, i.e. 3.9 m. As the toe is shallow, it will be exposed to wave breaking, therefore the armour stones need to be part of the falling apron rock material protecting the quarry run. This example could well have been designed similar to Figure 14.48, but the contractor preferred to avoid any seabed excavation.

A different example is shown in Figure 14.50 where the seabed is composed of caprock covered with a relatively thin layer of sand. Additional sand under the toe comes from the structure's core as it is made of hydraulically placed sand. In this type of sand placement, the sand will normally flow outside its design limit, hence increasing the sand layer thickness under the toe. The scour hole depth in this example is of course limited to the depth of the caprock layer, but the sand layer thickness is not known due to the flowing sand from the core. The depth has to be

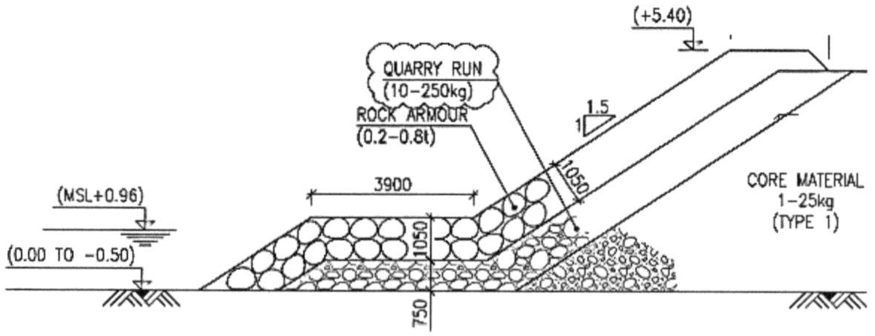

Figure 14.49. A small falling apron example in breaking waves zone.

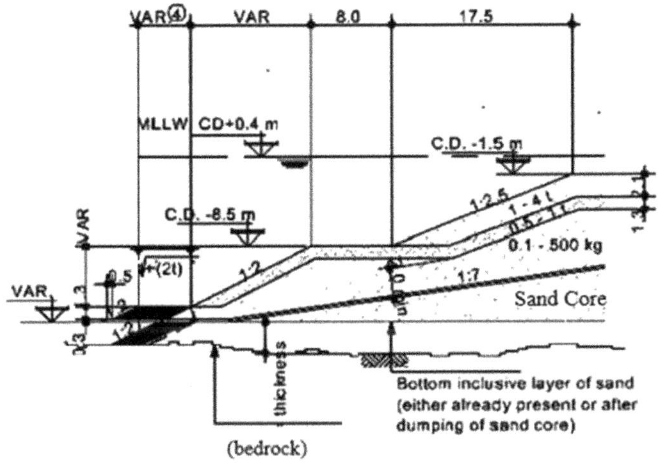

Figure 14.50. A falling apron with limited scour hole depth.

measured along the structure, say every 50 m, to determine the required length of the falling apron for that stretch, which is shown on the design drawing as "variable". Such a variation in the design, as construction progresses, will require special clauses in the construction contract.

The wave action at this toe is very small due to its depth and the section configuration of the armour layer face, which has a relatively flat and stepped slope. Accordingly, the main cause for scouring will be the currents. Therefore, quarry run rock material has been used for the falling apron with no need for larger rocks' protection as in the previous examples.

Falling aprons can provide significant protection against scouring, however, periodic monitoring is necessary to ensure that it has behaved as expected and to determine if maintenance is required by dumping of extra material.

14.8. Inferior Toe Design

Few toe design examples are given in this section to illustrate some issues which could be improved or avoided while developing the design to be more cost effective, more robust or practically constructable.

The toe of a CAU armour face must be secured as firmly as possible to avoid any possible settlement, as discussed in Section 14.3. However,

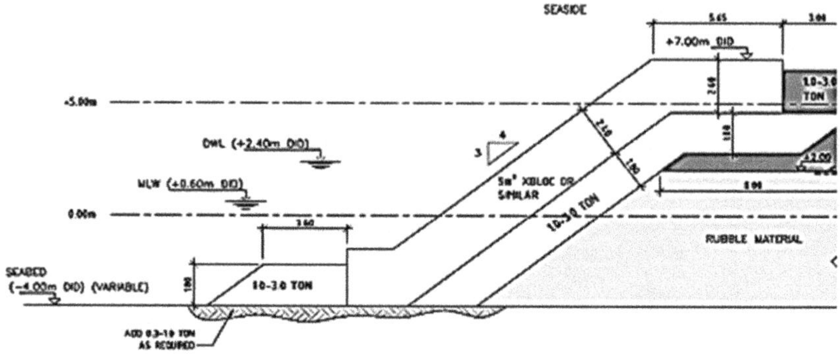

Figure 14.51. Insecure armour foundation layer.

a number of designers are taking this issue rather lightly. An example is shown in Figure 14.51 where the seabed is made of caprock with some scattered patches of sand with variable depths. The designer in this example recommends removing the sand, if it comes under the toe, and fill the pit with 0.3–1.0 t rocks. This is of course not the correct solution to this sand problem, as the sand could be of irregular thickness and this could lead to an irregular bed under the CAU because the extent and thickness of the fill has not, and cannot, be fixed in such a design and need to be finalised during construction. All these complications leave many doubts regarding the final integrity of the toe. It would have been better to indicate that if the sand thickness is more than 0.40 m, then it should be removed and replaced with selected quarry run with no rocks heavier than 100 kg. The CAU underlayer of 1.0–3.0 t rocks is to be extended under the CAU toe to form a proper foundation layer.

A second example for a CAU toe is shown in Figure 14.52, but on a sandy seabed and not bedrock. The toe is placed on a 1.0 m thick quarry run layer with an extra-long reinforcement rock bund. This type of a foundation layer, although used by many designers, is not actually preferred as the toe unit will not be fully restrained from rattling or rocking movement, and this will result in the armour face gradually sinking into the quarry run layer. Extending the underlayer of 1.0–2.0 t rocks under the toe will provide a much better solid and restraining foundation layer to the CAU. The second comment on this design is on the reinforcement rock bund in front of the CAU toe unit. This bund is only added for an extra safety as the CAU toe unit should be stable on its own. Therefore, a 3.0 m length (3 rocks of 2–4 t rock class) is considered sufficient.

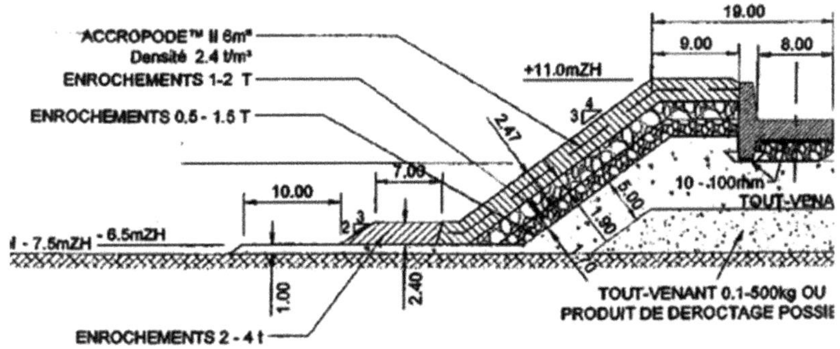

Figure 14.52. Insecure armour foundation layer and long reinforcement rock bund.

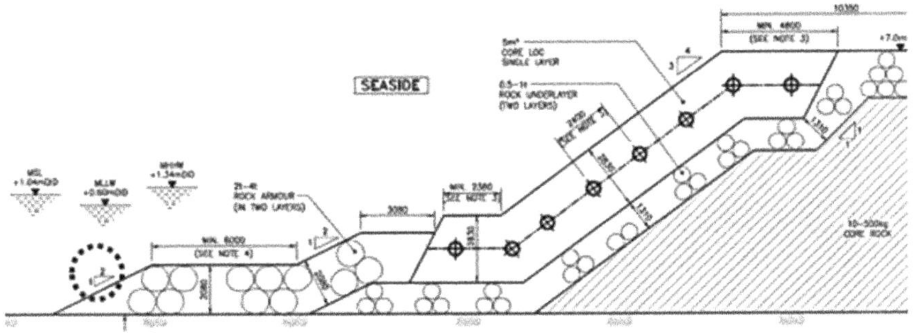

Figure 14.53. Not required apron of armour rocks.

A third example for a CAU toe is given in Figure 14.53. The toe unit has been placed on a foundation layer of the underlayer rocks. A reinforcement rock bund has been used and extended for further 6.0 m in what look like an apron. The design report did in fact describe this extension as a “falling apron” without offering any explanation for its necessity on a bedrock where it could not fall. The design should have extended the foundation layer a little more and used a smaller reinforcement rock bund, similar to Figure 14.11 (without the quarry run layer), resulting in significant saving in rock material.

Figure 14.54 shows another example of using excess material, for no reason, in the 6.0 m long foundation layer on a rocky seabed. Probably the designer was aiming to push away the tip of the layer from possible wave downrush or breaking waves turbulence. If there is such a risk, then at least one layer of the primary armour should have been used as a toe on top of the foundation layer.

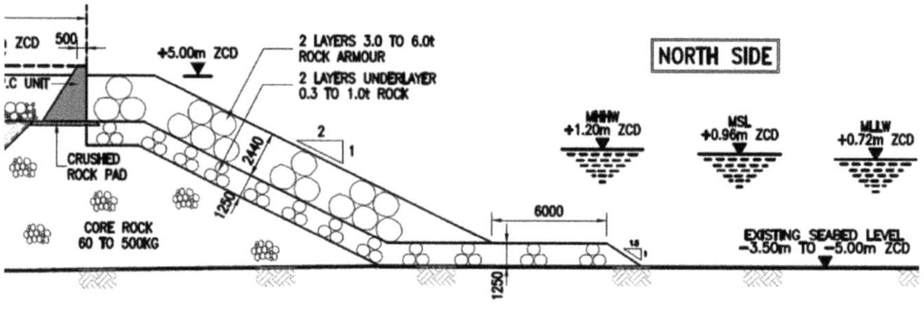

Figure 14.54. Long foundation layer on a rocky seabed.

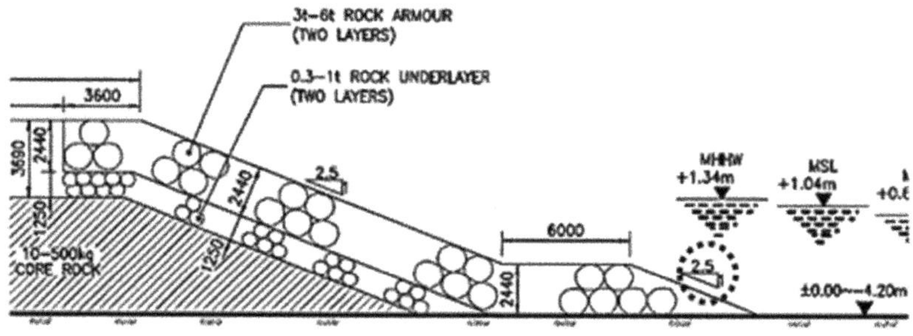

Figure 14.55. Long toe and incorrect toe side slope.

A frequent design flaw appears where the designer uses a relatively flat side slope for a two-layer toe, namely 1:2.5 as in Figure 14.55, or 1:2 in Figure 14.53. Practically the tip of the toe will have a slope of almost 1:1.5, hence most of the designers use this slope for rocks in two layers. Flatter slopes cannot physically be achieved from two layers, unless of course the bottom layer is extended as a single layer, and this is something which has never been specified by the designers who used such slopes.

One further comment on Figure 14.55 regarding the 6.0 m long toe, which is five rocks wide at the top layer, being excessive and not required. If the designer's intention is to provide more resistance to possible slippage, then using the underlayer as a foundation layer, as in Figure 14.54, and a normal toe would have worked better.

One of the common slips of designers is to give unnecessarily too much information/requirements on the drawings, which could cause confusion or contradiction. Figure 14.56 is an example where in (a) there is a

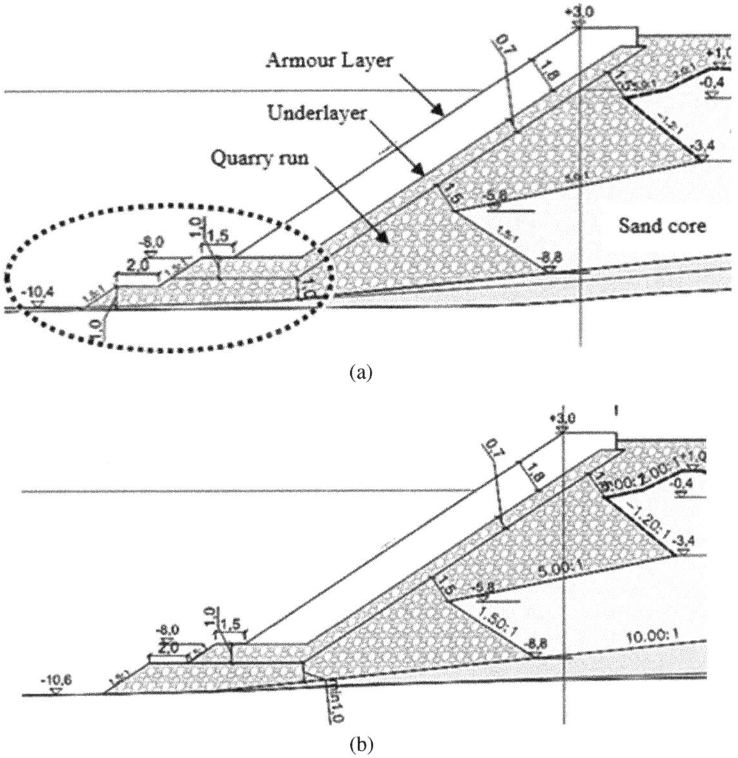

Figure 14.56. Toe level and thickness requirements.

1.0 m quarry run to be provided between the underlayer bottom corner and the sand fill, at the same time the quarry run of the scour protection layer is to be 1.0 m thick. Furthermore, the toe level is to be at –8.0 m. These many requirements led the designer to have a small step between the scour protection layer and the toe, and this means adding one more construction step to the toe. In (b), the same toe is shown with less requirements, which led to a simplified and a more practical design.

Chapter 15

Crest

15.1. Introduction

The crest design for a coastal structure involves determining and considering many aspects related to this element. These aspects would include: the height, the width, type and size of armour, access provisions (for maintenance or operational requirements), allowable overtopping discharge, possibility of using a crown wall (size and fabrication method, i.e. precast/cast-in-situ), wave agitation criteria on the leeside, aesthetic requirements and construction methodology.

Breakwaters and revetments without superstructures are considered in this chapter, those with a superstructure (crown/crest wall, wave wall, access concrete slab or even just a small separation wall) will be presented in Chapter 16.

15.2. Breakwater Crest

The required protection to the crest relative to the sea-side slope protection basically depends on the crest free board, Rc, relative to the design wave height. Width of the crest could also be taken into consideration, and also the construction methodology and feasibility of using different armour blocks. Based on the aforementioned factors, the armour block size (CAU/rock) for the crest protection could be smaller, the same or even larger than the seaside slope protection.

Breakwaters are generally categorised into two categories (EAU, 2012), depending on their freeboard (Rc), as follows:

Low crested breakwaters, with Rc < 1.0 Hs (with considerable wave overtopping), and,
High crested breakwaters, with Rc > 1.5 Hs (with little to negligible wave overtopping)

The above is to be considered as an appropriate indication for the expected overtopping discharge. To be more accurate, the overtopping depends also on the wave run-up level, which in turn depends on the water level, the structure slope configuration, armour type (i.e. roughness) and wave characteristics, specially the wave steepness. The above categorisation could then be more refined, based on Figure 15.2, into the following more representative categories:

Low crested breakwater, with Rc < 0.8 Hs
Medium crested breakwaters, with 0.8 Hs < Rc < 1.2 Hs
High crested breakwater, with Rc > 1.2 Hs

In general, the crest is normally not to be high enough to mitigate wave overtopping from taking place at design storm conditions. Therefore, the loading on the crest and the rear-side will depend on the percentage of the wave energy allowed to pass over the crest. This is to be carefully assessed during the design, and, in many cases, to be verified by the physical model testing. It is however highly recommended to carry out a joint probability analysis to select the design event (water level and wave height) for the wave overtopping as per the project design return period.

Although the crest freeboard (Rc) is the main factor in determining the amount of wave overtopping across the sea-side edge of the crest, the crest width also plays a part in reducing the overtopping discharge which crosses the rear-side edge of the crest and runs down on the lee-side slope.

Crest width is defined as the distance between the two design edges. The ***minimum*** recommended width to ensure an acceptable degree of support to the crest armour blocks is as follows:

Rock armour: 3–4 D_{n50} (CIRIA-CUR-CETMEF, 2007)

Accropode/Core loc: 4 Dv (CLI, https://www.concretelayer.com/en), where Dv = 0.55H (H is the unit height)

Xbloc: >2.28 D (Xbloc, 2018, https://www.xbloc.com/.),
where D is the unit height.

The above requirements will ensure at least three rows at the crest (at the same horizontal level), plus the two crest edge rows. Note that the recommended number of rows, for rock and CAU, are for crests with Rc > 0.8 Hs.

The construction methodology of the breakwater, i.e. land-based or marine equipment, is the deciding factor on crest width being as narrow as possible or wide enough to allow for the movement of the construction equipment on a construction access road on the core material, see Figure 15.1.

The presented examples in Figure 15.2 cover the commonly used range of crest heights in breakwaters, i.e. the low/medium/high crested

(a)

(b)

Figure 15.1. Narrow and wide breakwater crests: (a) Narrow crest; (b) Wide crest.

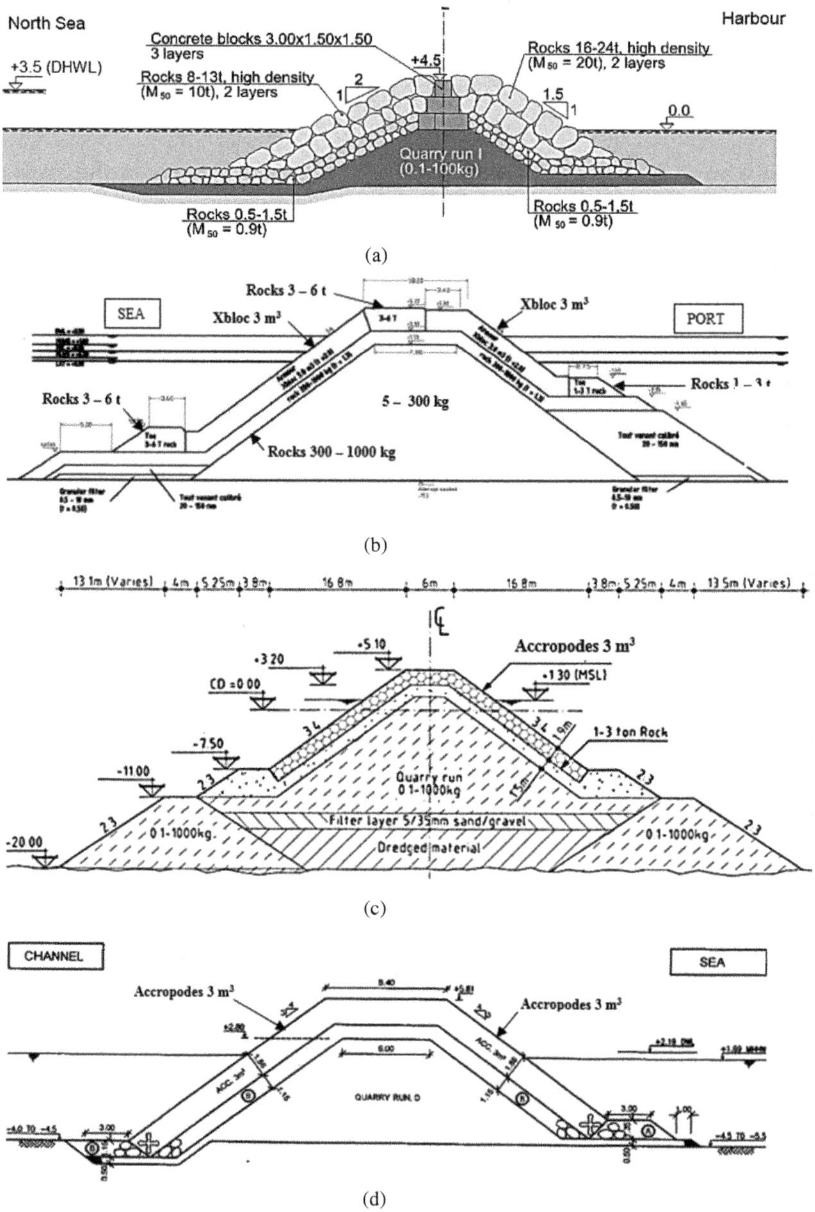

Figure 15.2. Breakwaters with different freeboard (*Rc*) values: (a) $Rc = 0.2$ Hs (Gierlevsen, *et al.*, 2013); (b) $Rc = 0.5$ Hs (Van Swicht, 2013); (c) $Rc - 0.62$ Hs; (d) $R_C = 0.8$ Hs; (e) $Rc = 0.82$ Hs; (f) $Rc = 0.9$ Hs; (g) $Rc = 0.9$ Hs; (h) $Rc - 0.9$ Hs; (i) $Rc = 1.0$ Hs; (j) $Rc = 1.0$ Hs; (k) $Rc = 1.0$ Hs; (l) $Rc = 1.3$ Hs; (m) $Rc = 1.4$ Hs; (n) $Rc = 1.3$ Hs (Jensen, 1984); (o) $Rc = 1.6$ Hs (Jensen, 1884).

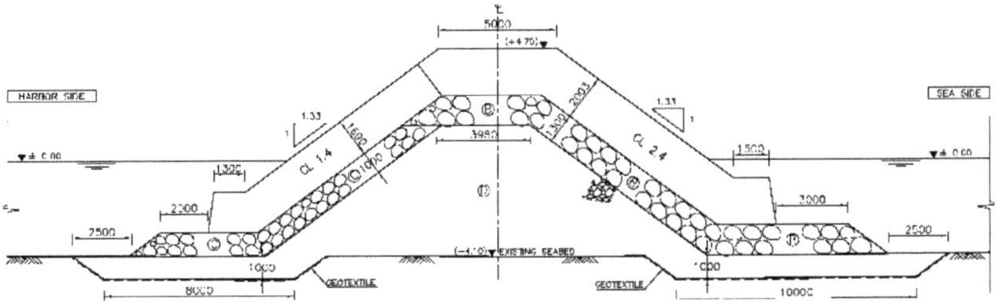

(e)

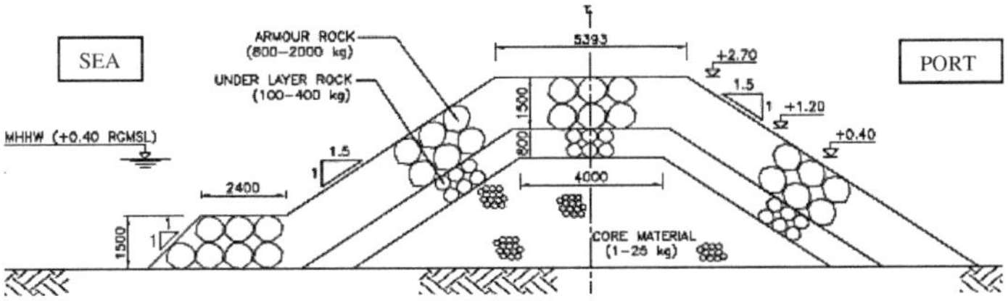

(f)

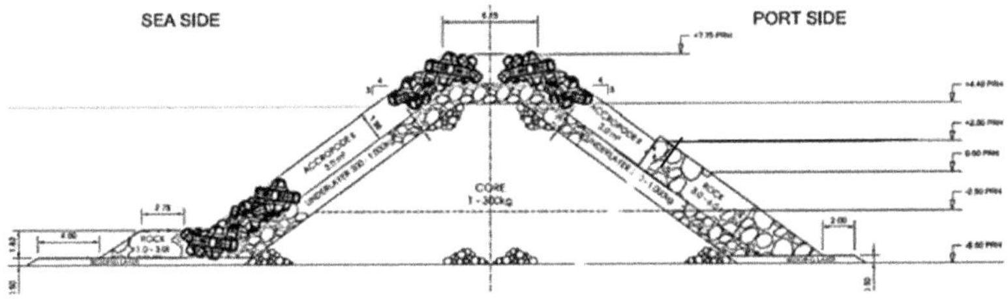

(g)

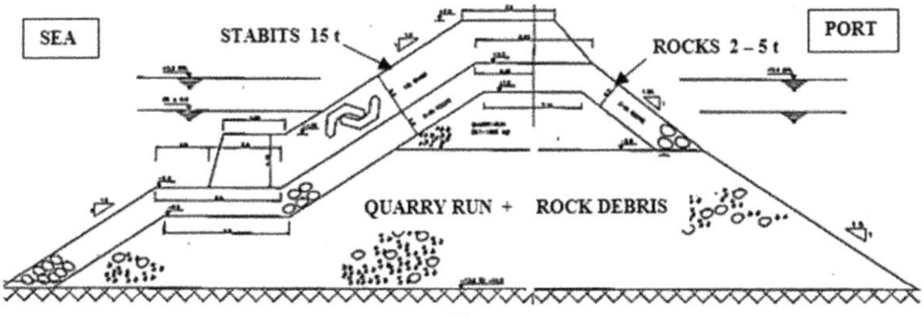

(h)

Figure 15.2. (*Continued*)

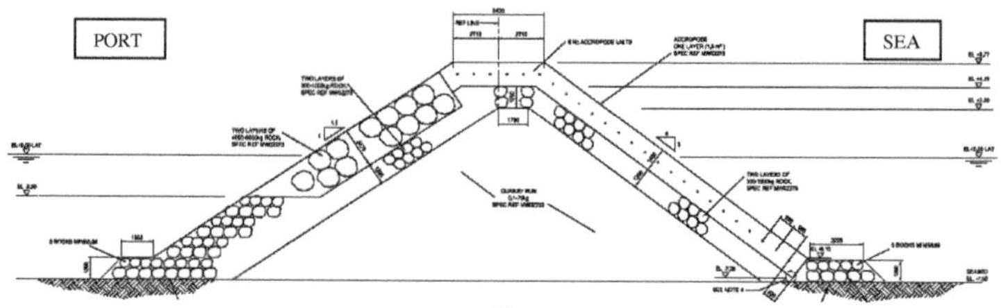

(i)

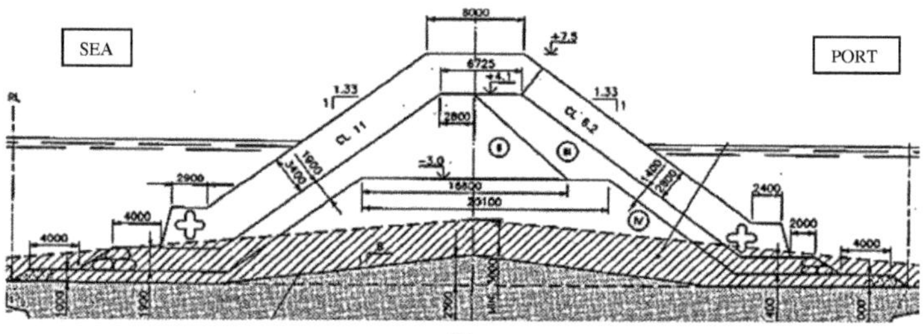

(j)

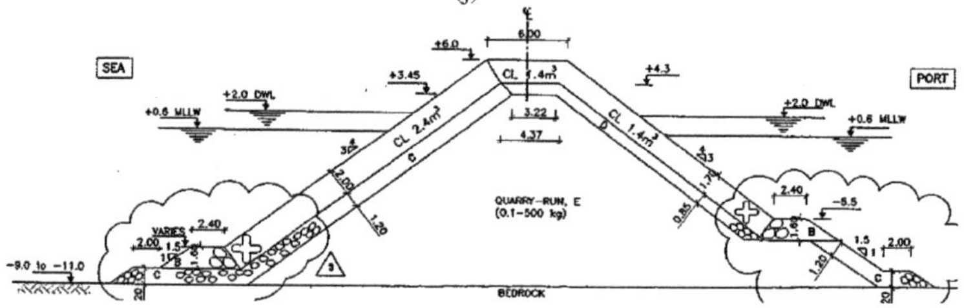

(k)

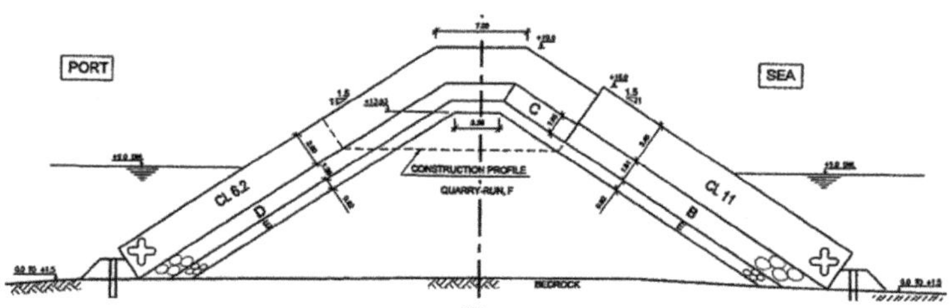

(l)

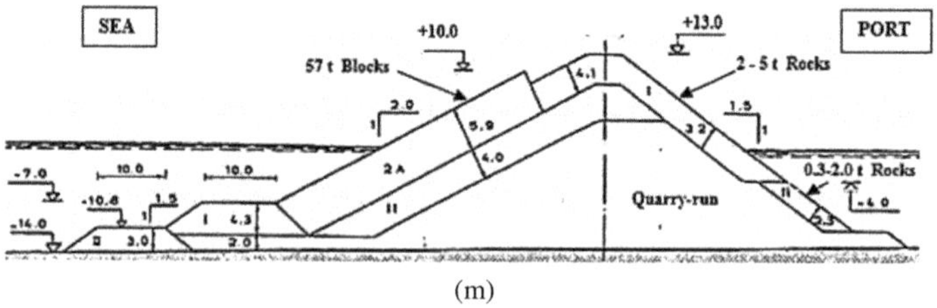

(m)

Figure 15.2. (*Continued*)

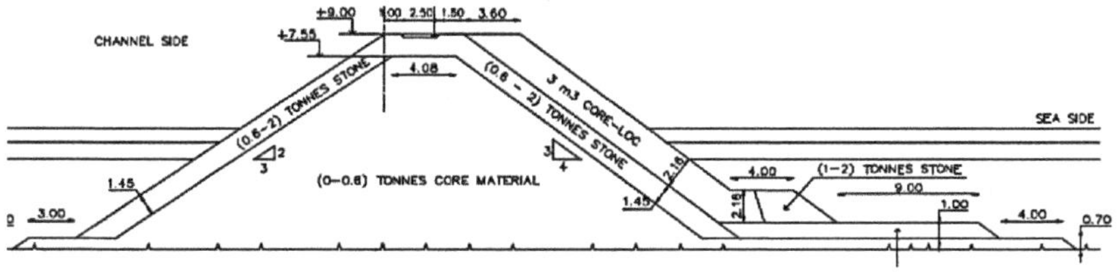

(n)

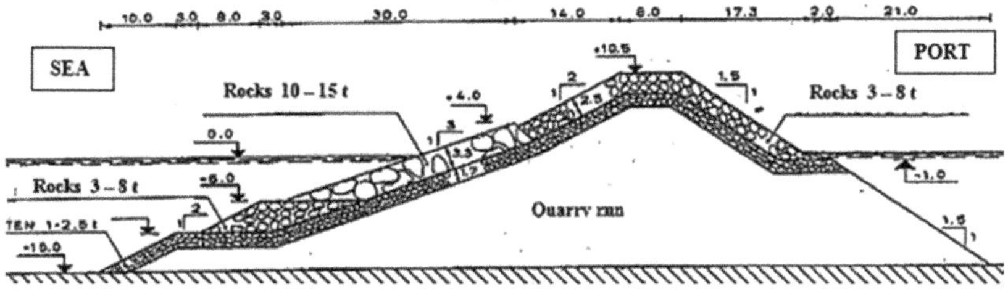

(o)

Figure 15.2. (*Continued*)

with Rc relative to Hs. Note that Hs is the design significant wave height and Rc is measured for the design high water level (DHWL).

The first category is the low-crested breakwaters, i.e. Rc < 0.8 Hs, as illustrated by the examples in (a), (b) and (c). The example in (a) has some concrete blocks forming a curtain at the crest centre. The blocks do not form a superstructure, but they are used to form an impervious barrier to sand carried by the overtopping waves, as the sand can easily pass through the large voids in between the rocks at the crest. Note also that the rock armour on the harbour side of the crest and the rear-side slope are larger than the sea-side slope and front part of the crest. This is as expected in such a low crested breakwater. In example (b), the 3.0 m^3 Xbloc unit size was shown to be unstable in the physical model if used alone to protect the crest, therefore it was used in combination with two layers of 3–6 t rocks to achieve a more stable crest. An Accropode unit of 3.0 m^3 size has been used in example (c) on seaside, the rear-side and the crest.

The next category of the examples, from (d) to (k), is for the widely used medium crest height, i.e. Rc = 0.8–1.2. A wide range of possible crest designs in this category are presented, which could be sub-grouped as follows:

One armour block size (CAU/Rock) for the three sides (front, rear and crest) – as in (d) and (f).
One CAU size for the seaside slope and crest, and a smaller size CAU for the rear-side – as in (e) and (j).
A large CAU size for the seaside slope and a smaller size CAU for the crest and the rear-side – as in (k).
One CAU size for the seaside slope and the crest with rocks for the full/ partial rear-side – as in (g), (h) and (i).

The final category is for high crested breakwaters, i.e. Rc > 1.2 Hs, which is illustrated by the examples in (l), (m), (n) and (o). In these examples, it has been possible to reduce the primary armour size on the upper seaside slope from a level approximately equal to Hs above the DHWL as in (l) and (m), or even lower as in (o), which also has a steeper upper part slope combined with a reduction of the rock size. In (n), the main armour protects the seaside face only and the underlayer uses the armour layer for the crest and the rear side.

15.3. Revetment Crest

There is a wider variety of revetment crest designs in comparison with the breakwaters as there are normally more demands to be fulfilled to meet the design criteria. Furthermore, there will be an interface between the revetment crest and the fill material behind, which is to be integrated in the structure, see Section 7.3.

The revetment crest level will most likely be the first item to have a value assigned to it at the preliminary design stage as the design of many other items will depend on this level. The end use requirements of the land behind the revetment will normally dictate the crest freeboard, Rc, as this will determine the overtopping discharge Q_1, as shown in Figure 15.3. The width of the crest, B_1, will determine the overtopping discharge, Q_2, which will pass beyond the revetment crest. The acceptable limits for this discharge are dictated by the type and extent (B_2) of surface protection and how much discharge is allowed (Q_3) to cross beyond B_2.

The width of the crest (B1) is similar to that of the breakwater, and is recommended to be at least three rocks/units.

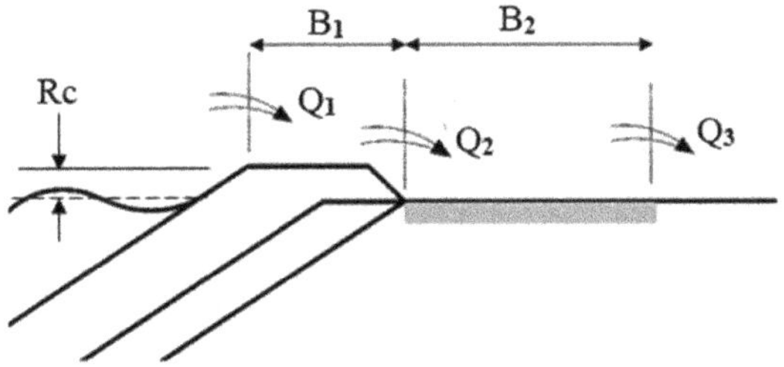

Figure 15.3. Wave overtopping discharge across the revetment crest.

Revetment crests, without a superstructure, could be categorised according to the level of the armour blocks relative to the land fill behind and the use of a kerb block or similar edge block to obtain a stable transition from the rock structure to the reclamation material.

15.3.1. *Flat crest*

A flat crest means that the top level of the armour blocks is the same as the land fill behind. Such a design criteria is normally required due to an aesthetic requirement. In order to meet such a design criterion, a careful study should be carried out on the magnitude and horizontal extent of the wave overtopping discharge as raising the crest level is not an option. Therefore, providing an adequate surface protection beyond the revetment crest may well be required. Otherwise, the overtopping might cause erosion and scouring of the land fill supporting the crest, and this in turn could cause the crest to collapse.

Rock armour crest examples are shown in Figure 15.4 to demonstrate various design configurations that could be adopted depending on the revetment design and the estimated wave overtopping discharge.

Two different examples for rock armour crests are shown in Figure 15.5, with an inverted armour interface slope in comparison with the examples in Figure 15.4. This design detail, although used by some designers, is not favoured in construction. This is because the inverted interface slope entails an extra construction activity. In Figure 15.4, the construction of each layer at the crest could be made to the required configuration before placing the next one until finally the last layer of the armour rocks is placed. Whereas this is not the case in the examples of

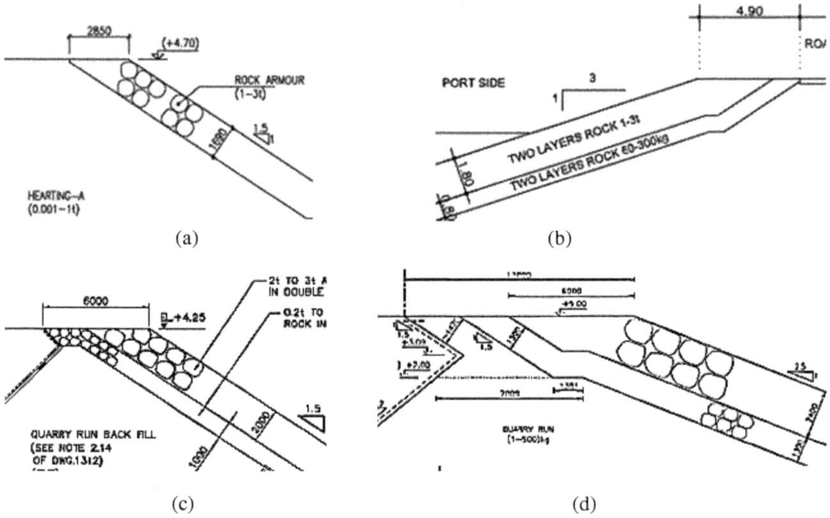

Figure 15.4. Rock armour flat top crest examples: (a) Armour + core; (b) Armour + underlayer + core; (c) Armour + extended underlayer; (d) Extended armour + underlayer + core.

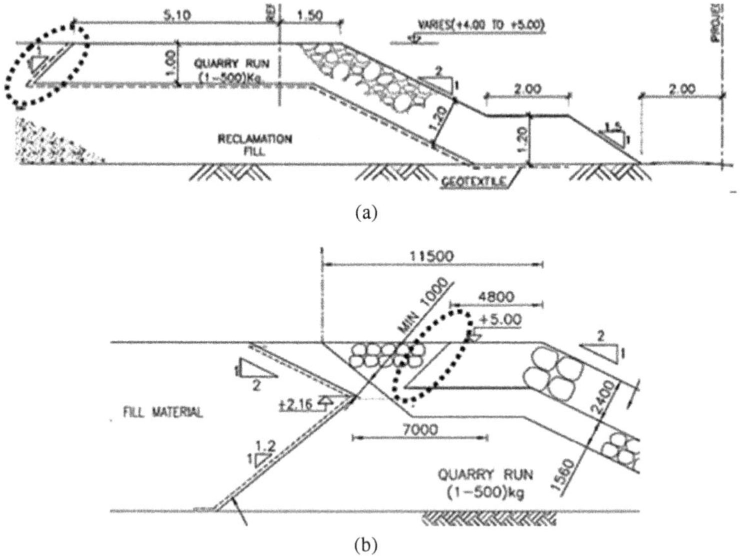

Figure 15.5. Crest examples with inverted interface slopes: (a) Extended armour; (b) Extended armour + underlayer + core.

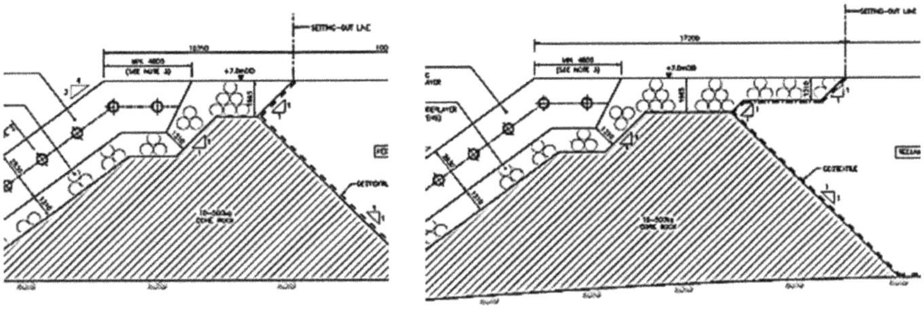

(a) (b)

Figure 15.6. CAU flat top crest examples: (a) CAU + underlayer with short extension; (b) CAU + underlayer with long extension.

Figure 15.5, as the armour layer needs to be placed first before closing up the interface with the preceding layer.

Two examples for CAU flat crest are shown in Figure 15.6, these are actually for the same revetment but at different locations, with low and high overtopping discharges, hence the difference in the underlayer surface extensions. It is however important to note that the underlayer must be continued to the top to provide support for the armour units, regardless of being extended or not.

15.3.2. *Elevated crest*

An elevated crest means that the armour block top level is higher than the land fill behind. This type of crest is more common in rock revetments where it can offer more protection against wave overtopping and also serves as a safety barrier at the water edge.

Rock armour crest examples are shown in Figure 15.7 to demonstrate various design configurations that could be adopted depending on the revetment design and the estimated wave overtopping discharge. The elevated part of the crest could be single- or double-layer thick, as shown in examples (a) and (b), respectively. However, in both cases the crest width shall be at least three rocks. Single- and double-layer crest examples are also shown in (c) and (d), respectively, with extended underlayer and quarry run to provide better protection to the landside against the wave overtopping. Example (e) illustrates the use of a larger stone as a kerb stone to separate the crest from the landside. This type of kerb can be made

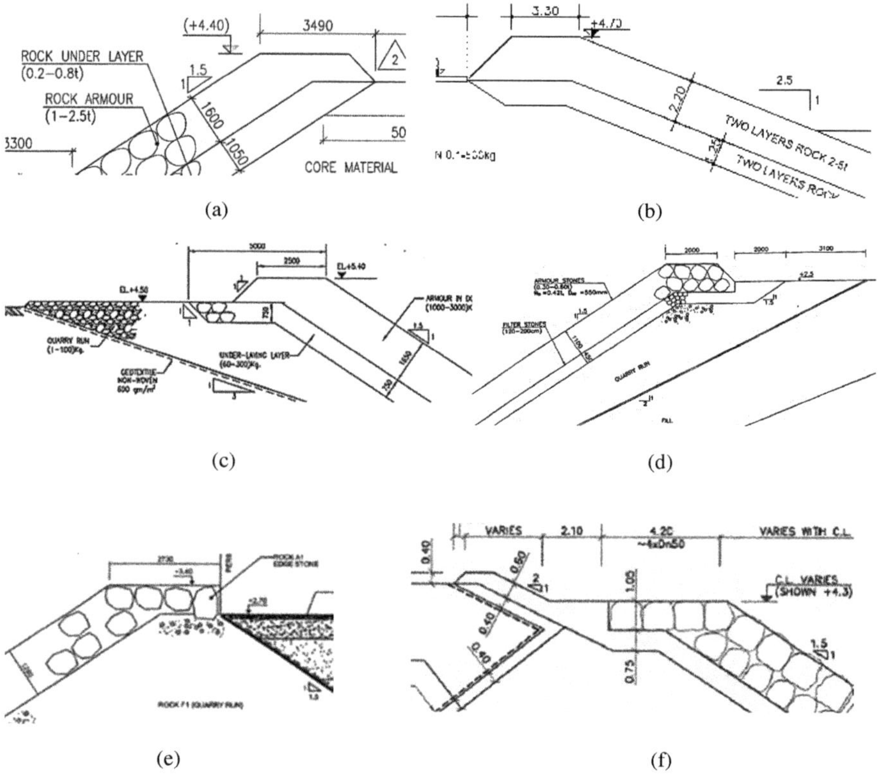

Figure 15.7. Elevated rock crest examples: (a) Single-layer armour; (b) Double-layer armour; (c) Single-layer armour with extended underlayer and core; (d) Double-layer armour with extended underlayer and core; (e) Single-layer armour with kerb stone; (f) Single-layer of underlayer.

more impermeable by pouring concrete in between the rocks to make a "masonry wall". A different concept is illustrated in example (f) where the primary crest level is below the land fill level and the underlayer is extended to the land level forming a secondary crest. Such a case of low crest level could be encountered where the wave overtopping is not large, and the land level is higher than what is required to be protected by the main armour. This last example with primary and secondary crests presents an optimised configuration for a low revetment and a high land level.

Five examples are shown in Figure 15.8 for various CAU armoured structures with elevated crests. The common design element between these five examples is the presence of sufficient rock material behind the CAU crests to provide support to the last unit, which is more crucial for the single-layer than the double-layer CAUs. A width equal to three armour rocks for the back rock crest could be considered adequate. Note that it is recommended to place the rocks in the back crest after placing the CAU armour layer, hence, having a reverse slope at the interface between the

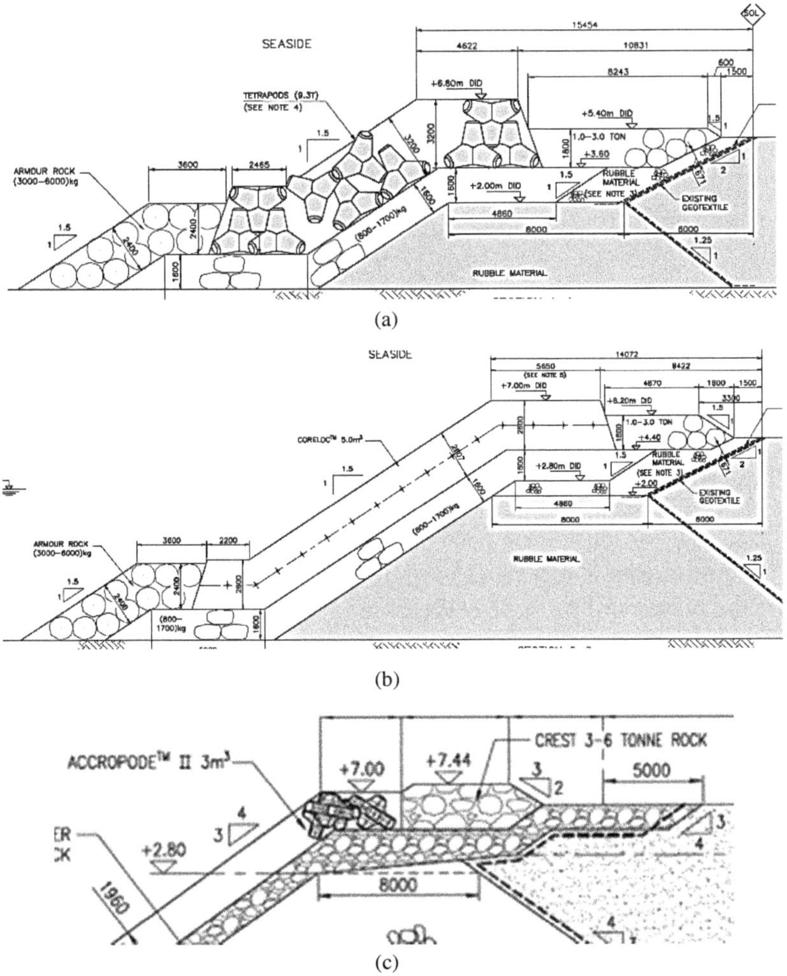

Figure 15.8. CAU elevated crest examples: (a) Tetrapod revetment crest; (b) Core-loc revetment crest; (c) Accropode revetment crest; (d) Xbloc revetment crest; (e) Concrete blocks.

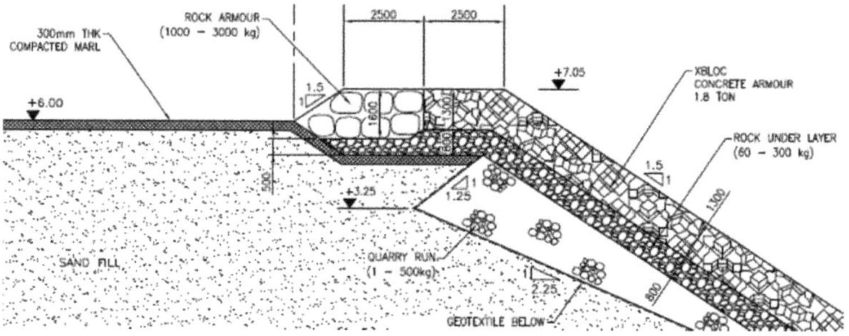

(d)

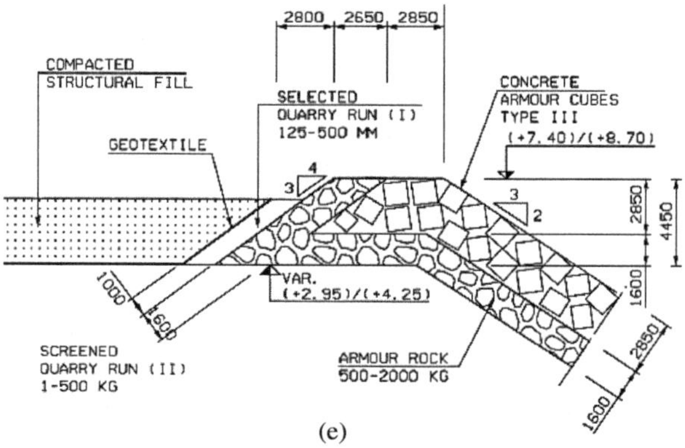

(e)

Figure 15.8. (*Continued*)

CAU and the rocks is recommended, unlike that for the rock armour crest, so that the rocks will be well-embedded in the CAU armour layer.

15.3.3. *Crests with Kerb blocks*

Where there is a road, a paved area or a high-profile development, there will invariably be a uniform and neat separation between the revetment crest and the landside by a kerb block. The kerb block will also help, though marginally, in reducing the wave overtopping which penetrates through the armour blocks. A proper foundation layer must be prepared for the kerb block as it cannot be founded directly on the land fill, and the fill below the block foundation layer must be well-compacted.

Different types of kerb blocks are presented in Figure 15.9, illustrating the possible various arrangements to be considered in securing the block in place. Note that most of the kerb blocks presented here are exposed to wave forces, either directly or through the armour layer.

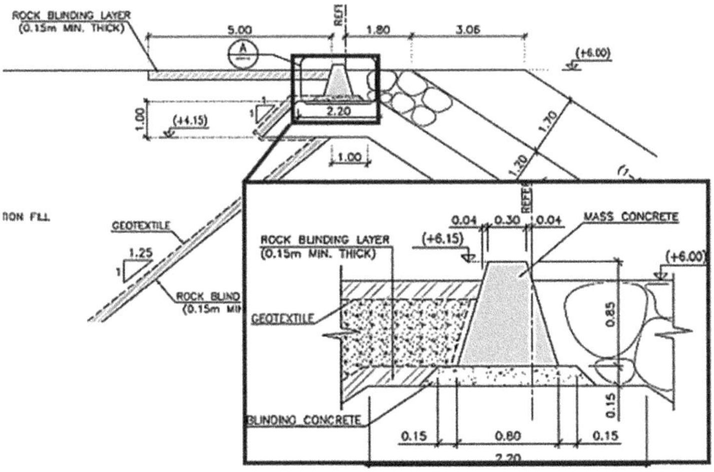

(a)

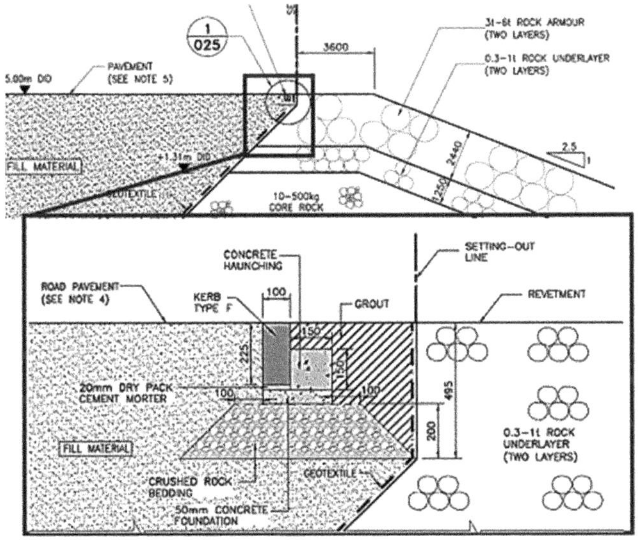

(b)

Figure 15.9. Examples of kerb block: (a) Kerb block placed on quarry run with a 150 mm mortar layer; (b) Kerb block placed on fill material and founded on crushed rock with 50 mm concrete layer with 20 mm mortar; (c) Kerb block placed on quarry run and founded on concrete bed with 15 mm mortar; (d) Kerb block placed on quarry run and founded on 150 mm blinding layer with 20 mm mortar; (e) Kerb block placed on fill material and founded on 100 mm crushed rock pad; (f) Kerb block on quarry run with 250 mm blinding and levelling layer.

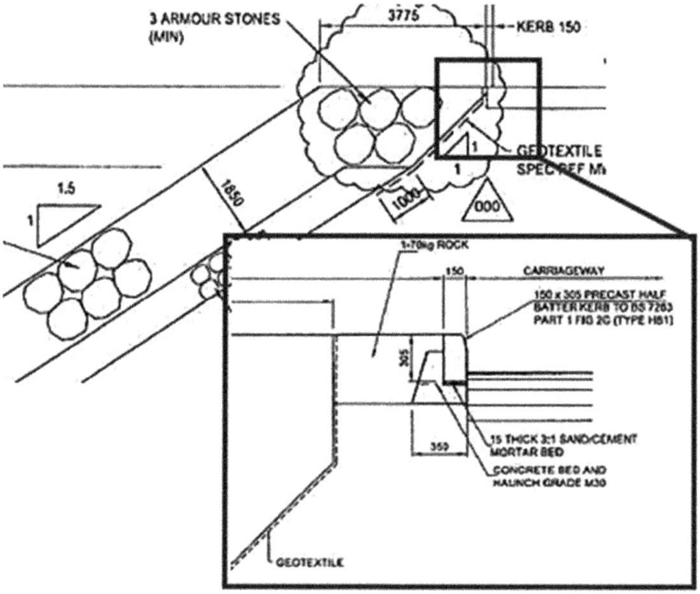

(c)

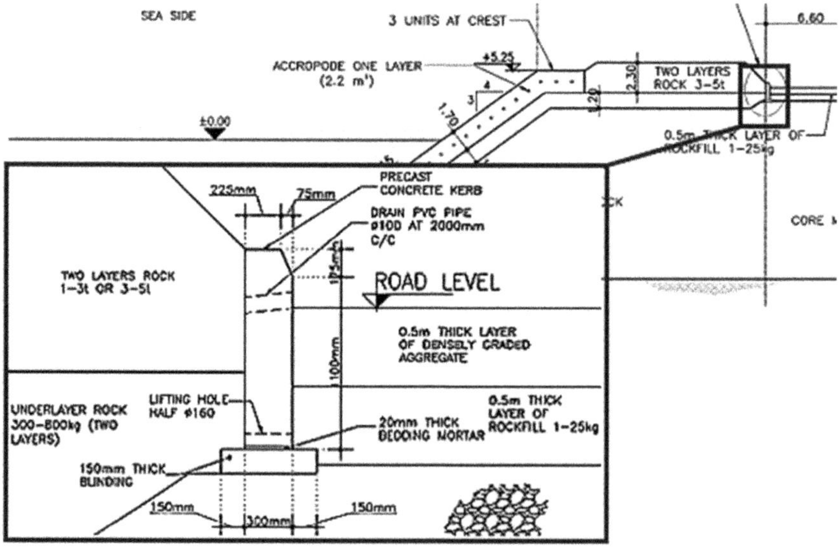

(d)

Figure 15.9. (*Continued*)

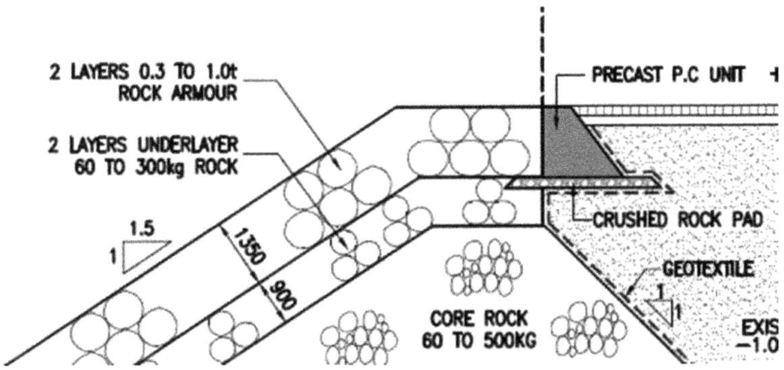

(e)

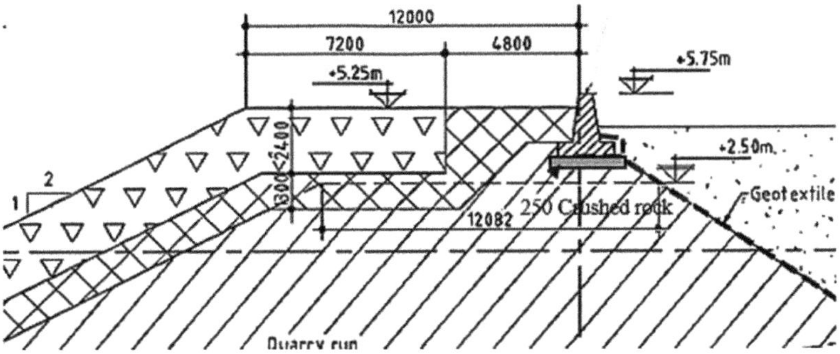

(f)

Figure 15.9. (*Continued*)

Therefore, the stability of these blocks is to be checked against design wave forces. The kerb blocks could be small, as in (b, c, and d), or relatively large, as in (a, e, and f).

By considering the kerb block top level aspect, two types could be differentiated. The first with a flat top level, i.e. at the same level of the crest and the land behind, as in (b, c, and e). This type's prime function is to provide a straight interface line between the crest and the landside. It will allow the surface water to pass both ways, from the sea to the landside and to be drained back to the sea. The second type of kerb block has an elevated top level, as in (a and f). This type will help a little more in stopping the wave overtopping from crossing to the landside, but it will also prevent the surface water from being drained from the landside back to the sea. Therefore this type will require some drainage system, perhaps simple drain pipes, to drain the landside surface water back to the sea.

15.4. Constructability Considerations

The designer should, if possible, consult the construction team regarding the planned construction methodology and the available equipment, as both could impose certain restrictions and have some requirements which

(a) (b) (c) (d) (e)

Figure 15.10 Examples of land-based armour placement equipment.

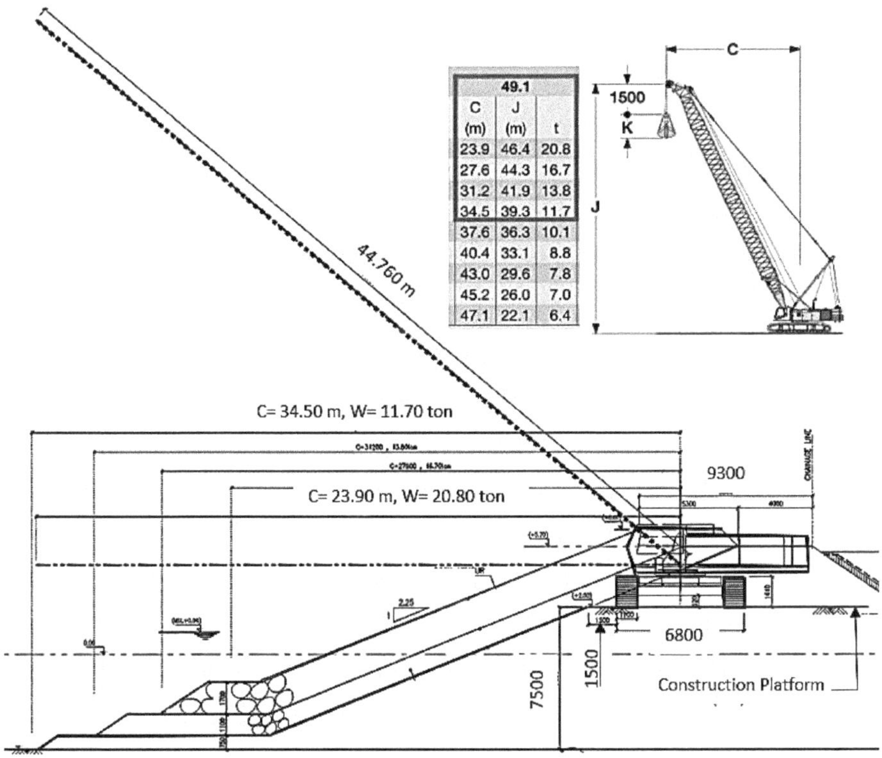

Figure 15.11 Crane lifting plan.

could affect the design in general and the crest in particular. This issue is of especial importance if the structure is in deep waters with heavy armour blocks. The lifting equipment, depending on its size and capacity, might require an intermediate platform to drive on to reach the toe of the structure, hence the section construction might be carried out in stages. Each stage needs to be checked for stability with the imposed loading of the construction equipment, thus some adjustments may be required to the design section profile.

Figure 15.10 presents few photos for rock placement activities with different equipment and methods. Examples for revetment construction are shown in (a, b and c) with a long boom hydraulic excavator, a short boom hydraulic clamshell and a sling crane with clamshell. Examples for breakwater construction are shown in (d and e) with a hydraulic excavator and a sling crane for placing Core locs. Noting that if marine-born equipment are used, i.e. cranes on barges, then it is not expected to influence the crest design as the land-based equipment might do.

The preliminary design of the section is better discussed with the construction team after they have prepared the lifting plan according to the available lifting equipment and the proposed design size of the armour block. An example is shown in Figure 15.11 for a sling crane with a clamshell. A similar diagram could be prepared for the long/short boom hydraulic excavator.

Chapter 16

Superstructures

16.1. Introduction

Superstructures include crown/crest walls, wave walls and access roads. These structures are commonly used in breakwaters and revetments, primarily for one, or more, of the following reasons:

- Cost saving in optimising construction material due to possible reduction in the structure height while maintaining an acceptable level for the wave overtopping discharge.
- Provide access on the crest for vehicles or pedestrians.
- In ports, berths could be constructed immediately behind the breakwater with the cargo-related facilities (conveyor belts, pipelines, etc.) could be placed on the crown wall base/slab.
- Serve as a separation block between the revetment slope protection and the landside.
- Provide support for the seaside crest armour units.
- Minimise wave downrush forces on the breakwater rear side.

The superstructures are either of bulky plain concrete or slender reinforced concrete, where both could be pre-cast units, cast in-situ or mixed. If the design can combine pre-cast and plain concrete to produce a satisfactory unit, then this will normally be the first choice for the designer. The advantage of plain concrete is to avoid the reinforcement steel and all its associated problems in harsh sea environments, and to have pre-cast units is to ease and speed up the construction and to be less affected by weather and sea conditions. However, as the crown wall gets larger, heavier and probably with thin sections, then cast *in-situ* cannot be avoided.

The use of crown walls could produce one, or more, of the following undesirable consequences:

- Rubble mound structures are flexible and can absorb minor settlements, however the crown wall is a rigid structure and can only tolerate very limited differential settlements.
- The reduction in wave overtopping over the breakwater means increasing the wave downrush force on the seaside armour layer.
- Higher cost and longer construction time are normally associated with larger crown walls in comparison with continuing the armour layer over the breakwater crest.
- A large crown wall with a long base tends to induce excessive pore pressure in the rubble mound.

The crown wall foundation, interface with armour layer, stability and integrity are some of the issues which do require special attention as these contribute to the crown wall and the rubble mound structure stability.

16.2. Breakwaters

Superstructures in breakwaters are mainly crown walls, access road slabs or a crown wall integrated with a berthing facility, see Figure 16.1. Crown walls are primarily utilised to minimise the wave overtopping and frequently in combination with providing access. Whereas crest structures are mainly to facilitate access of personnel or vehicles.

Superstructures could be categorised, or defined, by the shape, location, utilisation, interface, material and casting method, as will be explained and elaborated on in the following paragraphs.

The ***height of the wall*** could be used as one of the defining characteristics, as illustrated in Figures 16.1(a–d), and this will normally be related to the allowable rate of wave overtopping discharge.

The ***location of the crest structure*** and its use for access or other functions could also be another defining character, as in Figures 16.1(e, f), and also in Figure 16.1(g).

The configuration, and accordingly the construction, of the crest structure, are to be carefully considered as it will define the structural

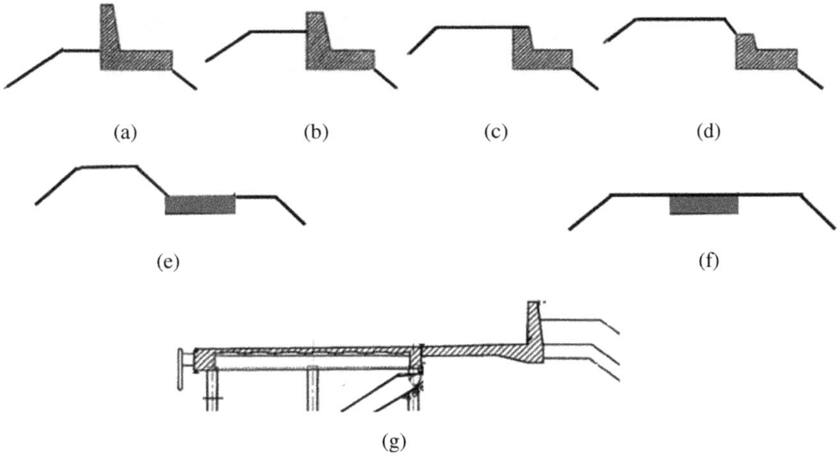

Figure 16.1. Typical breakwater superstructures: (a) High wall; (b) Medium wall; (c) Levelled wall; (d) Receded wall; (e) Sheltered crest slab; (f) Levelled crest slab; (g) High wall with extended slab.

design and material composition, i.e. whether it is to be ***mass plain concrete*** or ***reinforced concrete***, and if it is to be of ***pre-cast units***, ***cast in-situ*** or ***mixed***.

The interface between the crown wall base slab and the breakwater rear slope, as illustrated in Figure 16.2, is another character which is to be well-defined in view of the wave overtopping discharge and the stability of the rear-side armour layer. An overhanging base slab, as in Figure 16.2(a), is normally used with large discharges of wave overtopping to ensure some sheltering to the rear side from the falling water jet, which in many cases will hit the water first before the rear-side slope. Chute blocks may be used, as indicated by the dotted line, as energy dissipators. For less severe discharges, the bas-slab could be made flushed with the rear slope armour, as in Figure 16.2(b). For even less wave discharges, or for wider breakwater crests, the crown wall slab may not reach the rear-side slope, accordingly a dropped shoulder may be the result of such a case, as in Figure 16.2(c), or a rear crest berm, as in Figure 16.2(d). However, it is important to note that the arrangement shown in Figure 16.2(c) is not recommended as the shoulder will most likely get damaged by the falling water jet at extreme events.

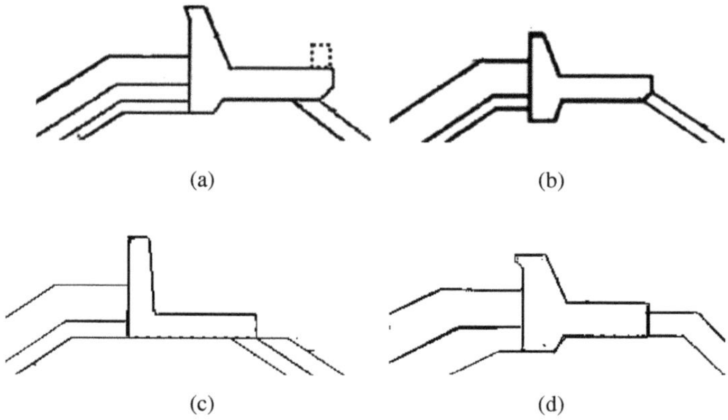

Figure 16.2. Crown wall interface with the breakwater rear-side: (a) Overhanging base-slab; (b) Flushed base-slab edge; (c) Dropped rear-side shoulder; (d) Rear crest berm.

On the seaside of the crown wall, the crest shoulder width is recommended to be 2–3 blocks (CAU/rocks) wide. The armour blocks should be placed tightly on the crest and be well-supported by the crown wall, no gap should be left between the last row of the armour and the crown wall, which is practically difficult to achieve with CAU or large armour rocks. Theoretically, the length of the crest shoulder could be adjusted, according to the slope length and the armour block size to fit the two or three blocks without leaving a gap to the crown wall. However, this is not always achieved in the construction, and in this case such a gap must either be filled with rocks of reasonably large size in case of CAUs, or by trying to rearrange the armour rocks at the crest or even replace the small ones with larger rocks. In both cases, the last row must be fully supported, and no rocking or other movement of the blocks is to take place. Illustrative photos for the correct and incorrect placements are shown in Figure 16.3.

It is preferred to place the superstructures (crown wall/crest slab) on the quarry run and not on the underlayer rocks. This is to help in avoiding the higher penetration of the uplift pressure under the superstructures because of the smaller voids and higher resistance to flow. The quarry run is usually covered with 0.20–0.30 m layer of filter size stones to provide a level surface for the pre-cast superstructure. However, for a cast *in-situ* superstructure it is preferred to place a 100 mm blinding layer of lean concrete instead of the filter layer to hold the water in the fresh concrete.

(a) (b) (c)

Figure 16.3. Placement of armour blocks against the crown wall: (a) Proper placement of Rock armour; (b) Proper placement of CAU (Core-loc); (c) Improper placement of CAU (Core-loc).

A wide selection of reinforced concrete crown walls is presented in Figure 16.4, which are generally associated with thin sections and a key/heel embedded in the foundation layer. The need for the key is primarily dictated by the reduction in weight, therefore, to compensate for the reduced frictional resistance to sliding a mechanical anchor is required, i.e. the key. The presented examples are with different configurations regarding the wall height, the key shape, the use of chute blocks and the rear-side interface with the armour layer. From the presented examples in Figures 16.4 and 16.6, a general deduction could be made regarding the crest shoulder height in relation to the wall height. The crest shoulder freeboard, Rc, tends to be relatively large (Rc > 1) for levelled or receded

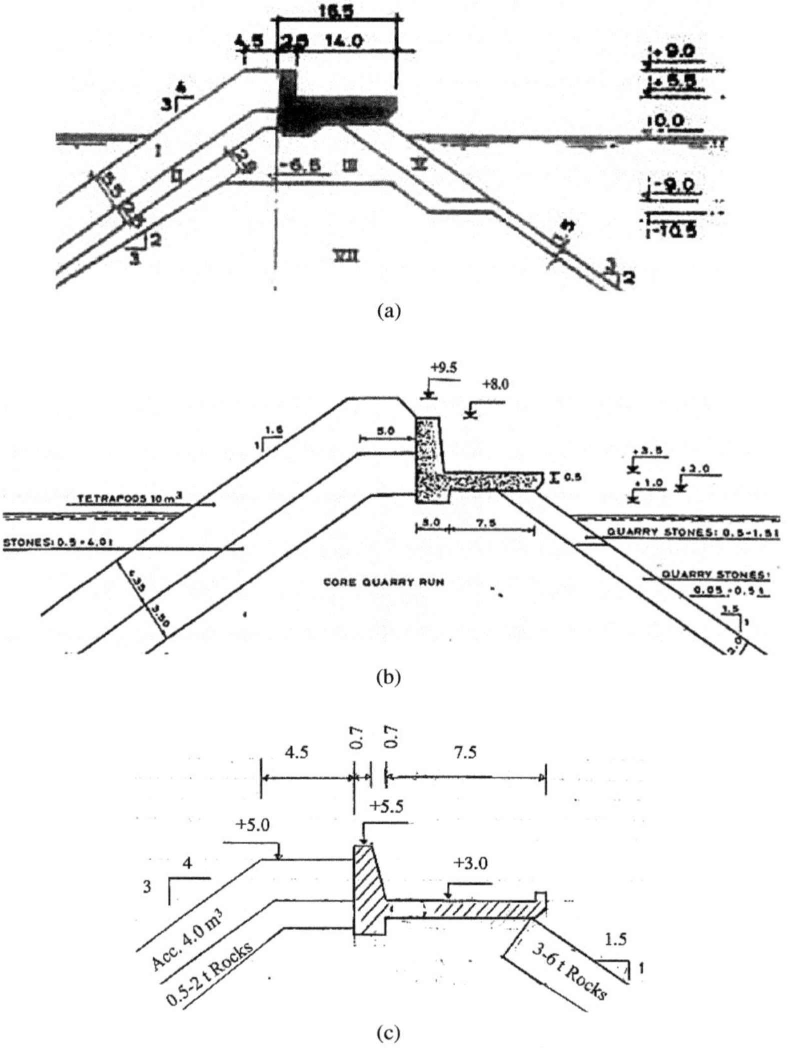

Figure 16.4. Examples of reinforced concrete crown walls with different crest shoulder freeboards (Rc): (a) Rc = 1.2 Hs (Jensen, 1984); (b) Rc = 1.14 Hs (Gravesen, 2008); (c) Rc = 0.9 Hs; (d) Rc = 1.43 Hs (Von-Rovers *et al.*, 2008); (e) Rc = 1.15 Hs; (f) Rc = 1.44 Hs; (g) Rc = 0.95 Hs; (h) Rc = 1.2 Hs (Coutos *et al.*, 2016).

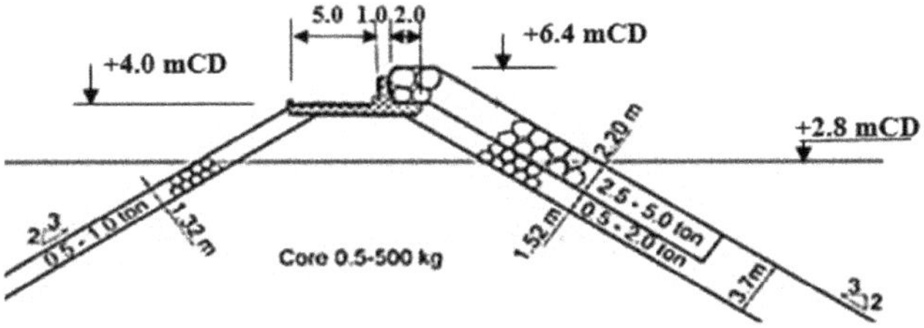

(d)

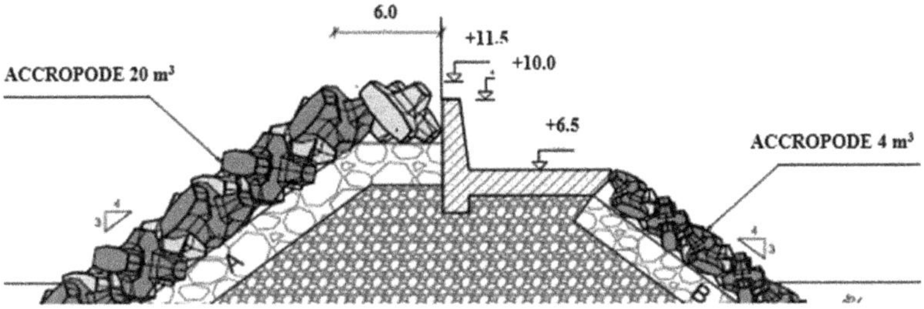

(e)

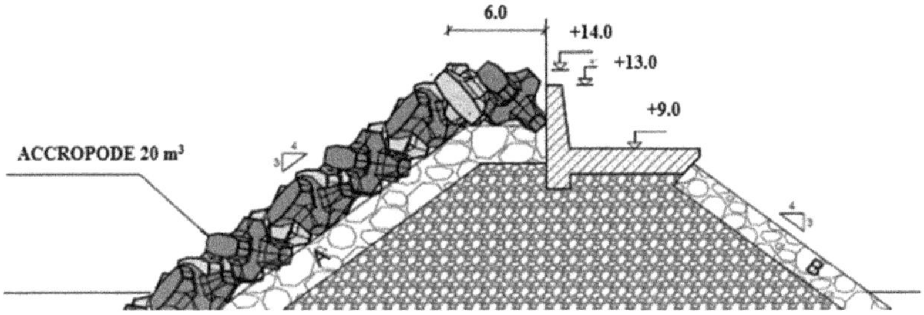

(f)

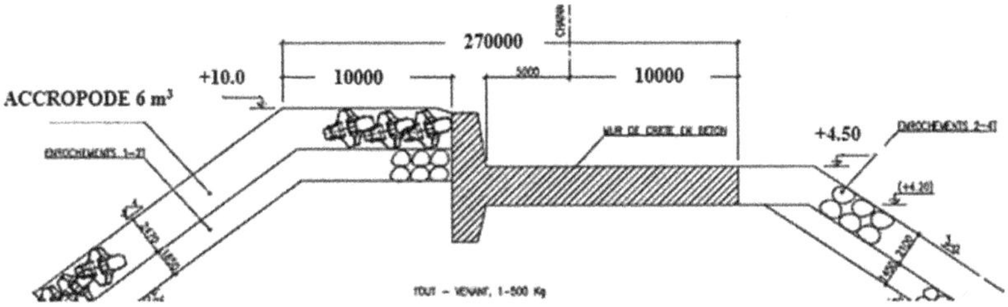

(g)

Figure 16.4. (*Continued*)

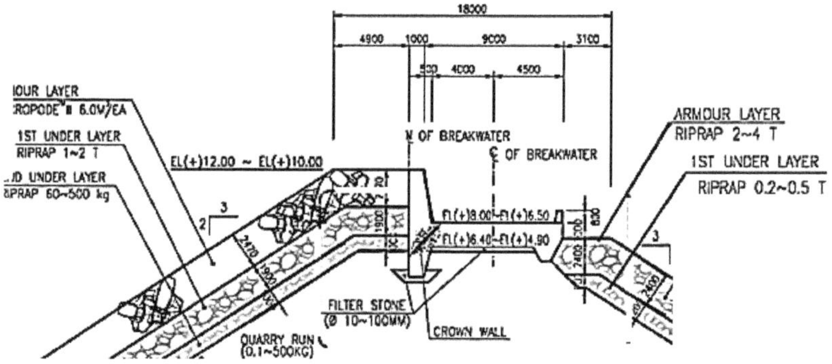

(h)

Figure 16.4. (*Continued*)

walls, and relatively small (Rc < 1) for medium and high walls (the Rc is measured from the design high water level (DHWL)). Of course, there will be exceptions to this general trend as in Figure 16.4(g), where the crest shoulder width is wide.

The construction methodology of crown walls in this category is mostly cast in-situ. However, few are made of pre-cast elements as the example shown in Figure 16.4(d), which included a toe in the front (instead of a key) to utilise the weight of the armour rocks to generate more frictional resistance. The cast *in-situ* crown wall usually cannot be completed in one stage, two or more stages could be required, specially between the base and the wall.

An important note regarding the stability of the pre-cast crown wall units needs to be mentioned, and it is regarding the width of the pre-cast element. Although the stability calculations, which are based on forces per meter run, may show that the crown wall section is stable, the actual width of the pre-cast element does in fact play a role in its stability. Narrow pre-cast elements, as normally required to reduce the weight, and thus minimising the required lifting capacity at the construction site, could be tolerated up to a limit, then the element becomes unstable at design condition. There is no calculation method that can predict the critical width limit, and the stability could only be verified by physical modelling. This behaviour could be attributed to the wave incident angle, which could concentrate the wave force on a short length of the crown wall.

Mass plain concrete crown walls are presented in Figure 16.5, and these are generally associated with bulky sections so that the plain

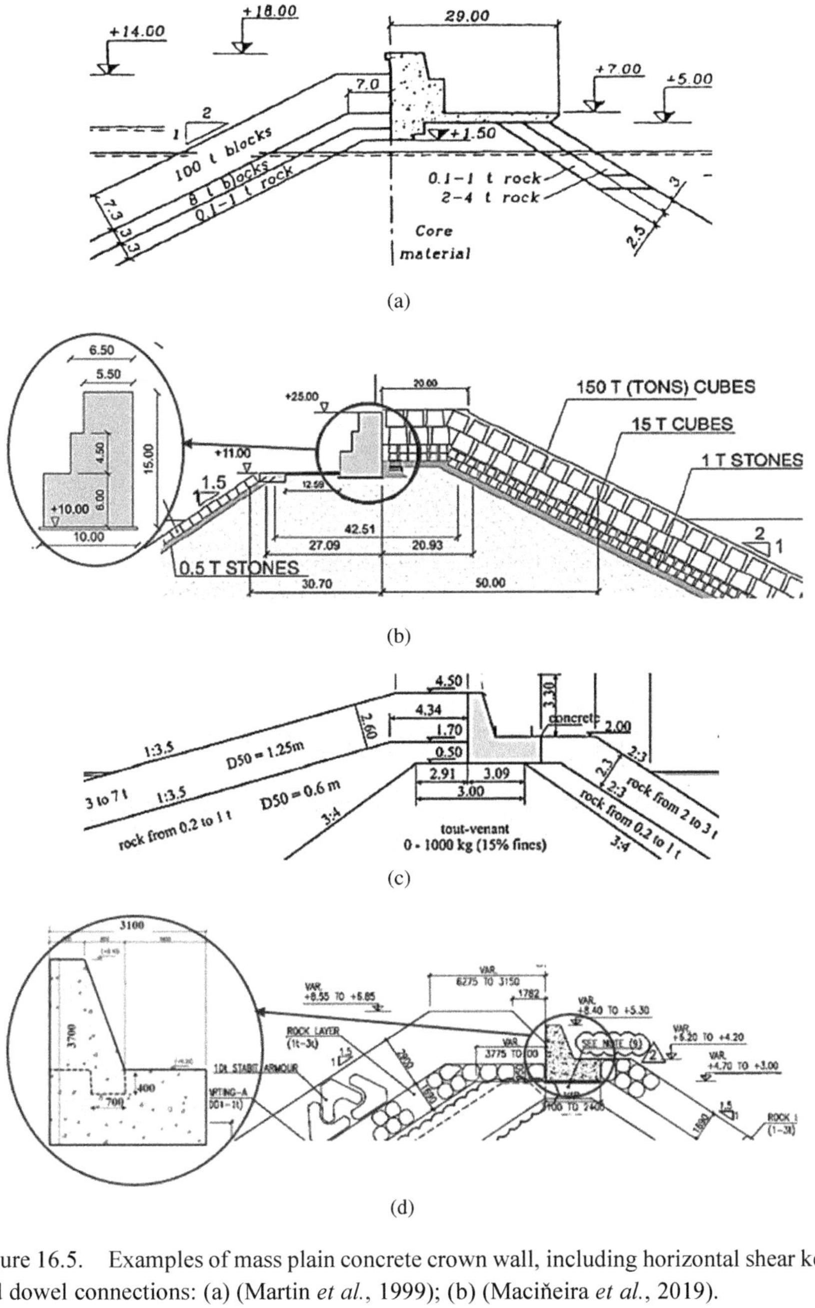

Figure 16.5. Examples of mass plain concrete crown wall, including horizontal shear key and dowel connections: (a) (Martin *et al.*, 1999); (b) (Maciñeira *et al.*, 2019).

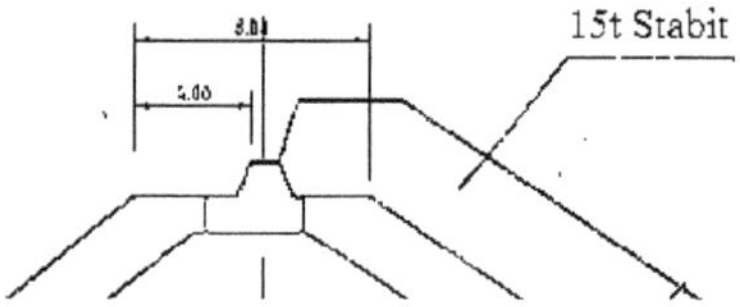

(e)

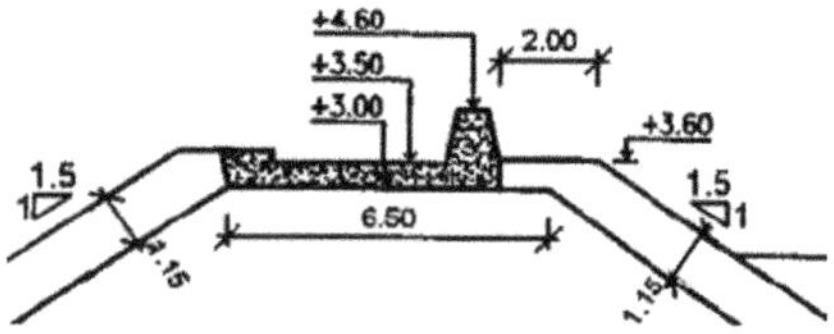

(f)

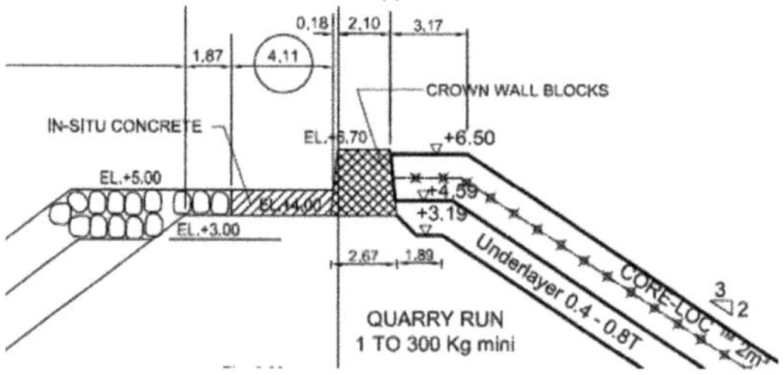

(g)

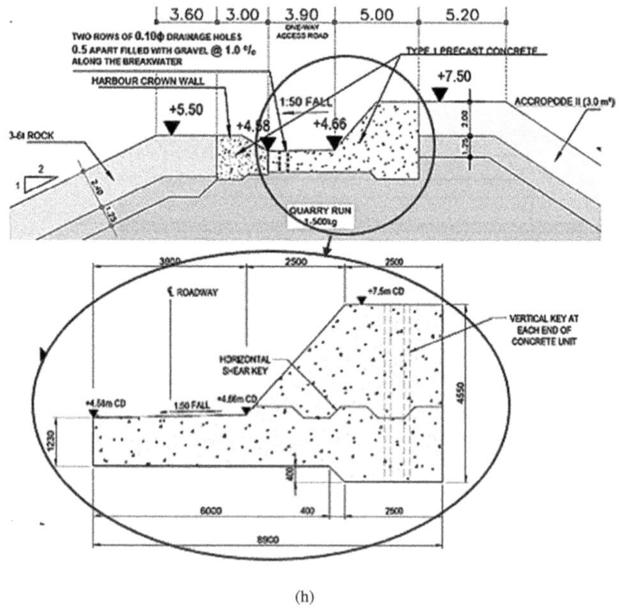

(h)

Figure 16.5. (*Continued*)

(i)

Figure 16.5. (*Continued*)

concrete can withstand all internal stresses without the need for steel reinforcement. The presented examples of the crown wall include a very bulky type for severely exposed sections, as in (a) and (b), a relatively smaller type for moderately exposed sections, as in (c), (d) and (e), and crown walls with access roads, as in (f), (g) and (h). The use of pre-cast elements in this type of crown walls, i.e. mass plain concrete, is very common, however, the entire crown wall section might become very heavy for the construction site lifting equipment capacity. Therefore, it could be divided into segments connected either by shear keys, as shown in examples (d), (g) and (h), or dowels, similar to the example shown in (i).

The crown wall could be an integrated part of the port structures performing and serving operational requirements other than just protecting the port from wave overtopping, as shown in Figure 16.6. The first two examples are shown from the same project where the crown wall is connected to the open quay deck, with (a) being a moderately exposed region, and (b) in a more exposed region. Whereas the crown wall in (c) is connected to the capping slab of a gravity wall quay. The Rc for these examples is measured from the DHWL to the crest level of the crown wall.

The crest slabs in some projects are required only to provide an access road over the breakwater crest. It will serve in breakwater maintenance works and/or reaching to navigation light beacons, which are normally mounted on the breakwater roundhead and along the breakwater, or for some other reason. Examples of such structures are shown in Figure 16.7.

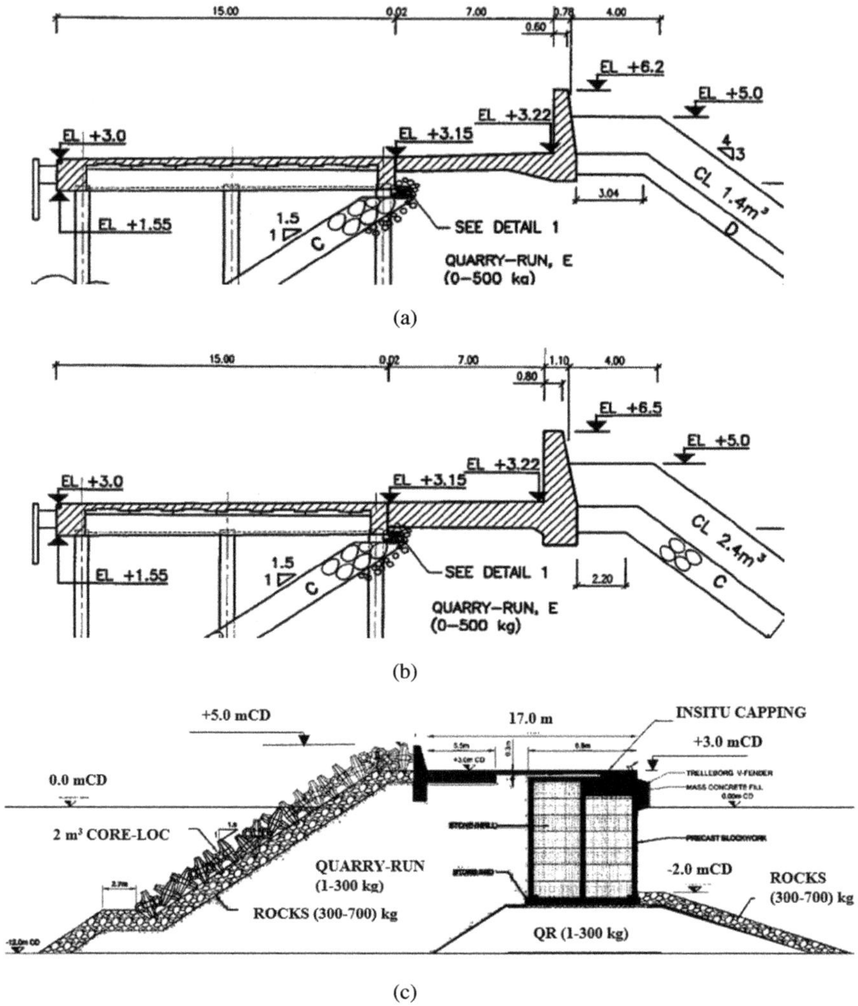

Figure 16.6. Examples of crown walls connected to berthing wharves: (a) Rc = 0.9 Hs; (b) Rc = 0.8 Hs; (c) Rc = 0.85 Hs.

Crest slabs, as shown in the presented examples and unlike the crown wall, could be placed on the underlayer rocks as their stability is not as critical as that of the crown wall. However, they are exposed to wave forces and their thickness will subsequently depend on their exposure.

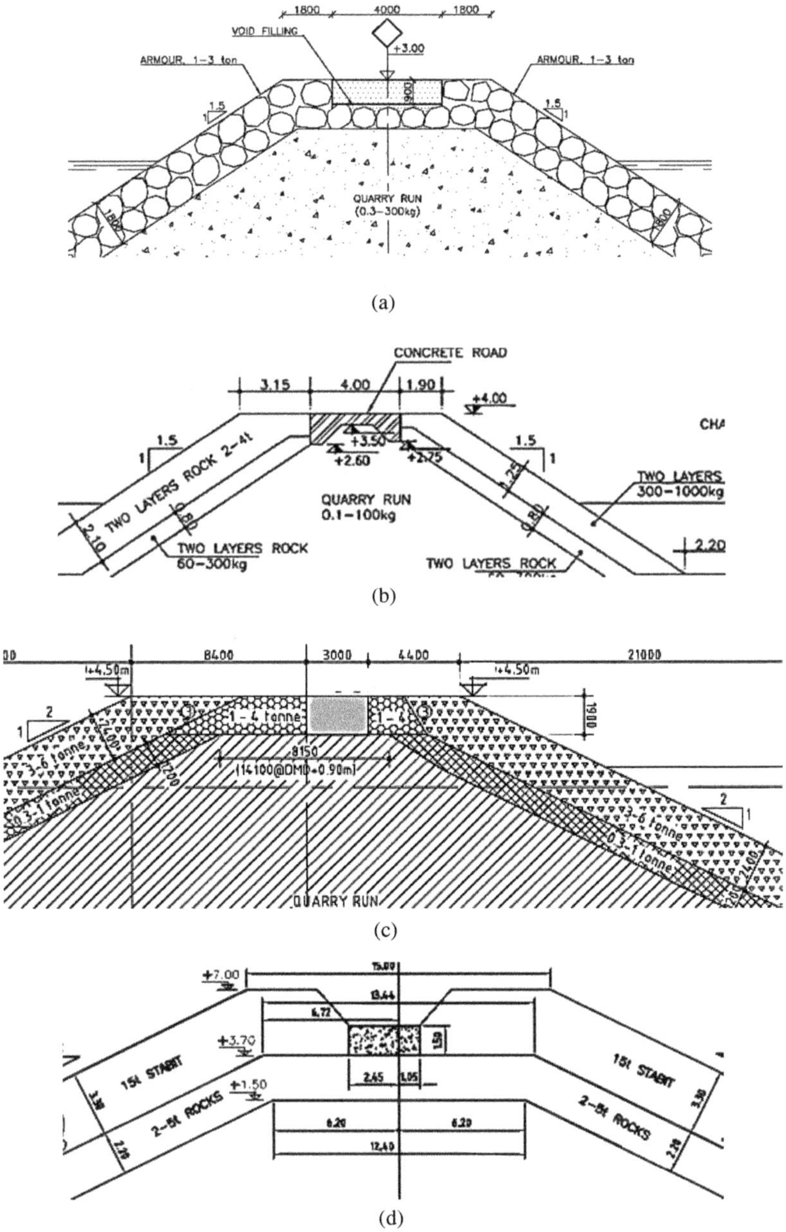

Figure 16.7. Examples of crest slabs: (a) Levelled crest slab; (b) Levelled crest slab — with keys; (c) Levelled crest slab; (d) Sheltered crest slab; (e) Sheltered crest slab (Feng *et al.*, 2016).

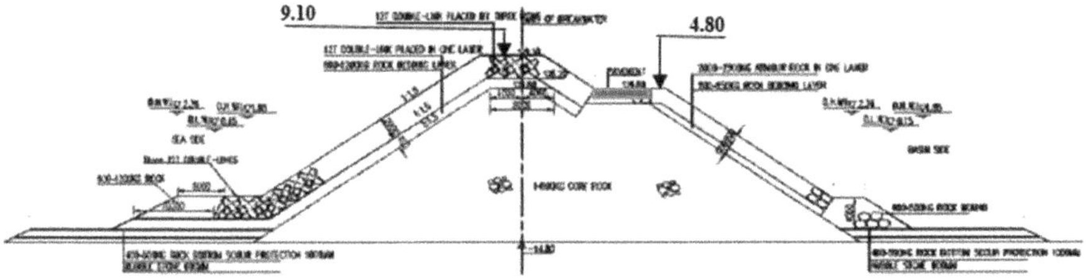

(e)

Figure 16.7. *(Continued)*

16.3. Revetments

The use of crown walls is very common in revetments and for many reasons as they could be for reducing/preventing wave overtopping, seeping water through armour layer, providing support to the armour blocks and acting as kerb block forming a neat crest edge line.

With reference to Figure 16.1, a high wall type is rarely used in revetments as the most common are the medium to levelled walls. This is because, except from revetments in industrial ports, the view of the sea is not to be obstructed in most cases and for different reasons. The size of the crown walls in revetments is generally smaller than that in the breakwaters, and in most cases, this will very much promote use of pre-cast elements.

Revetment crown walls are usually partially buried by the landside fill, leaving part of the wall as a kerb block. However, there are situations where they will be fully buried or fully exposed. Being partially or fully buried, crown walls will significantly improve their stability against turning and sliding, and this means a lighter crown wall section could be achieved. The landside fill level is usually fixed by others; however, the revetment crest level, width, configuration and the crown wall level will normally be fixed according to the design criteria of the wave overtopping discharge. Different illustrative examples are shown in Figure 16.8. A fully buried wall is shown in (a) and a partially buried in (b) and (c). Also, a partially buried type is shown in (d). However, it is only the base slab of the crown wall which is buried. Two different types are shown in (e) and (f), where these two crown walls are partially buried but the wall level is the same as that of the crest shoulder. Two examples for not buried crown walls are shown in (g) and (h), and consequently these two are expected to be a little bulkier than if partially buried.

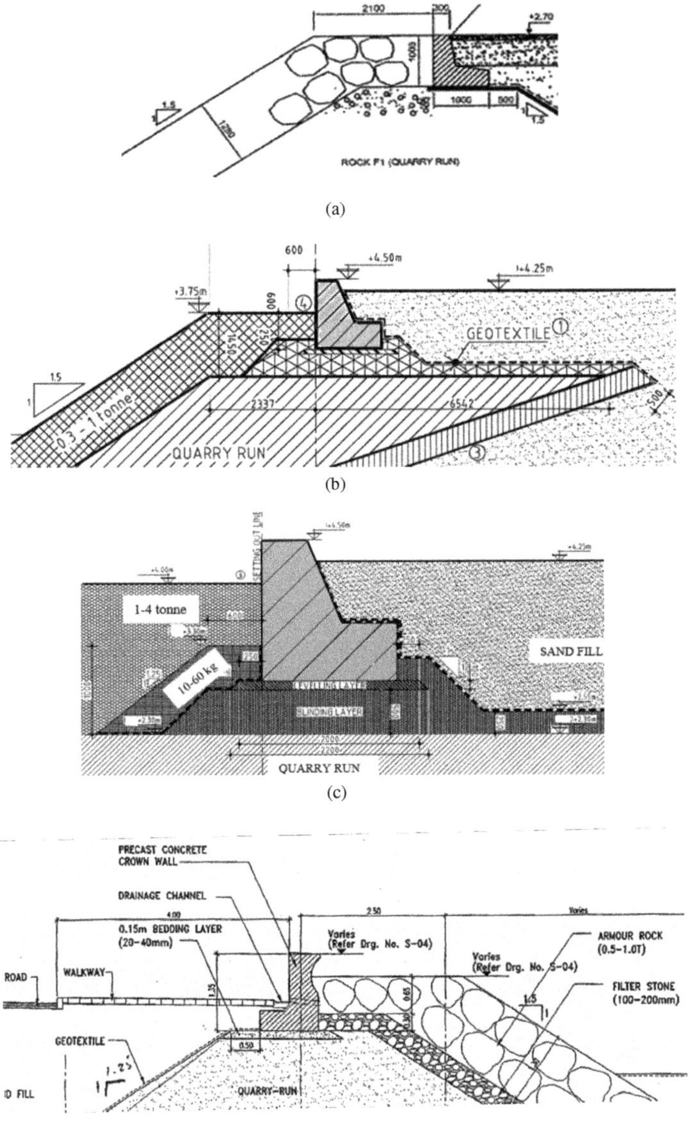

Figure 16.8. Revetment crown wall examples with different burial levels: (a) Fully buried; (b) Partially buried wall — small armour rocks; (c) Partially buried wall — large armour rocks; (d) Buried base with recurve wall; (e) Buried base with levelled wall; (f) Partially buried inverted crown wall with levelled wall (Smith *et al.*, 2013); (g) Not buried — flat crest shoulder; (h) Not buried — compound crest shoulder (Ceuppens, 2007).

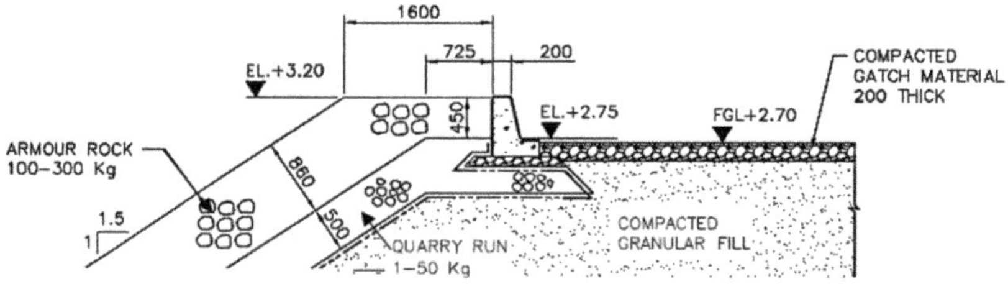

(e)

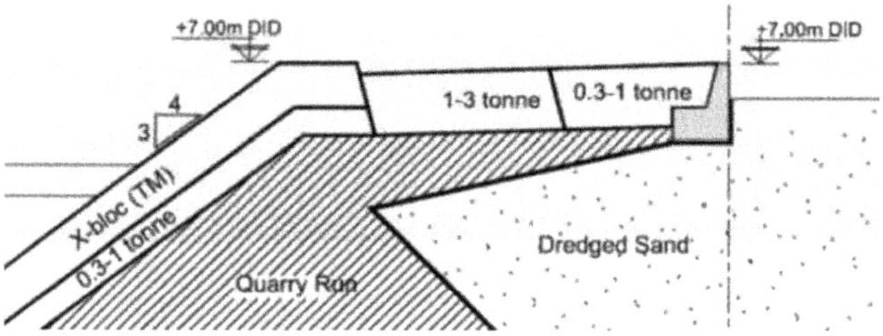

(f)

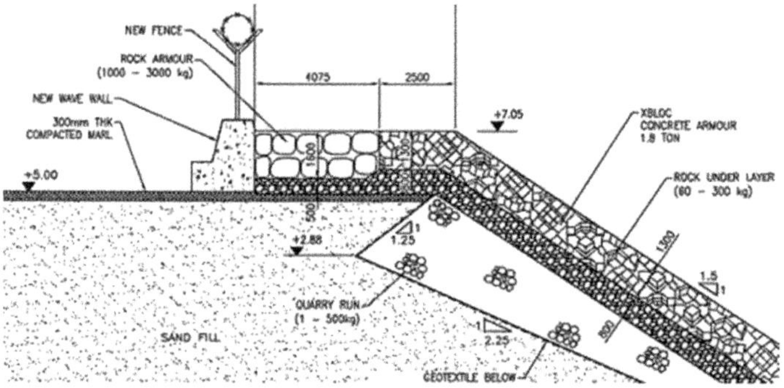

(g)

(h)

Figure 16.8. (*Continued*)

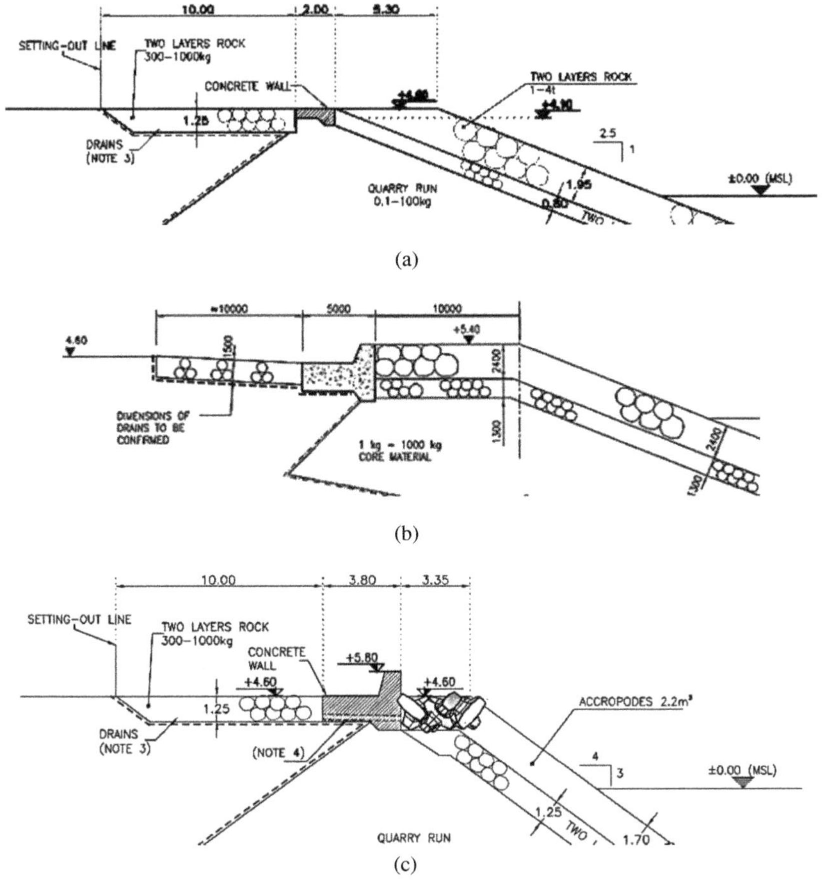

Figure 16.9. Revetment crown walls with extended landside protection.

In some projects the amount of the wave overtopping discharge may be excessive, and that could be attributed to one or more of the following: low crest level, adverse wave climate, low crown wall height, simple slope revetment design (i.e. no wide berms to reduce the wave overtopping) and short crest shoulder. Accordingly, in such a situation provisional measures are to be taken to protect the revetment from damage and to prevent the wave overtopping discharge from reaching the unprotected landside area. Examples for such cases are shown in Figure 16.9, where the landside, as in (a), (b) and (c), has been protected by an extended layer of larger rocks, which can also drain the overtopping discharge. There are different reasons for using larger rocks, which include: preventing vehicles from driving on this area which could damage the revetment or damage the geotextile, larger rocks will not easily get moved by wave overtopping like quarry run, and it will create larger voids, which are not easily blocked by dust or other material that could obstruct draining the overtopping discharge.

A different arrangement for the crown wall is shown in Figure 16.10, and that is by using a drainage channel, either on the seaside or the landside of the crown wall, to collect the wave overtopping discharge. For the seaside channel, it has been found that instead of raising the height of the crown wall, leaving a gap between the crown wall and the crest shoulder is very effective in reducing the discharge crossing over to the landside. Two examples, from the same project, are shown in (a) and (b), where in (a) the crown wall was kept at the same height as the crest, however in (b) as there is no wide berm as in (a) it was required to raise the crest level to maintain the same wave overtopping discharge and keeping the crown wall at the same level. The gap, i.e. the channel, is to be filled with relatively large rocks, say 0.1–0.4 t, to give support to the crown wall and the armour layer, and to ensure large voids for quick drainage. The other channel arrangement is to have the crown wall, as usually, adjacent to the crest, and to have the collecting drainage channel on the landside, as shown in (c) and (d).

It is preferred to place the crown wall on quarry run and not on larger rocks. This is to help in avoiding or reducing penetration of the uplift pressure under the crown wall, and to give a better overall stability. The quarry run is usually covered with 0.10–0.20 m layer of filter size gravel stones as shown in Figure 16.8(a, d), or even two layers could be used, a blinding layer (fx. 50–150 mm) over the quarry run and a levelling layer (fx. 5–50 mm) under the crown wall, as shown in Figure 16.8(b, c). In (b), no geotextile has been used on the seaside foundation layers because filter requirements are fulfilled. Whereas in (c) the filter requirements cannot be achieved with the 1–4 t armour rocks, this led to the use of a geotextile, and a separation layer of 10–60 kg to avoid placing the large rocks directly on the geotextile.

The crown wall could be placed, though not recommended, on well-compacted fill material after providing a foundation layer, as shown in Figure 16.8(g). A better arrangement is as shown in Figure 16.8(e) where a 0.30 m layer of quarry run is used as a foundation layer on top of the compacted sand. Similarly, it is also not preferred to place larger rocks

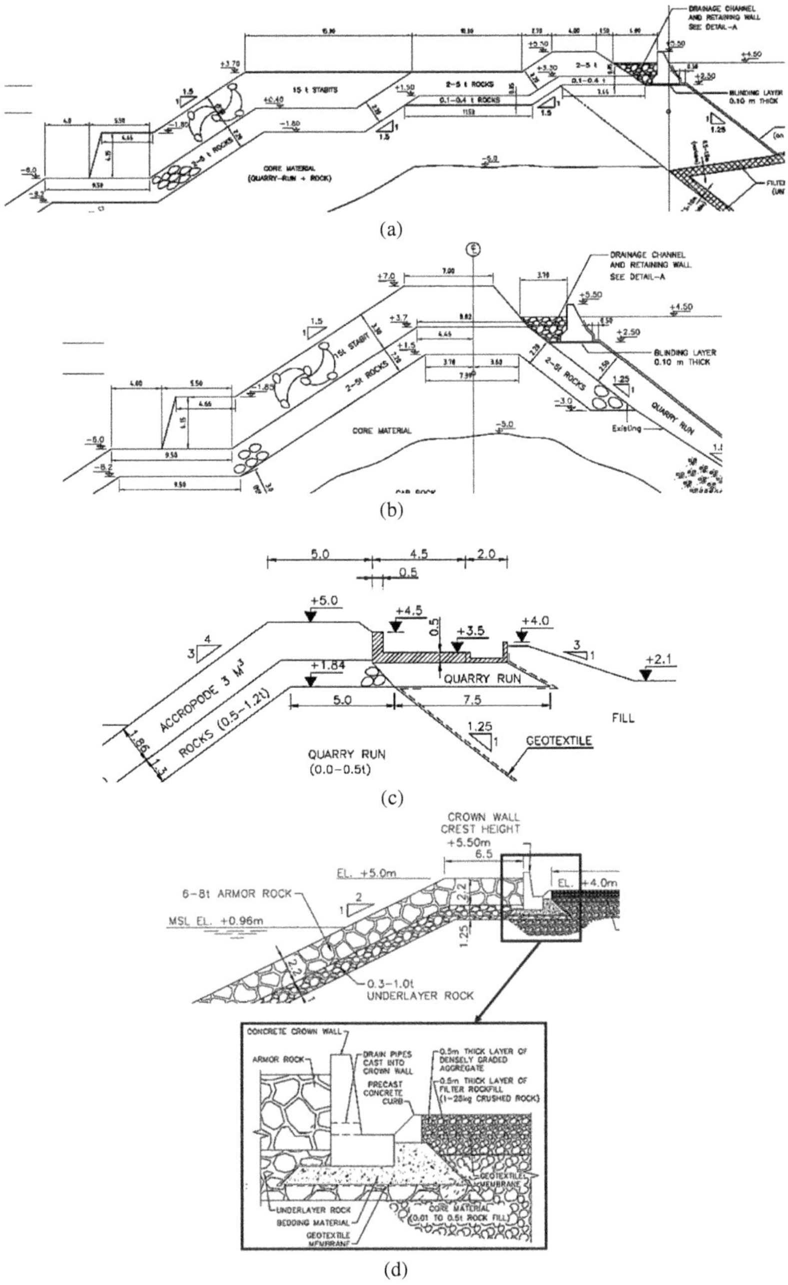

Figure 16.10. Revetment crown walls with drainage channels.

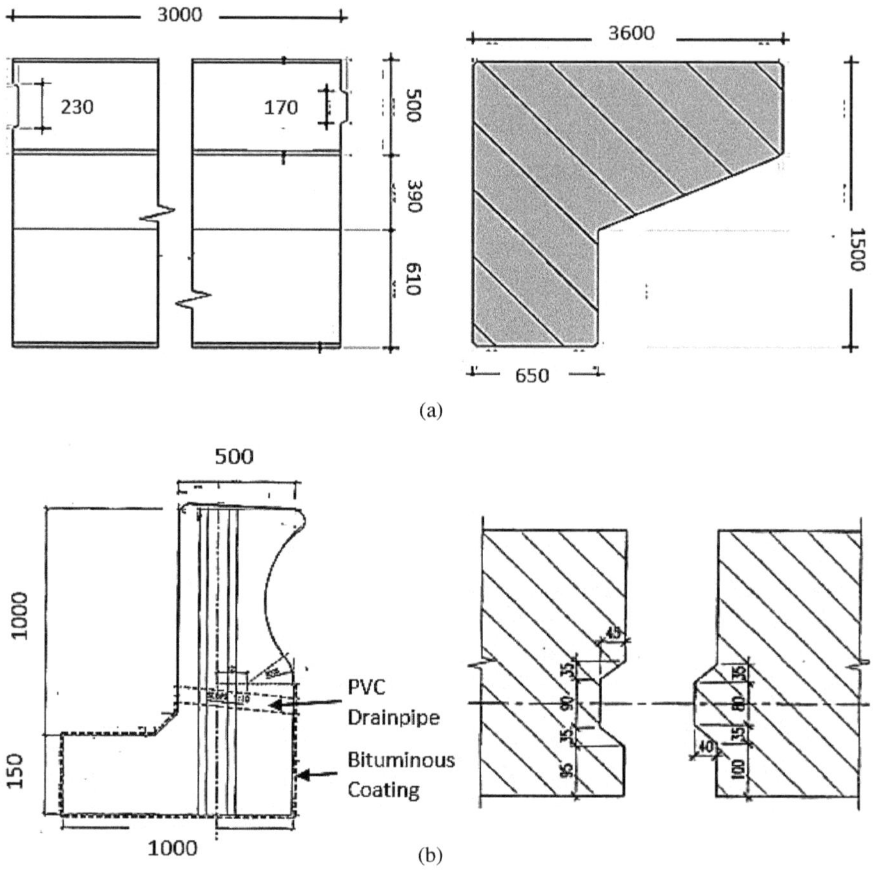

Figure 16.11. Examples of vertical shear key connection between crown wall units.

directly on the geotextile founded on compacted sand. Therefore, it is recommended to replace the two layers of 0.3–1.0 t used for protection and as a drain, shown in Figure 16.10, with three or more layers of smaller rocks like 0.1–0.4 t. This size of rocks is usually stable enough against possible damage due to overtopping and with large enough voids to maintain good drainage.

Pre-cast crown wall units are to be provided with vertical shear keys; examples are shown in Figure 16.11. These will help during the installation by keeping all the units aligned, stop wave overtopping from passing through from between the units and will prevent individual units from horizontal movements under wave pressure, i.e. provides a better overall crown wall integrity.

Chapter 17

Breakwater Rear Side

17.1. Introduction

The stability of the armour layer on the rear side depends on the type and size of the armour block (rock/CAU), the slope and certainly on the wave overtopping discharge which crosses the rear edge of the crest and how this discharge falls down and hits the rear side. The wave overtopping discharge depends on the water level, wave characteristics (height, period and direction) and the breakwater configuration, which includes the seaside slope (steep or flat, simple or compound, with/without a berm) and the breakwater crest (high or low, wide or narrow, with/without a crown wall). The armour block stability on the rear side is actually more sensitive to external forces than it is on the seaside of the structure. On the seaside, the block weight, i.e. gravitational force, will act against the main wave forces during the wave uprush as it tries to dislodge the block, whereas on the rear side the block weight assists the falling wave overtopping discharge in dislodging the blocks. Thus, a smoother surface of the armour layer is beneficial in reducing the acting forces from the falling wave overtopping on the armour blocks. This is unlike the recommendation for the armour layer surface on the seaside, where a rougher surface helps in reducing the wave runup and consequently wave overtopping.

Other factors, though minor, are also to be considered that might affect the design of the rear side. These are the wave climate on the rear side of the breakwater, as it might be partially exposed to open sea waves, and also the propeller jet wash of closely passing vessels in a port entrance channel or in a port basin.

17.2. No Superstructures

In relatively low crested breakwaters, the primary armour block on the seaside face is usually to be continued over the crest and the upper part (i.e. slightly above to a little below water level (WL)), or even the whole length of the rear side, as shown in Figures 15.2(c, d, f, g). As the ratio (Rc/Hs) gets larger, either the crest or the exposed rear-side slope, or both, will require smaller size armour units or a different type of armour, as shown in Figures 15.2(h, i, j, k).

The exposed part of the rear-side slope, i.e. above WL and a little below, will receive the full impact of the falling wave overtopping discharge, therefore it will require a protection corresponding to the applied downward pressure. Whereas the immersed part of the rear-side slope will receive a reduced impact force, a reduction related to the immersion depth. Accordingly, the armour block size could be reduced after a recommended minimum depth that corresponds to $3D_{n50}$ of the seaside armour block (Burchart, 2007), as illustrated in Figure 17.1. However, Burchart has left to the designer to select the relevant water level to be used as a reference for measuring the depth from. Wave overtopping discharge is strongly related to the water level; therefore, the highest discharges are taking place at high water levels. Accordingly, measuring the depth $3D_{n50}$ should be from the DHWL, but for added safety the MWL is recommended to be considered as the reference level.

Further to the given examples in Figure 15.2, more examples are presented in Figure 17.2 for breakwaters with CAU and in Figure 17.3 for breakwaters with rock armour. Various arrangements of breakwaters with concrete armour units (CAU) could be designed to meet the relevant design conditions, stability requirements, cost and construction-oriented

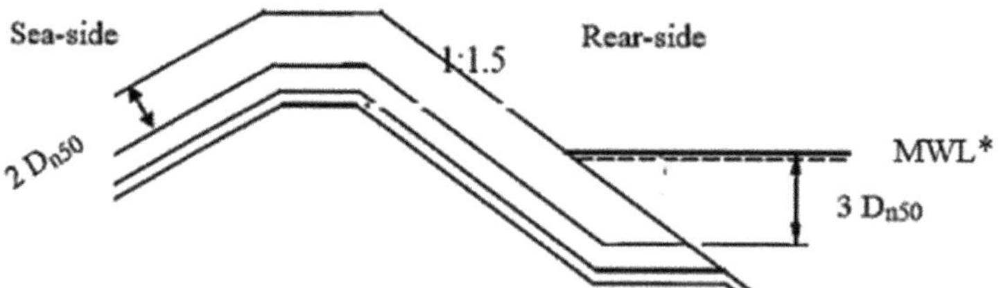

Figure 17.1. Recommended primary armour extent on the breakwater rear side (Burcharth, 2007).

Note: *MWL is added by the author.

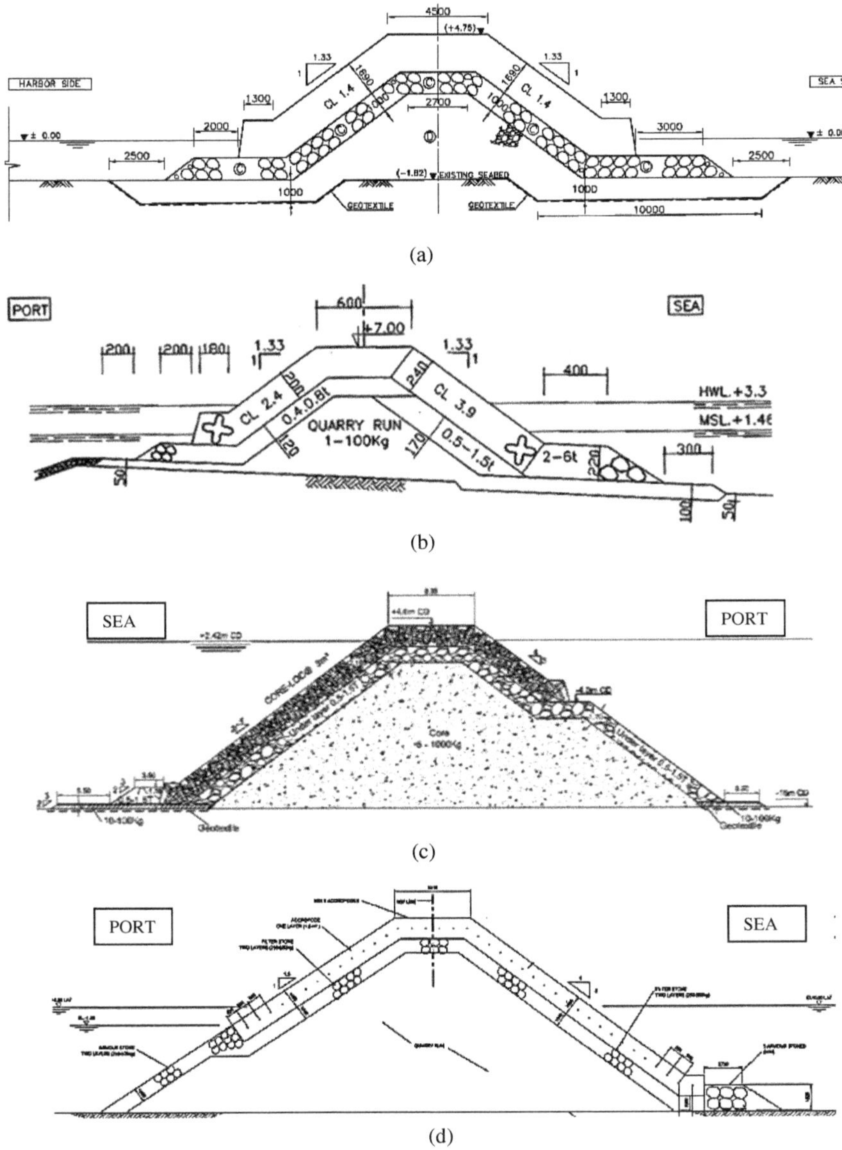

Figure 17.2. Rear side for Breakwaters with CAU seaside armour and no superstructure: (d) (Glennerster *et al.*, 2013); (f) (PIANC, 1992).

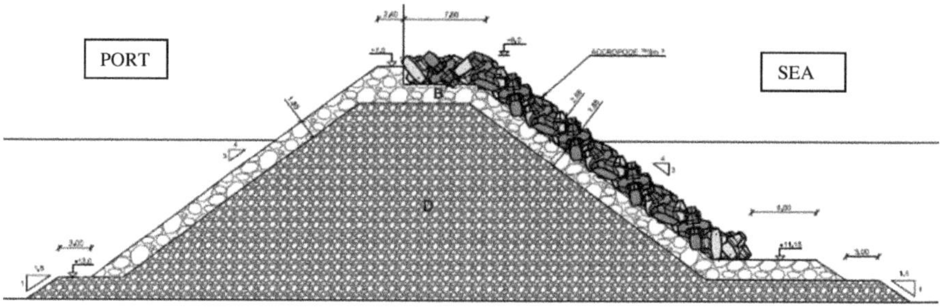

(e)

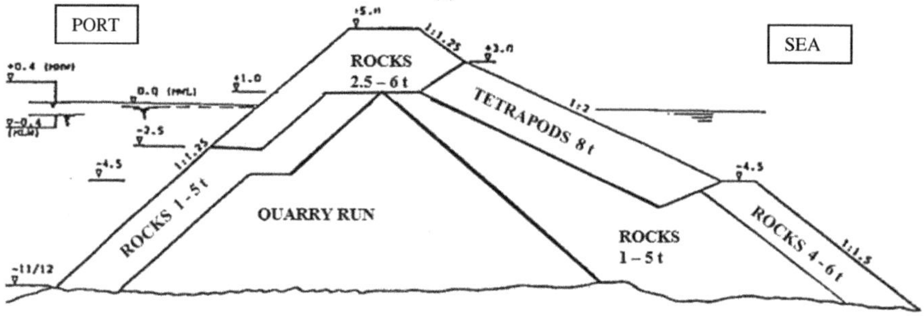

(f)

Figure 17.2. (*Continued*)

optimisations. The appropriate CAU size shall be selected for all the three sides of the breakwater, that is for the seaside, crest and the rear side. The CAU may only be required to protect the exposed part of the rear-side slope and to be replaced by rock armour for the immersed part, as in Figures 15.2(g) and 17.2(c, d), or for the full length of the rear-side slope as in Figures 15.2(h, i) and 17.2(e), or even for the crest and the rear-side slope as in Figure 17.2(f).

Figure 17.3, together with Figure 15.2, shows diverse examples for breakwaters with rock armour, starting from using the same rock armour layer on both sides, with the same slope as in Figure 15.2(f) or with a steeper slope for the rear side as in Figure 17.3(a). In Figure 17.3(b), the seaside primary armour layer has been used for the rear-side exposed upper part only, whereas the seaside rock armour size and same slope has been used for the rear-side but with one layer instead of two layers as in Figure 17.3(c). The examples in Figure 17.3(d), e also show the rear side with one layer of the primary rock armour while the second layer is of a smaller rock size. A smaller rock size has been used in Figure 17.3(f) by

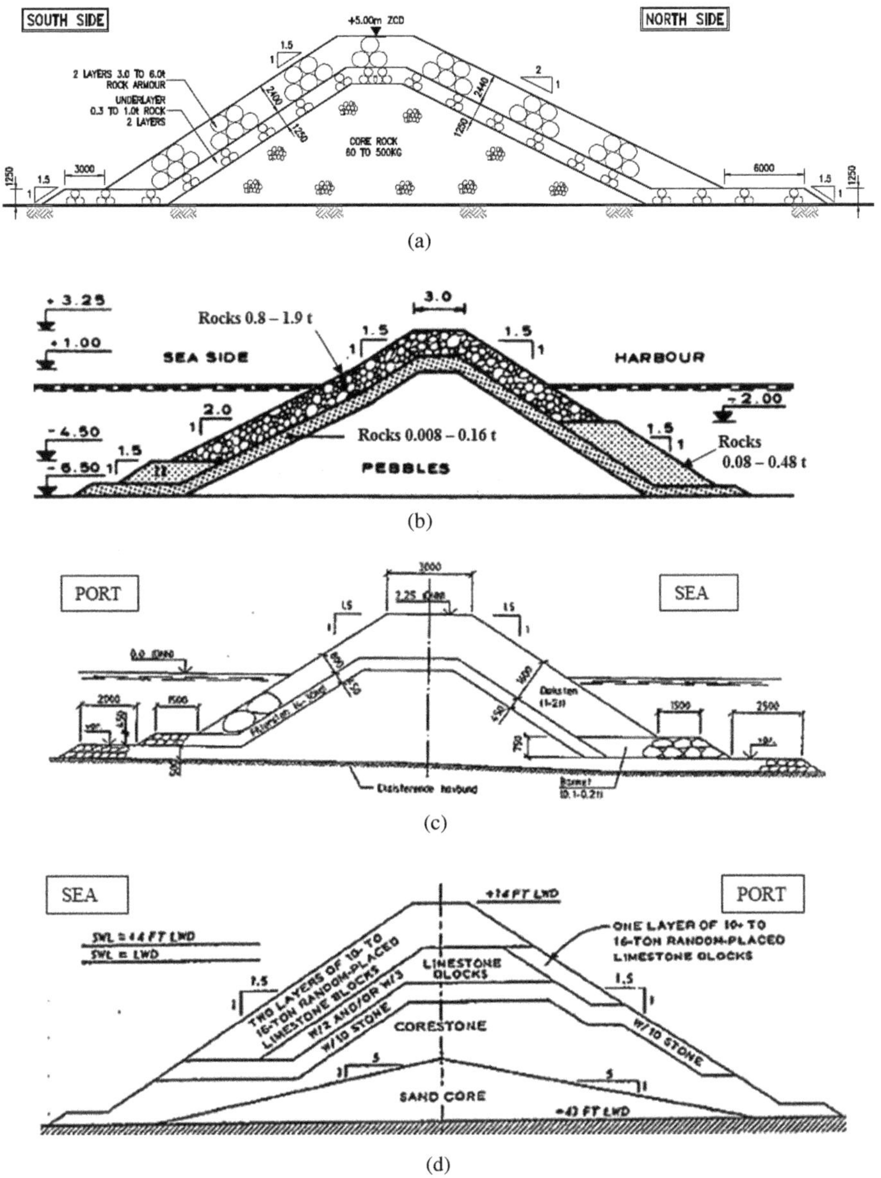

Figure 17.3. Rear side for breakwaters with rock armour and no superstructures: (b) (Jensen, 1984); (e) (Jensen, 1984).

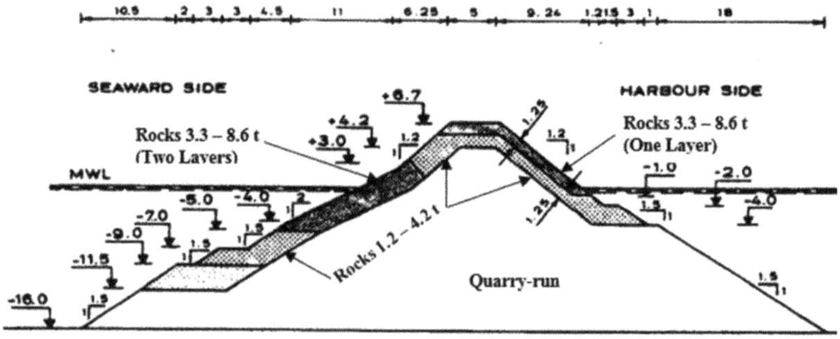

(e)

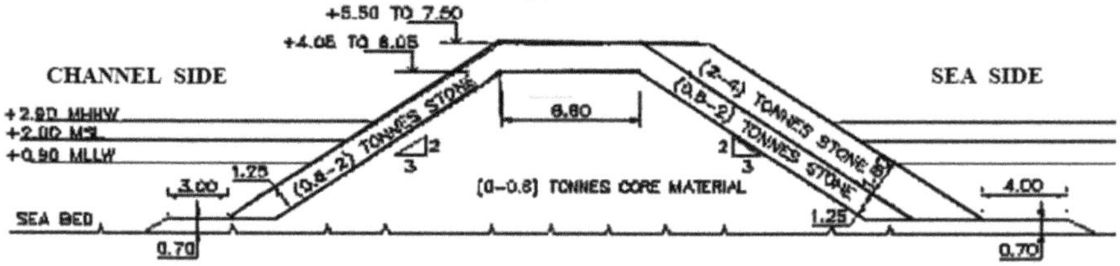

(f)

Figure 17.3. (*Continued*)

extending the underlayer of the seaside primary armour to be the armour layer of the crest and the rear side.

It is worth noting that a simple section design with fewer rock classes, not many different side slopes and fewer construction stages is always preferred by construction teams. Intricate design sections, similar to Figures 17.2(f) and 17.3(d, e), are only developed because of very special local circumstances.

17.3. With Superstructure

The rear-side armour stability for breakwaters with crown walls is only partially dependent on the Rc/Hs ratio, nevertheless this ratio is still a good indication for the expected wave overtopping discharge. The armour layer stability is however very much influenced by the configuration of the interface with the crown wall base slab. Different arrangements for this interface are presented in Figure 16.2 with the recommendation to avoid the interface of Figure 16.2(c), also shown in Figure 17.4(b), as the possibility of damage to the rock shoulder is very high. Whereas the

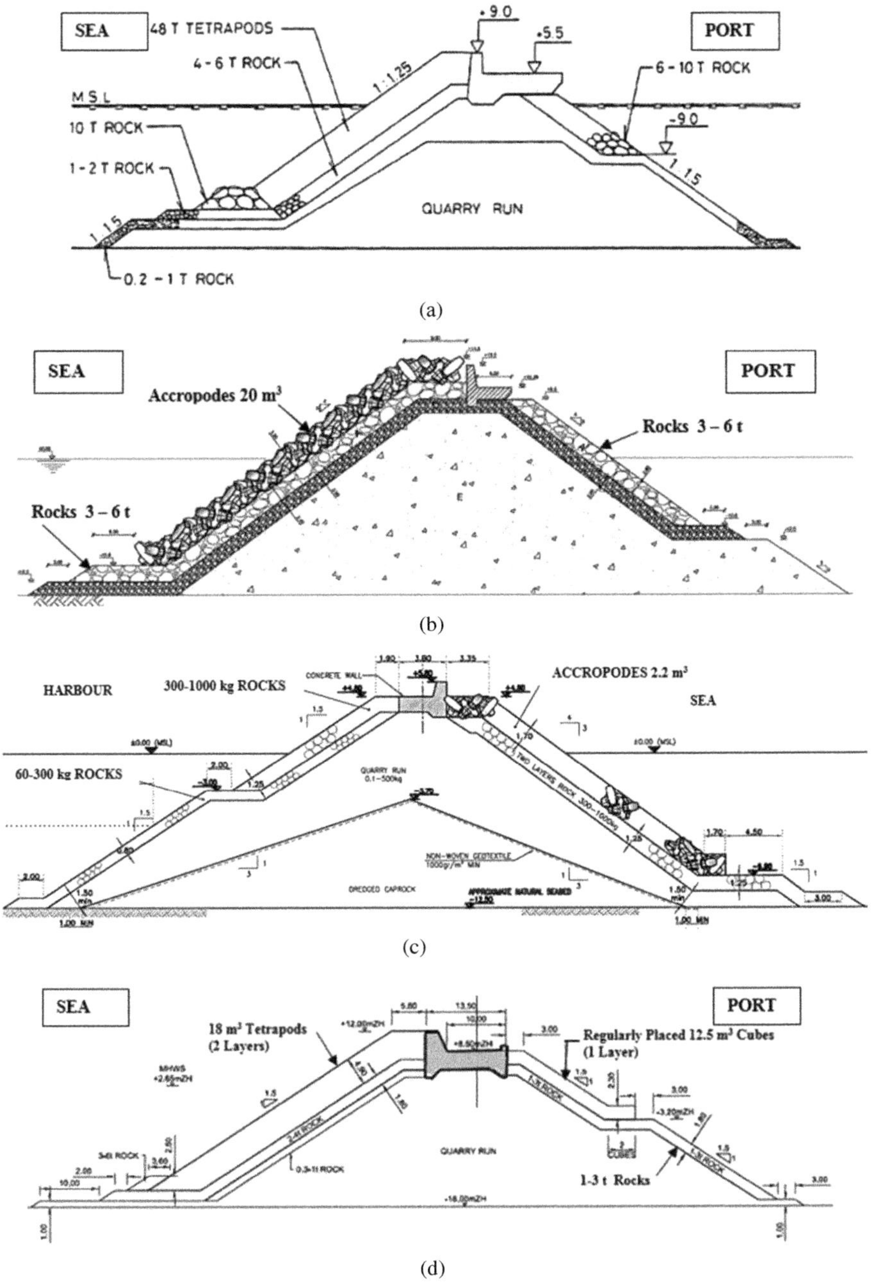

Figure 17.4. Rear side for breakwaters with crown walls.

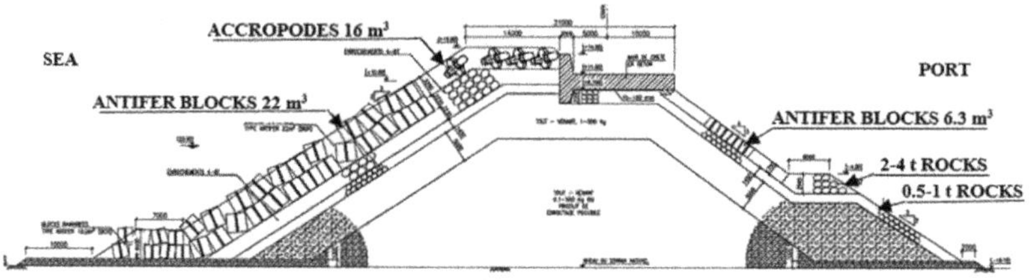

(e)

Figure 17.4. (*Continued*)

interface shown in Figure 16.2(d) could be used with relatively low wave overtopping discharge due to high Rc/Hs ratio as in Figure 16.4(h), or with wide crest and wide crown wall base, as in Figure 16.4(g). This type of interface is also shown in Figures 17.4(c, d). The examples in Figures 17.4(a, e) have an interface that could tolerate high wave overtopping discharges.

As mentioned in Section 17.1, a smoother rear-side surface is beneficial in reducing the forces on the armour block, examples of such designs are shown in Figures 17.4(d, e).

Breakwater crests may have road slabs, as shown in Figure 17.5, and the rear-side slope in these breakwaters is to be treated as a breakwater without a superstructure.

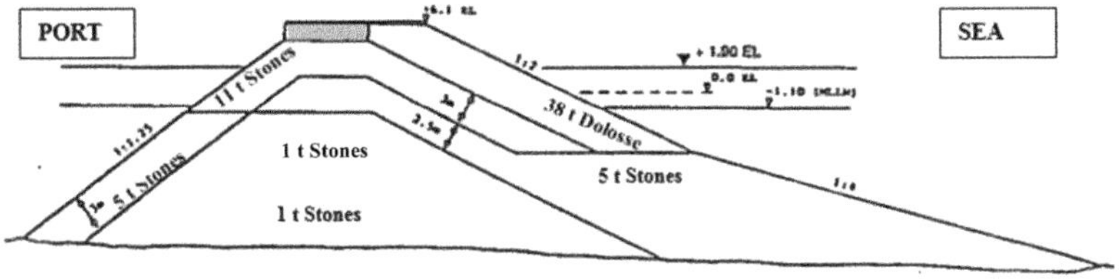

(a)

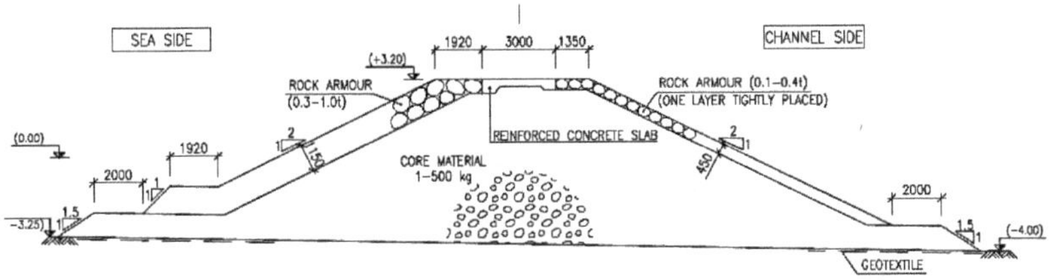

(b)

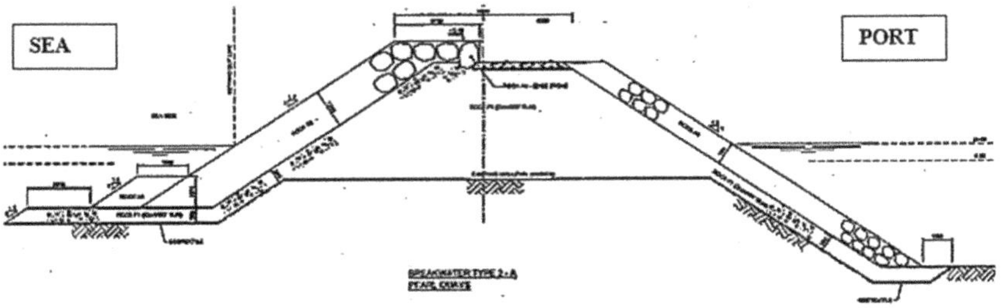

(c)

Figure 17.5. Rear side for breakwaters with crest slabs: (a) (PIANC, 1992).

Chapter 18

Slope Protection Under Open-Piled Structures

18.1. Introduction

Open-piled structure could be an open quay, a bridge, a jetty or a gangway. Open quay structures typically comprise a deck supported by vertical, or a combination of vertical and raking, piles, overlying an embankment with a slope protection. The interface between the embankment and the landside could be of different forms and shapes, as shown in Figure 18.1. A reinforced concrete abutment structure is widely used to form the interface and it could have different configurations, in (a) is an independent concrete retaining structure, in (c) is a friction slab linking the deck to the embankment, and in (d) a curtain wall integrated with the pile cap. A different type of retaining structure is also common and that is by using sheet piles, as in (b). However, there are situations where neither an abutment nor a sheet pile is used, as in (e). Practical examples are given in Section 18.4 for each of these types.

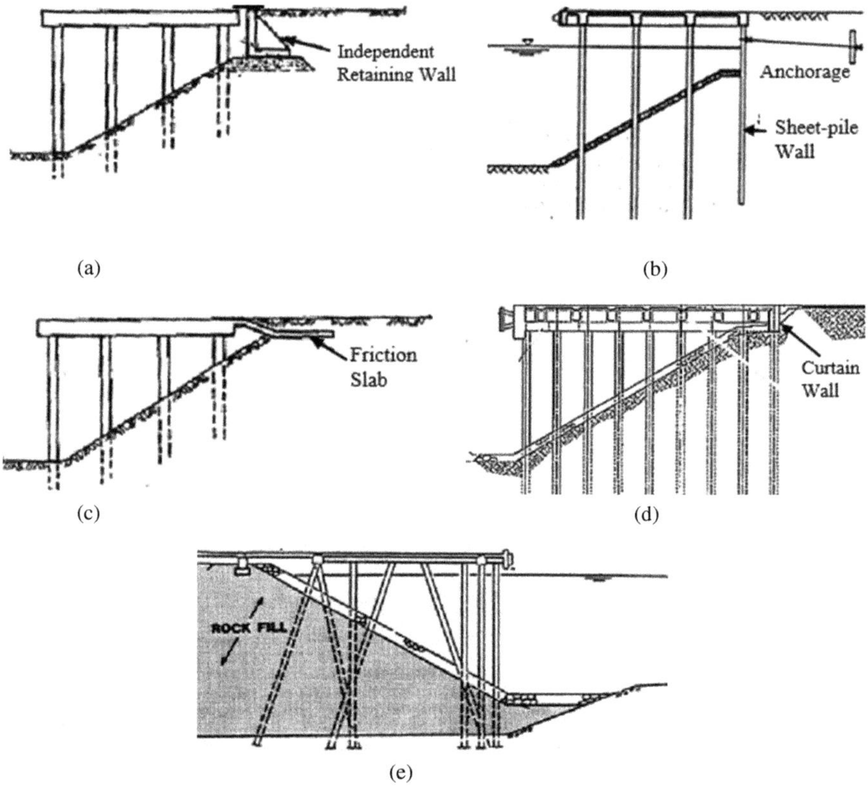

Figure 18.1. Typical land interface arrangements for open-piled deck: (a) Independent retaining wall; (b) Sheet pile retaining wall; (c) Friction slab; (d) Curtain wall; (e) Embankment (no abutment).

18.2. Slope Protection

The slope protection must be stable mainly against vessel propulsion; however, although waves and currents are normally of a lesser importance for these slopes, they are also to be considered. Different propulsion types are in use, as illustrated in Figure 18.2, and these have different hydraulic loadings on the slope protection during the berthing and unberthing operations (PIANC, 2015). Velocities at propeller jets and side thrusters at the exits could be as high as 12 m/s. The jet velocity will be partially dissipated due to turbulence depending on the dispersion pattern and/or deflection by a rudder before reaching the slope or the bed. However, the

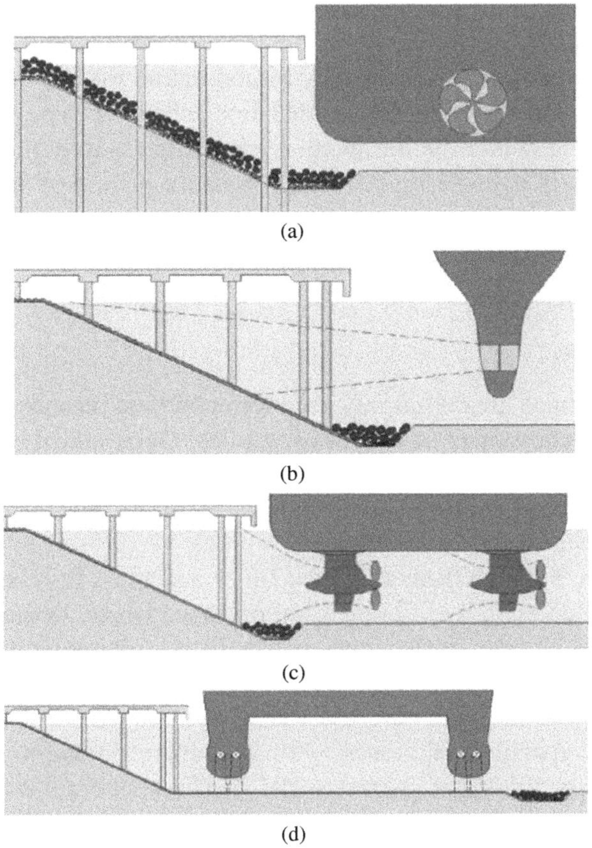

Figure 18.2. Basic types of ship main propulsion and manoeuvring jets (Hawkswood *et al.*, 2014): (a) Stern propeller; (b) Side thruster; (c) Podded propellers; (d) Water jet.

impact velocity will depend on many factors including the distance from the slope and bed, the slope angle, type of propulsion and direction of the jet (BAW, 2010).

Different slope protection material could be used, including rock armour, CAU, grouted rock with liquid asphalt or concrete, concrete slabs, concrete block mattresses and concrete geomattresses, see PIANC (2015) and Pilarczyk (2003). The interlocking and integrity of the protection layer requires special attention and needs to be carefully assessed and supervised during the design and the construction. It is also important to ensure that the underlayer/filter layer is fully covered by the main protection layer in all interfaces/transitions around the piles, at the crest and at the toe.

The slope protection is normally placed before the construction of the deck, and as such it is recommended that whatever type of protection material is used it is to be designed as maintenance free, in other words, to be moderately over-designed. This is because in case of damage due to overload, age or exposure to aggressive substances, it is extremely difficult to maintain/repair the slope protection.

18.3. Piles

Tubular steel piles are commonly used for economic reasons, which could be used as friction piles or end-bearing piles. Open-ended steel piles are used as friction piles and are usually equipped with either an external or internal reinforcement ring shoe, see Figures 18.3(a, b). This type of pile is to be driven through soft strata, i.e. no rock fill or bed rock.

If the pile must go through a rock fill or a shallow thin bed rock, then it will be fitted with a different type of shoe, which makes it an end-bearing pile. The piles of this type are usually of a diameter 500–800 mm. Steel piles can be driven into harder strata than can concrete piles. For example, a 700 mm diameter tubular steel pile can penetrate a 20 m thick rubble fill consisting of stones with diameters as large as 500 mm (approximately 500 kg), (Thoresen, 2014). These steel piles are normally fitted with a conical pile shoe with dowel, as shown in Figure 18.3(c). This type of pile shoe is used specially to prevent the pile from skidding on the sloping rock surface and to get a better grip in the rock by easing its penetration into the quarry run rock fill (few stones of 600–700 kg in the quarry run class of 1–500 kg could be tolerated, however, such large stones are better to be removed). The top part of these piles could be filled with concrete to enhance the rigidity of the pile. Example projects are shown in Figures 18.4, 18.7, 18.10 and 18.12.

Cast in-situ reinforced concrete piles with steel pipe casings are normally used in wharves for large vessels where larger piles are required. The tubular steel casing is to be fitted with a ring shoe and to be driven until it penetrates the seabed to a stable depth. Then drilling will commence and continues beyond the steel casing until reaching the required design depth. An example project is shown in Figure 18.9. It is, however, important to note that steel pile casings in this example are to be driven

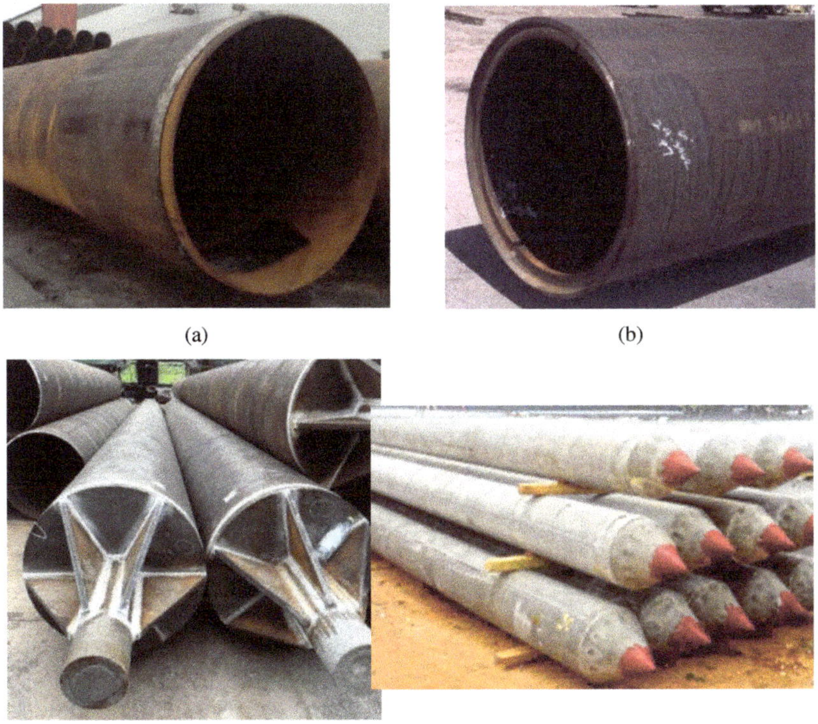

(a) (b) (c) (d)

Figure 18.3. Examples of pile shoes: (a) Steel pile — outside ring shoe; (b) Steel pile — inside ring shoe; (c) Steel pile with conical shoe with dowel; (d) Pre-cast-pre-stressed concrete pile with steel shoe.

after placing the (1–500 kg) quarry run rock embankment, which is not possible with such large rock sizes. Therefore, the rock material for the outer part of the embankment is replaced with a smaller grade, (2–20 kg), to facilitate driving the piles. As the embankment is exposed to sea waves, it needs to be protected soon after driving the pile casings, however the casings must be protected against possible denting and damage. A special wooden jacket is used to protect the pile from damage during the under-layer rocks and the armour units placing operation, see Figure 18.9(b).

A different use for the pile casing is shown in Figure 18.11, where it is used as a sleeve for a steel pile to be driven inside it. The construction methodology in this example requires that the piles are to be installed after

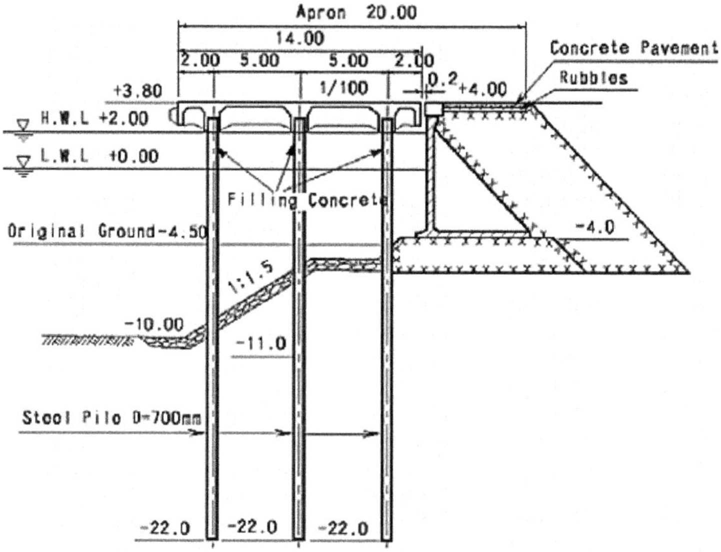

Figure 18.4. Large pre-cast R.C. retaining wall (Nozu *et al.*, 2004).

completing the construction of the breakwater. Therefore, pile sleeves are to be installed first before the construction of the breakwater, and to be well braced to hold its position and alignment. After the completion of the breakwater construction, the steel piles can then be installed through the sleeves.

Pre-cast-pre-stressed concrete piles could represent a more attractive option for smaller wharves. These are to be fitted with steel shoes for driving the pile to avoid any damage. An example of a steel pile shoe is shown in Figure 18.3(d), and an example project is shown in Figure 18.6.

The pile type and size must be carefully selected together with their spacing and distribution in view of the following:

(a) Seabed condition
(b) Seabed depth
(c) Static and live loads on the deck
(d) Tidal levels
(e) Construction work requirements and preferences
(f) Arrangement of bollards
(g) Arrangement, shape and dimensions of fenders
(h) Crane rail alignment

Practical examples of different types of piles are shown in Section 18.4.

18.4. Interface with Landside

The interface with the landside is actually an interface between three elements: the deck, the embankment and the landside fill. The interface must ensure the stability of the embankment and its protection, the stability and the confinement of the landside fill material, and the stability of the interface structure itself.

Basically, there are two categories for the interface, the first is with a retaining structure/wall where the embankment, in most cases, does not reach the underside of the deck. The second category is by having a full embankment reach the underside of the deck with a simple arrangement to retain the landside fill material. The selected type for a particular project depends on many factors including the nature of the seabed, the type of the embankment fill, the height of the deck from the seabed, construction methodology, cost, live loads, berthing and mooring loads and seismic loading.

Different practical examples for the interface arrangements are presented in the following sub-sections.

18.4.1. *Partial embankment with a retaining structure*

This type of interface is widely used in many projects as the embankment slope does not reach the underside of the deck. Therefore, a retaining structure is required as an interface closing the gap between the deck and the embankment while retaining the landside fill material.

The retaining structures could be a pre-cast/cast-in-situ R.C. or a sheet pile. The type and height of the retaining structure is project-specific, depending on many factors including construction methodology, design criteria, exposure to wave and hydraulic loading, seabed condition, availability of rock material and cost.

A vertical retaining wall is characterised by a high wave energy reflection coefficient. This is expected to induce relatively more turbulence under the deck, which might affect the berthing/unberthing operations. Therefore, high retaining walls are normally used in well-sheltered basins, i.e. not exposed to sea waves to minimise the unwanted disturbances.

A relatively large pre-cast R.C. retaining wall, about 7.0 m high, is shown in Figure 18.4. Whereas smaller R.C. retaining walls, which could

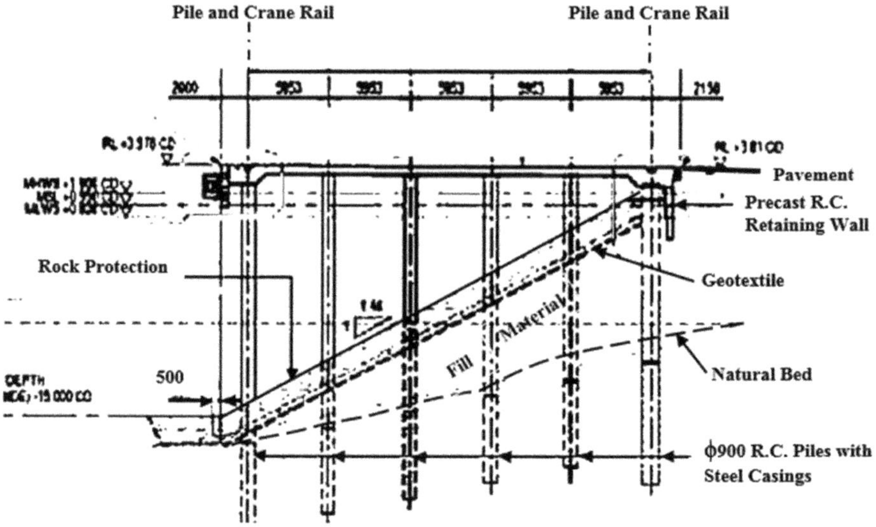

Figure 18.5. Small pre-cast R.C. retaining wall connected to the deck (Port of Napier, Wharf 6, 2020).

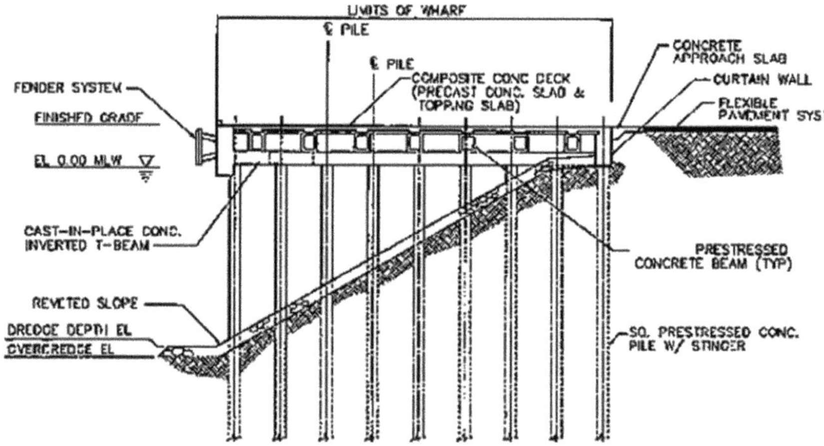

Figure 18.6. Small cast-in-situ retaining/curtain wall.

also be described as curtain walls, with almost full embankments are shown in Figures 18.5 and 18.6 for pre-cast and cast-in-situ types, respectively.

A different example is shown in Figure 18.7 for a jetty with the embankment being part of a breakwater exposed to sea waves. It is almost a full embankment with two types of small R.C. retaining walls being used, one above the pile cap and the other in between the piles, as shown in Figures 18.7(b) and 18.7(c).

The use of sheet piles as abutments are shown in Figure 18.8 for a larger height with an anchoring block, and in Figure 18.13(a), for a smaller retaining height with a back sheet pile anchor.

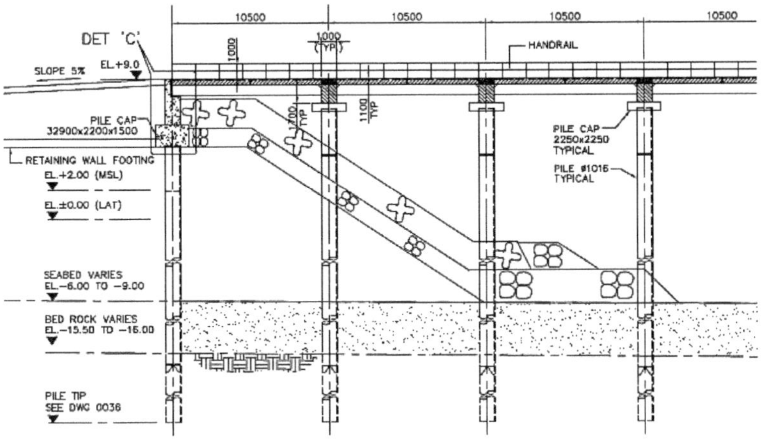

(a)

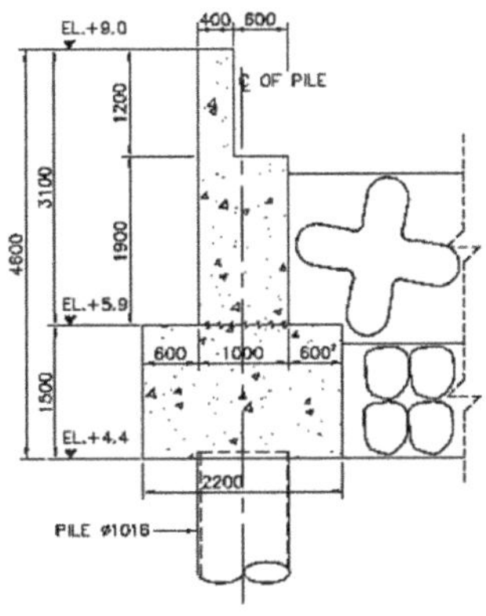

(b)

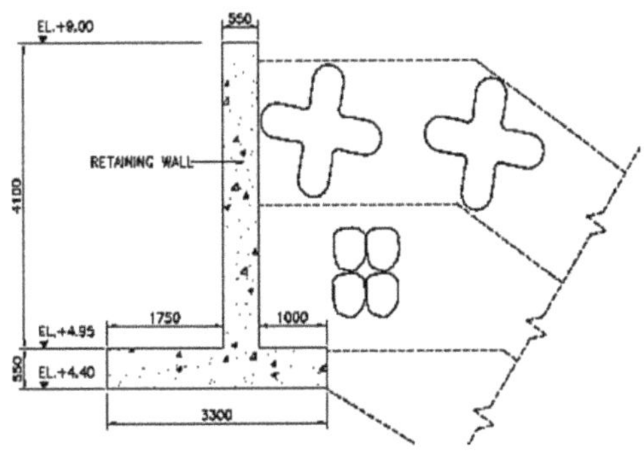

(c)

Figure 18.7. Small pre-cast R.C. retaining wall for a jetty: (a) Embankment and jetty cross-section; (b) Retaining wall over the pile cap; (c) Retaining wall between pile caps.

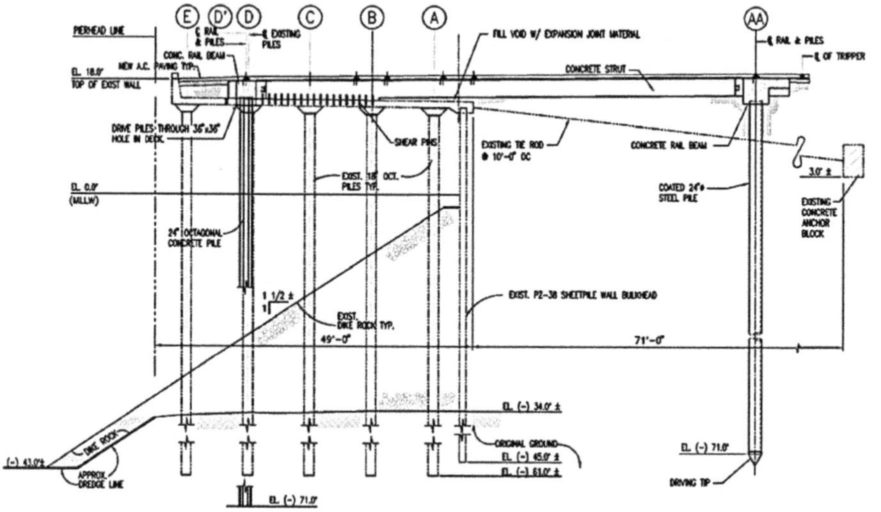

Figure 18.8. Long sheet pile retaining wall with anchor block (Mansour *et al.*, 2004).

18.4.2. *Full embankment*

A full embankment means that the embankment slope extends to, or very close to, the soffit/underside of the deck with no retaining wall structure. The construction sequence in such projects usually starts with completing the embankment including the placement of the armour units, thereafter the construction of the deck. The final construction step is to close the small remaining gap between the embankment and the deck. Gunny bags, geobags or something similar are normally used for this purpose. These bags are small, easy to handle and could be filled with cement mortar or similar grout. The bags will then be stacked to provide a stable and durable construction element. Two examples are given in what follows, one with CAU armour units in Figure 18.9, and the other with rock armour in Figure 18.10.

A different type of structure is shown in Figure 18.11 where the embankment is on the lee side of a breakwater, with a crown wall on the crest. There is no gap to be closed by bags and no need for a retaining wall, as was the case in Figure 18.7(c). The quarry run is well-covered and protected by the crown wall and the underlayer rocks from being washed out.

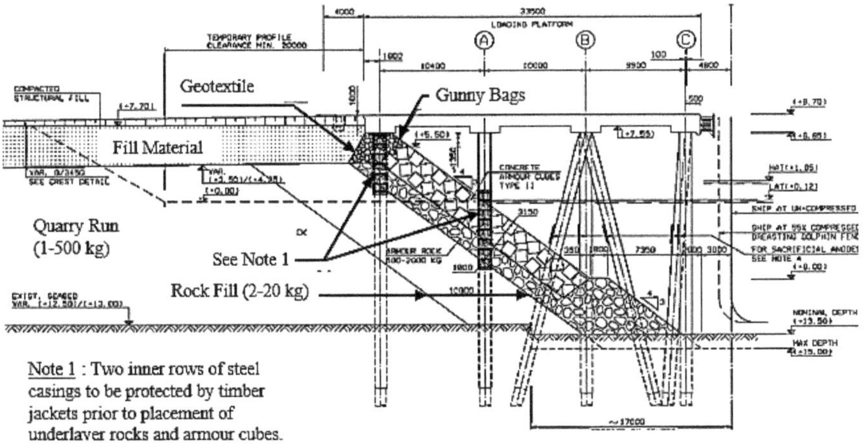

(a)

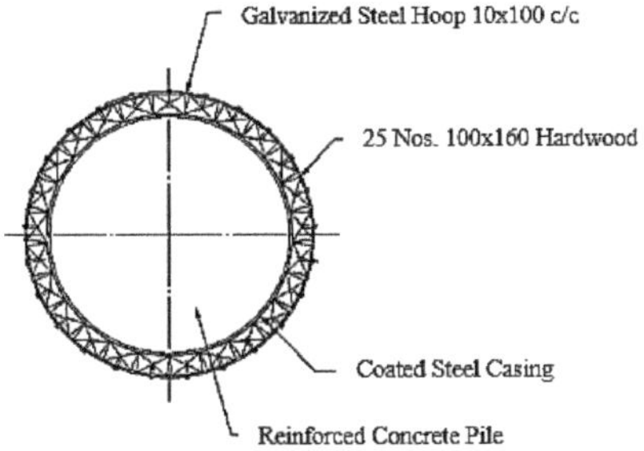

(b)

Figure 18.9. Embankment and deck cross-section with R.C. piles and steel casings: (a) Embankment and open berth cross-section; (b) Pile timber jacket details.

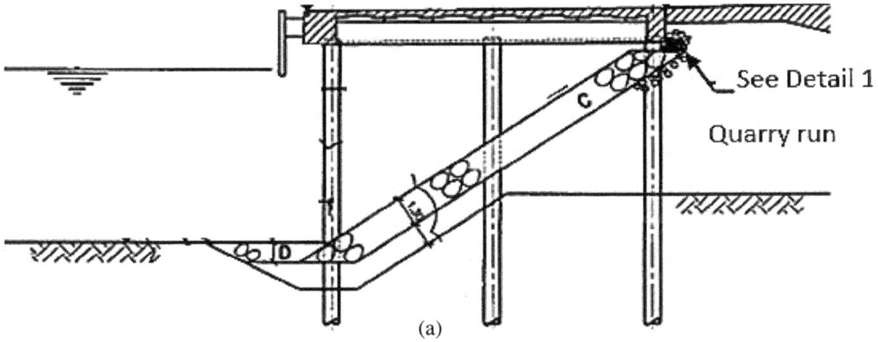

(a)

Figure 18.10. Embankment and deck cross-section with steel piles: (a) Embankment and open berth cross-section; (b) Embankment crest closure detail.

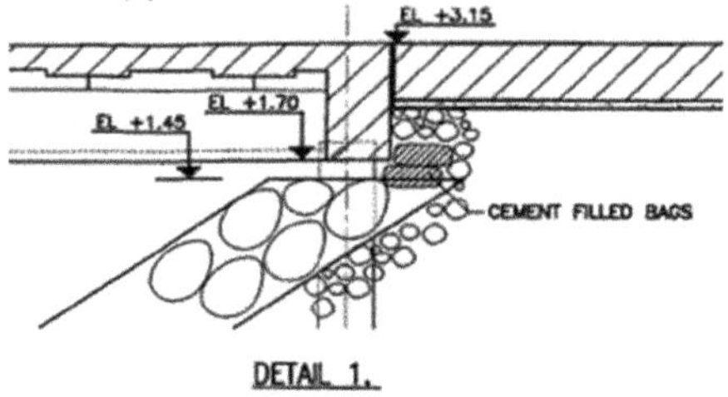

(b)

Figure 18.10. (*Continued*)

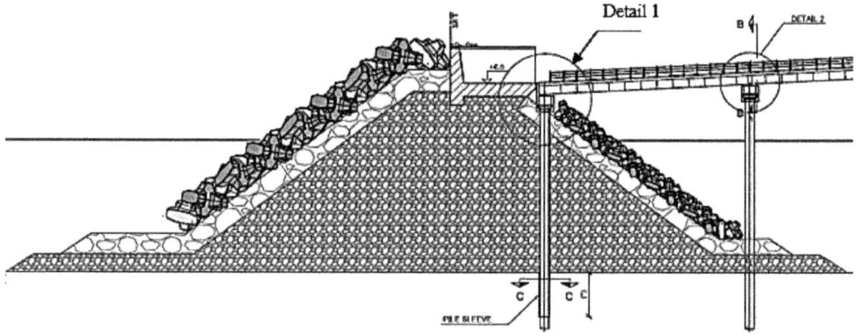

(a)

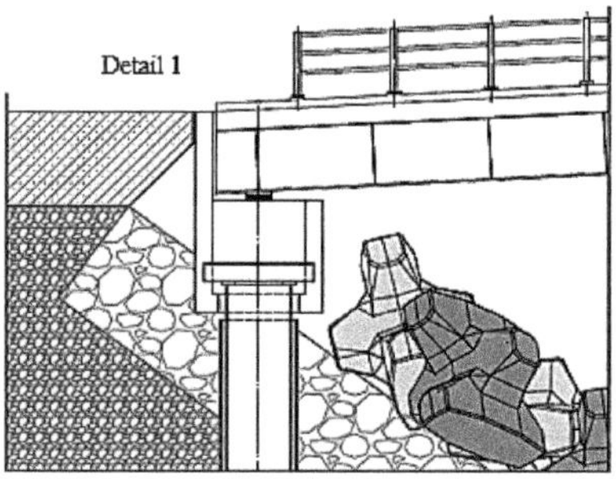

(b)

Figure 18.11. Breakwater and jetty cross-section with steel piles in pile sleeves: (a) Breakwater and jetty cross-section; (b) Embankment crest and crown wall details.

Another type of project characterised by a long deck and/or the quay is constructed on a rocky shoreline that does not require any protection, the embankcd in either case will invariably be small and detached from the berthing line, as shown in Figure 18.12.

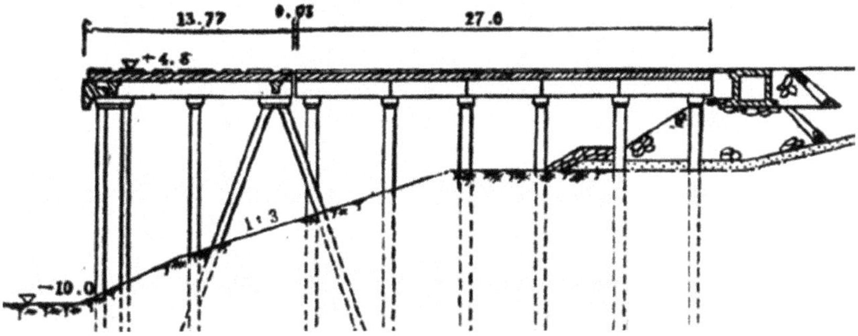

Figure 18.12. Long deck with landside small embankment (Zhaohua *et al.*, 2018).

18.5. Bed Scour Protection

The material type, size and extent (width and length) of the bed scour protection depend on:

(a) Size and draft of the vessel,
(b) Type of vessel's propulsion and side thrusters,
(c) Berthing/unberthing procedure of the vessel (i.e. using the vessel's own propulsion or a tug-assisted operation),
(d) The nature of the seabed,
(e) The geotechnical stability of the embankment (i.e. the width of the passive soil volume in front of the embankment).

The used material for the bed protection is almost the same as that for the slope protection apart from the CAU. A combination of different materials is also used, for example a concrete mattress with the edge being of rock fill forming a falling apron to secure the edge of the mattress. In bed protection design, the water level is obviously an important design input and the proposed levels by Hawkswood *et al.* (2014) seem very reasonable to be adopted. These proposed design levels are (MLW) and (LAT), depending on the local tidal range, being large (>1.0 m) or small (<1.0 m), respectively. The start of the bed protection, or the end of the toe to the embankment slope, is recommended to be at least 1.0 m behind the berth line (i.e. landward from the berth line), to minimise the effect of the propeller turbulence on the slope.

Few examples are presented to demonstrate the required extent of the bed protection width. Starting with no protection being required as in Figures 18.8, 18.12 and 18.13(a), to a narrow protection width of 7.5 m in Figure 18.13(b), it increased to 14.0 m in (c), to 23.0 m in (d) and 40.0 m in (e).

For a soft seabed material with vessels carrying out the berthing/un-berthing operations by its own propulsion and side thrusters, the extent of the bed protection is to cover the affected area by the jets (side thruster and main propulsion jets). The starting design dimensions for the scour protection are recommended to be as follows: the length is to be equal to the design vessel length plus extra 50 m in front of the bow and another 50 m behind the stern, whereas the width is to be as wide as the design vessel width. However, these dimensions are to be refined in the detailed design as per the recommendations of PIANC Report no. 180 (2015).

Concrete mattresses are becoming more popular than the rip-rap rock protection due to many reasons, primarily to the increases in vessel dimensions and the installed engine power. Propulsion and manoeuvring jets from roll-on/roll-off, ferries, cruise and container ships have a wide flow influence area as these vessels can change the jet direction. Possible erosion by strong flows has made protecting the bed and the embankment, specially around the piles, an issue of great importance.

Mattresses can offer high resistance to high velocities and shear stresses with easy and fast installation, and usually have a cover thickness less than the rock protection. Mattresses used for bed protection are basically either articulated concrete mattresses or geofabrics filled with fine aggregate concrete, and both come in different forms and shapes. It is worth noting that gabions are also used as a possible efficient bed protection. However, the final choice of the bed protection scheme shall be selected after close consultation with the construction team and potential suppliers.

In using mattresses, it is important to consider the uplift forces which are due to the reduced pressure zone at the top because of the high-water velocities or wave troughs, and the excess pore water pressure from underneath. These forces could either be reduced by slowing the velocities through creating more turbulence or provide sufficient weight in the mattress to counteract the uplift forces. However, in concrete articulated mattresses, and in some geofabric ones, there will be enough surface roughness to disrupt the surface flow and enough porosity to dissipate the pore pressure.

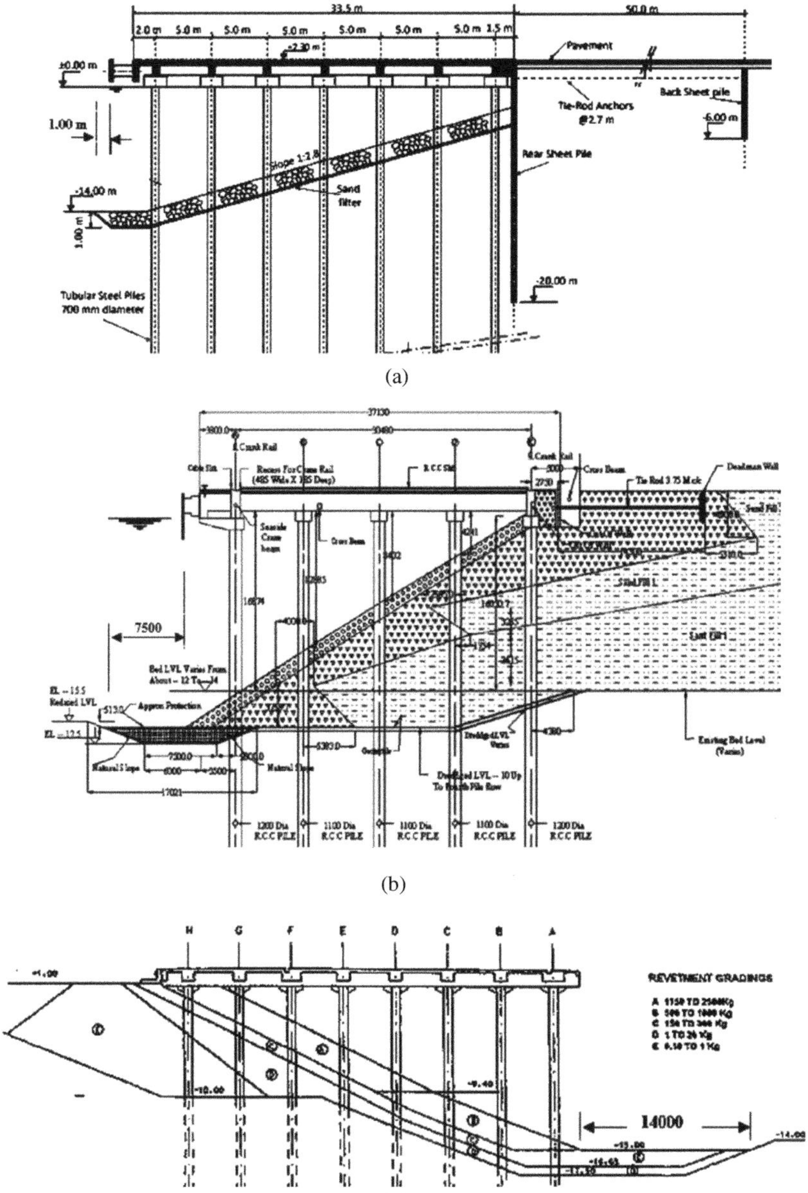

Figure 18.13. Examples of bed rock protection with different widths: (a) (Galal, 2017); (b) (Premalatha *et al.*, 2017); (c) (Hallett *et al.*, 1995).

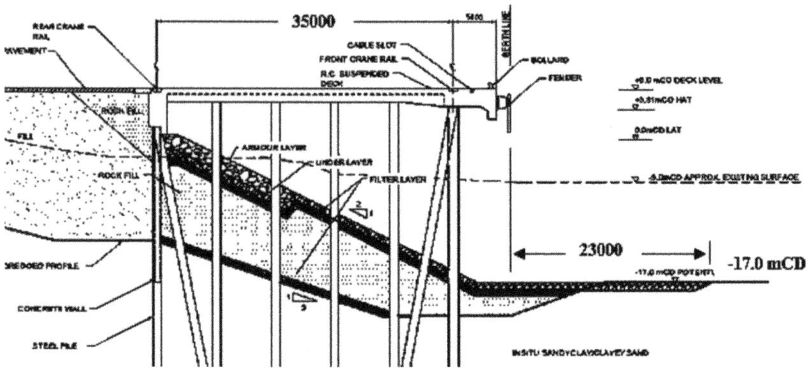

(d)

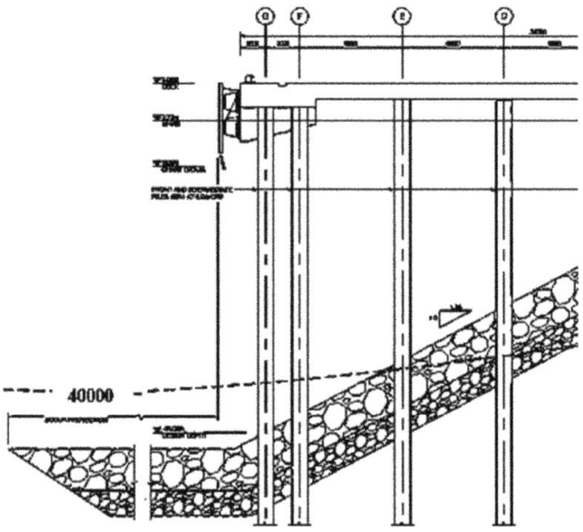

(e)

Figure 18.13. (*Continued*)

An example of an articulated concrete mattress is given in Figure 18.14, showing a section in the berth with the design details of the mattress. The gaps between the concrete blocks are sized according to the handling requirements and they are large enough to allow the bedding filter material to escape. Therefore, the selected geogrid must act as the connector between the blocks and as a filter preventing the loss of the bedding material. Accordingly, the geogrid must satisfy the strength requirement and must also have an aperture size smaller than the D_{n50} of the bedding material to prevent loss of material.

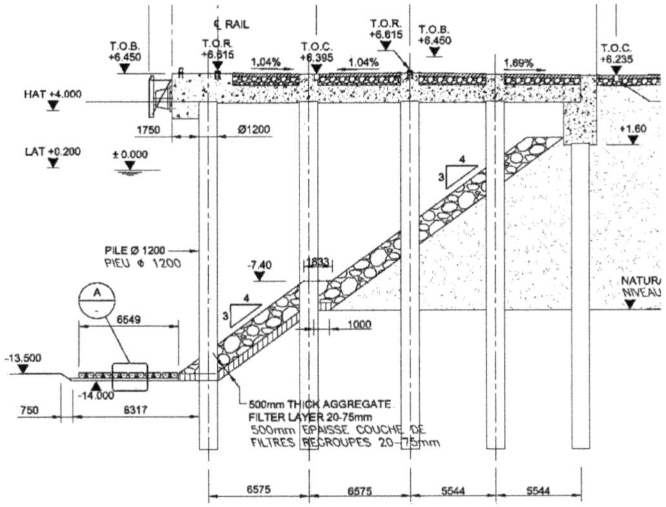

(a)

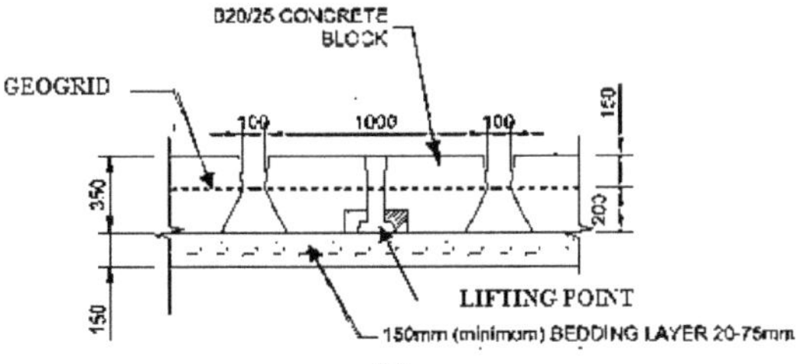

(b)

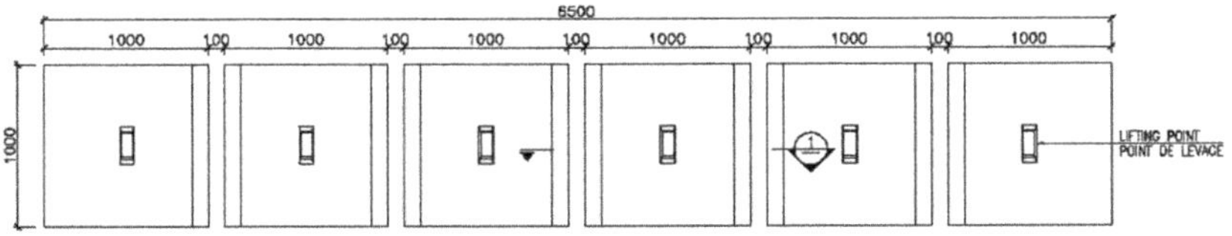

(c)

Figure 18.14. Concrete mattress bed protection: (a) Embankment and open berth cross-section; (b) Concrete mattress section details; (c) Concrete mattress plan details.

18.6. Small Boat Berths

Embankments under gangway bridges to boat berths and floating docks are usually part of a longer revetment along a shoreline or a riverbank. At the embankment crest is often used a heavy pedestal block to support and anchor the gangway.

An example is shown in Figure 18.15 for two options of a gangway, with guide piles and without. In Figure 18.16, two examples of gangways to a floating deck with guide piles are shown.

In most cases the revetment toe arrangement will continue as for the rest of the revetment stretch without any requirement for additional bed protection. However, if larger boats are using the berths, with shallow water depths, and the boats are getting close to the revetment toe area, then some bed protection will be recommended to avoid scouring problems.

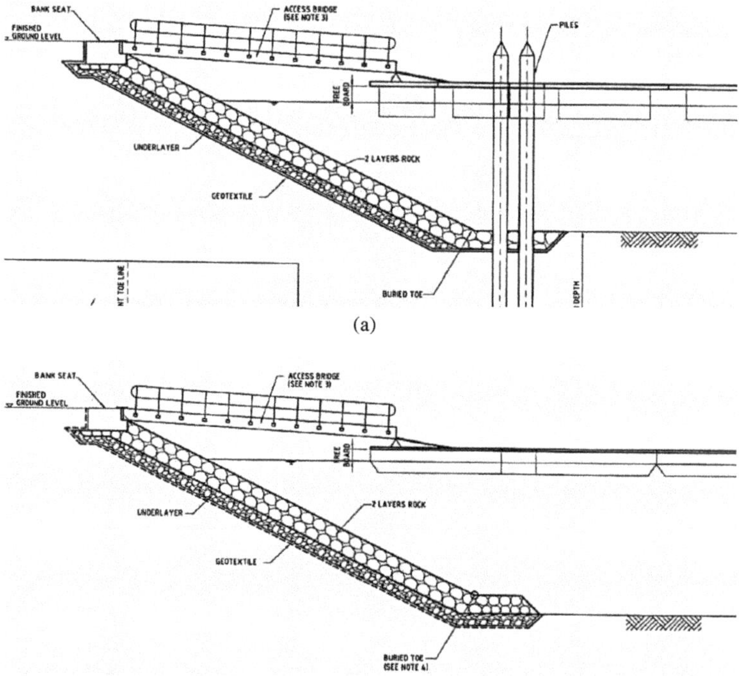

(a)

(b)

Figure 18.15. Embankment examples to access gangways — floating berths.

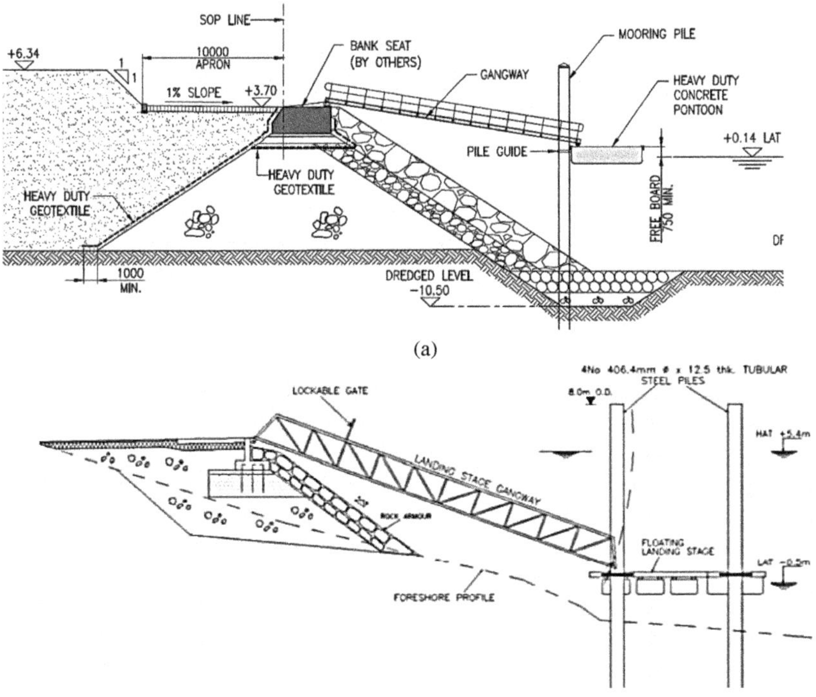

Figure 18.16. Embankment examples to access gangways — floating decks.

Chapter 19

Shoreline Outfalls, Pipelines and Cables

19.1. Introduction

It is frequently required to have outfall pipelines discharging to the sea either directly on the shoreline or through a revetment. The pipelines could be carrying storm drainage water or effluents from inland industrial activities.

In many coastal structures, surface drainage is adopted to drain the wave overtopping and rainwater where the water will simply flow over or percolate through the rock material towards the sea. However, this arrangement is not always applicable due to many reasons. These could be because the revetment section configuration is obstructing or retarding the free flow of water towards the sea, or due to other requirements depending on the nature of the project, or because the surface is fully paved and surface water has to be collected by drainage pipes.

The drainage pipes usually penetrate different materials before reaching the revetment side slope surface, starting with the usual sand fill material at the landside, then rock material (quarry run, underlayer rocks and possibly armour rocks). Therefore, the pipes need to be protected in the rock material against possible damage, and they are to be resting on a good foundation layer throughout the pipeline alignment ensuring the pipe design slope.

The outlet opening of the outfall pipeline at the revetment slope needs careful consideration and evaluation as different factors may affect the flow and the stability of the pipe and also of the armour layer. These factors include the wave exposure, the height of the pipe opening relative to the sea level, prevention of return flow, type/size of the armour blocks and the possible need for an outlet structure with a proper interface with surrounding armour rocks/CAUs without affecting the integrity of the armour layer.

Examples for three categories shall be presented in the following sections covering the slopes with Concrete Armour Units (CAU), Rock Armour and a third category for pipes and cables on the seabed and passing underneath the revetment slope.

19.2. Slopes with Concrete Armour Units (CAU)

As illustrated in Figure 10.1, most of the CAUs depend, in various degrees, on interlocking, therefore it is highly recommended to avoid breaking the "fabric" of the units, especially for the single-layer units. Therefore, it is not advised to have an outfall outlet structure on the slope face and above the toe level. Accordingly, if the outfall pipeline is at a level higher than the toe, then it is best to have it terminated before reaching the armour layer, i.e. not to go through it. Such an arrangement is practically possible after considering that the armour layer is very porous, with a porosity of $P \geq 50\%$ (see Table 10.5). There are however some limitations and requirements, as listed in what follows, to be observed for not affecting the integrity of the armour layer. These requirements are actually more relevant to single-layer armour units and could be relaxed for the double-layer armour units (excluding the Cubes and Cubipods, where this arrangement is practically not possible).

A. The maximum size of pipe/sleeve (i.e. the hole) shall not be larger than H/6 (where H is the CAU height). This is to ensure proper placing and good resting of the CAU on the underlayer, as the H/6 is slightly smaller than the Dn of the ELL gradation rock size limit of the relevant recommended underlayer rock grading.

   This requirement means that if the flow discharge is larger than what could reasonably be accommodated by one pipe, then a multiple outfall pipeline system is to be adopted with a manifold pipe.
B. For a multiple outfall pipeline system, the spacing between the pipelines should approximately match the horizontal spacing between the units, D_H. This is to approximately ensure one opening per unit.

C. The flowrate from the outfall pipeline must be evaluated to avoid excessive hydraulic loading on the unit. This is to safeguard against partial dislodging of the unit and a reduction in the interlocking.

Note that the flow rate is normally not an issue of concern with gravity flows, but it is if the flow is pumped out. Similarly, point (b) could also be relaxed for this reason.

A typical example for an outfall pipeline system with a single-layer CAU is shown in Figure 19.1, with all the pipes in the system are HDPE pipes. The plan view is shown in (a), with a manhole connected to manifold pipe (OD 1.40 m) and thereafter 14 outflow pipelines (OD 0.56 m) branching from the manifold pipe towards the sea. The main armour in this example is 5.0 m^3 Core-loc. This Core-loc unit size has a height (H) of 2.83 m, with a D_H of 3.14 m (see Figure 10.4). Therefore, the pipe hole size should not exceed an approximate diameter of 0.47 m, say 0.50 m, (i.e. H/6), and the spacing between the pipelines is to be selected in line with the D_H distance.

Figure 19.1(c) shows that the hole size, i.e. pipe inner diameter, is 0.49 m, which could be considered fulfilling the requirement of point (A) above, and the spacing between the pipes is 2.56 m c/c as shown in Figure 19.1(a). The spacing is slightly less than the recommended, though this could be tolerated as the flow is a gravity flow, i.e. low velocity.

Figure 19.1(b) is a cross-section through the revetment showing the connection between the manhole, the manifold pipe and one of the outflow pipes. The outflow pipes are protected by pre-cast concrete casings as they enter the rock fill. A duckbill valve, i.e. a non-return valve, is provided at the discharge pipe from the manhole, i.e. the feeder pipe to the manifold pipe. It is placed at this location for easy access from the manhole for maintenance or replacement. A section in the pre-cast concrete casing protection is shown in Figure 19.1(c) with the required appropriate bedding arrangement (geotextile, small rocks and lean concrete) to achieve the design longitudinal slope.

During the development of the outfall system shown in Figure 19.1, other potential designs, presented in Figure 19.2, were also investigated. The two alternative potential designs were based on letting the discharge water freely drain through the rocks without using any outfall pipes. This is to be achieved by having the exit pipe from the manhole ending in a vertical/horizontal discharge outlet that allows the water to seep through the rocks. In order to accommodate the outlet structure and the rock fill

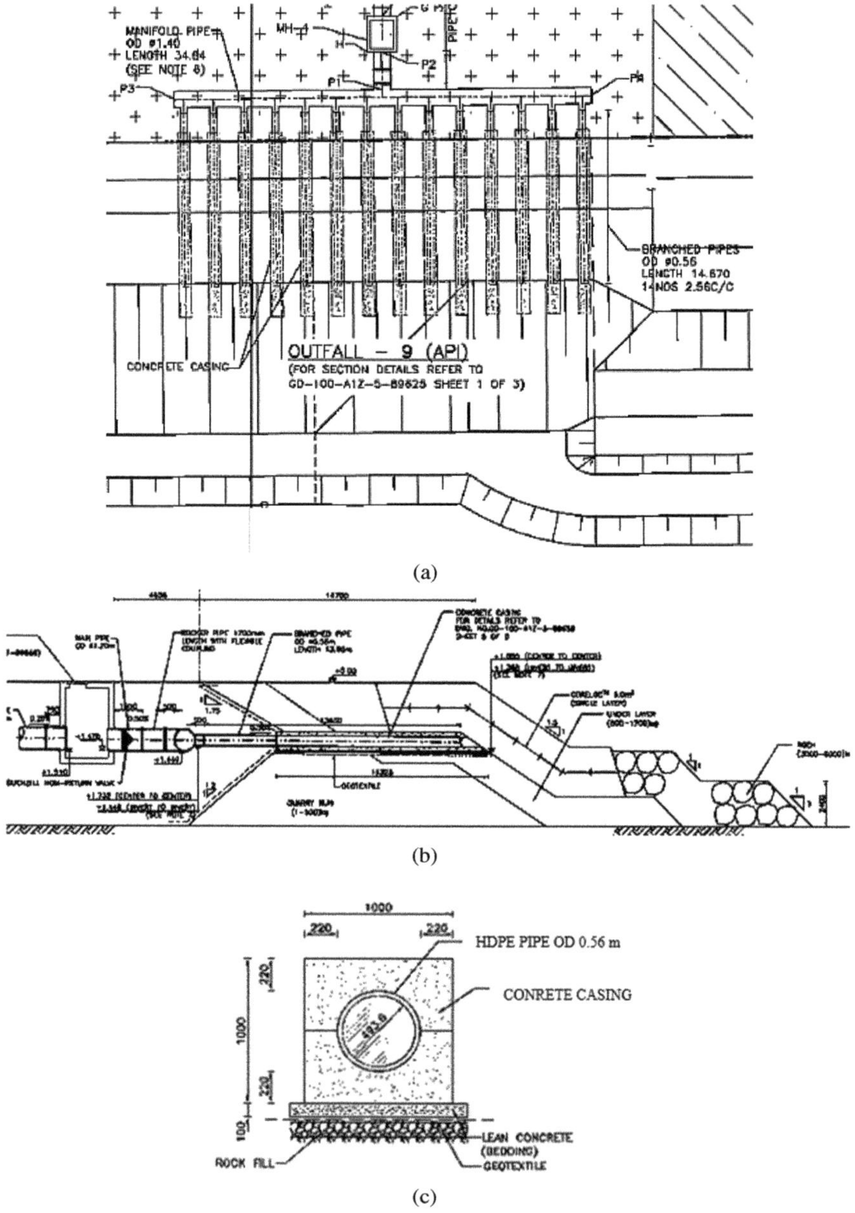

Figure 19.1. Multi-pipeline outfall system in a revetment with Core-loc armour: (a) Plan view; (b) Longitudinal section; (c) outflow pipe — concrete casing and bedding.

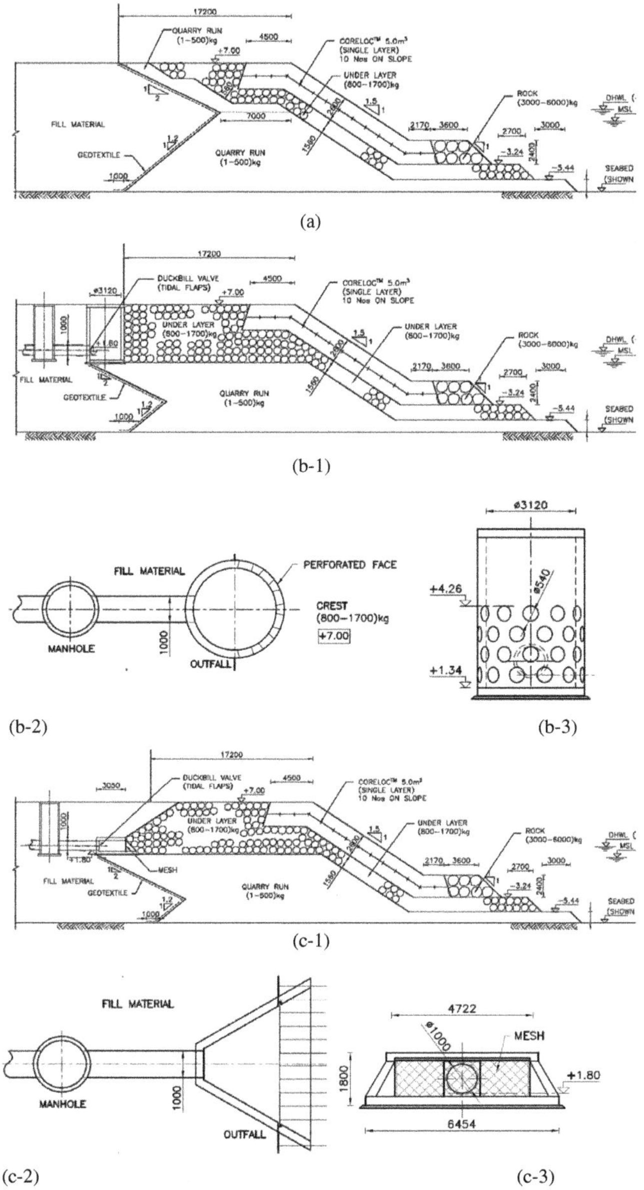

Figure 19.2. Alternative outflow systems to the multi-pipeline system in Figure 19.1: (a) Revetment section without an outfall system; (b-1) Alternative-B: Revetment section with a Tank Outfall; (b-2) Manhole and perforated tank outfall; (b-3) Perforated tank details; (c-1) Alternative-C: Revetment section with a divergent outfall; (c-2) Manhole and a divergent outfall; (c-3) Outfall structure — front view.

for the water passage, the revetment section would require some modifications at every manhole along the revetment.

Figure 19.2(a) presents the original revetment section, and Figure 19.2(b) illustrates Alternative-B with a perforated drum, i.e. vertical outlet, as the outflow structure that allows for the free gravity flow of water through the rocks. Figure 19.2(c) illustrates Alternative-C with a divergent structure, i.e. horizontal outlet, which allows for the water to seep through the rocks.

The design presented in Figure 19.1 is assessed to be more reliable as the two design alternatives in Figure 19.2 basically have the same disadvantage, i.e. the dependence on the porosity of the rock material. The flow of water could be greatly reduced due to variations in the rock grading, and consequently the porosity, and also the eventual accumulation of dirt and dust in the relatively long passage to the sea.

Another example of pipe outfall system is shown in Figure 19.3 with 2.2 m^3 Accropode armour. This Accropode unit size has a height (H) of 1.86 m, with a D_H of 2.32 m (see Figure 10.4). The hole size of the sleeve pipe (OD 0.30 m) satisfies the H/6 requirement, and the 3.0 m c/c spacing

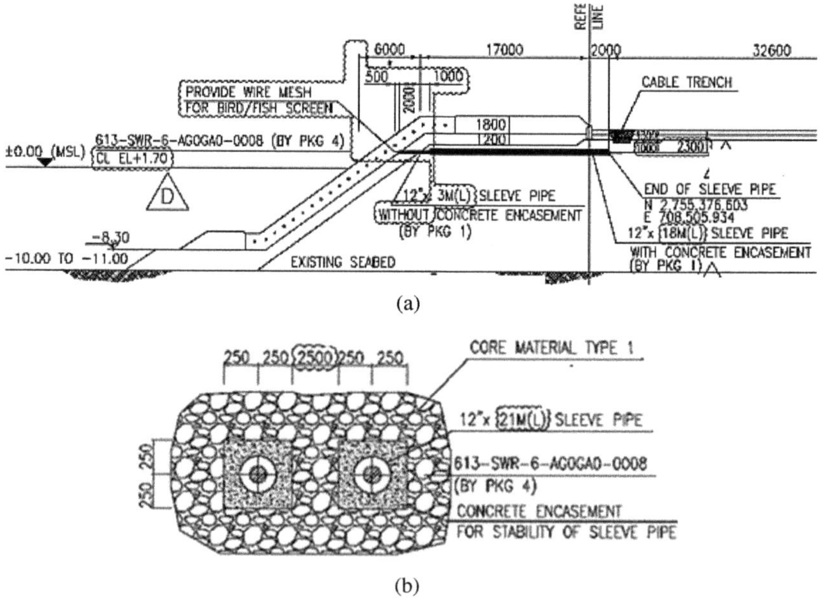

Figure 19.3. Two outflow pipelines in a revetment with Accropode armour: (a) Longitudinal section; (b) Outfall pipes — concrete encasement.

between the pipes is greater than the D_H. In this example, the spacing between the pipes has more importance than it did in the example of Figure 19.1 as the outflow is pumped water from a small desalination plant. The outflow has more momentum, therefore, it is important to ensure the adherence to this requirement. Moreover, it is also important to stipulate that the pumped discharge is not to exceed a certain operational limit, which is to be carefully assessed in order not to dislodge the armour units.

A different example of an outlet structure is shown in Figure 19.4, where the structure is placed high up in the Core-loc armour slope, extending over the crest. This type of an outlet structure, as discussed above, is definitely not recommended, specially with a single-layer CAU. However, an outlet structure at the toe level, as shown in Figure 19.5, could be acceptable and used as it replaces the first row and the toe units.

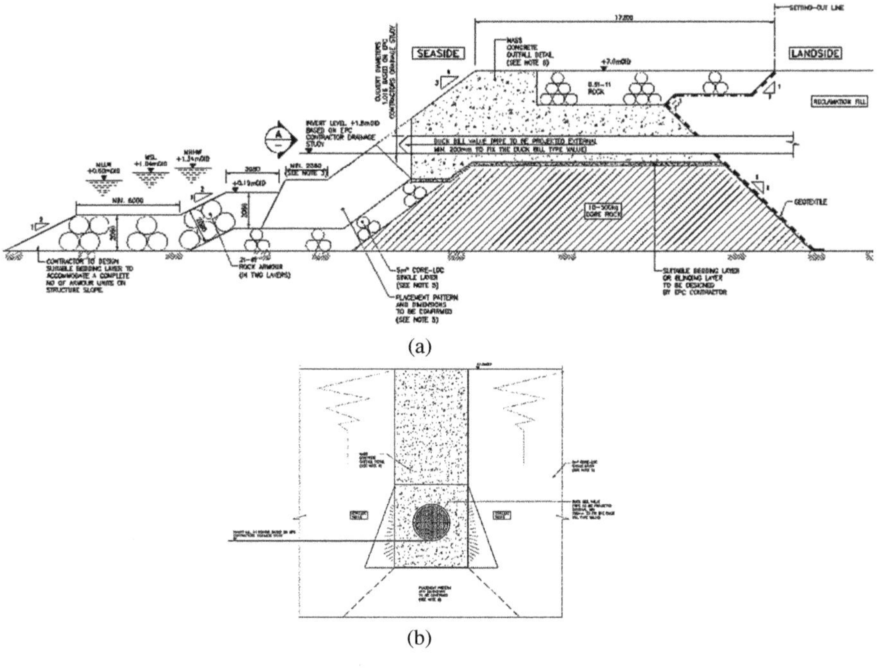

Figure 19.4. An outlet structure in a revetment with Core-loc armour: (a) Longitudinal section; (b) Outlet structure — front view.

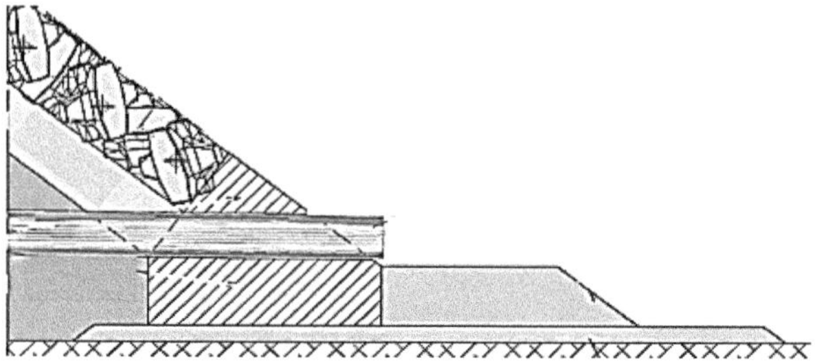

Figure 19.5. An outlet structure at toe level in a revetment with Accropode armour.

19.3. Slopes with Rock Armour

Rock armour, unlike most of the CAU, depends primarily on its own weight to achieve stability, though contact between adjacent rocks is still a requirement. Therefore, having an outfall pipeline or an outlet structure could be integrated relatively more easily in the rock armour slope than in the CAU slope.

Many examples are presented in this section to demonstrate the different possible arrangements that could be used to accommodate an outfall pipeline or an outlet structure. Figure 19.6 presents an outlet structure, similar in principle to the one shown in Figure 19.5, and could be used at any required level. The advantage of this type and configuration of an outlet is that it gives a good support for the upper part of the armour layer. The pipe could protrude outside the structure, as is the case here, or preferably be flushed with the surface of the outlet structure. A protective mesh at the hole could be used with a possible non-return valve.

A different concept for the outlet structure is shown in Figure 19.7, where the wing walls and the headwall protrude slightly outside the slope of the revetment. The headwall acts as a toe support for the armour layer above. This type of an outlet is not preferred where there is wave action as every abrupt change in the slope profile will induce turbulence and extra stresses. In this example, a galvanised grating is used to prevent floating debris from getting in, with a flap valve being used as a non-return valve.

Figure 19.8 shows an outlet structure with the same concept as the one in Figure 19.7 but with no protruding walls, all flushed with the slope profile. This has the advantage of reducing the turbulence, although the toe of the upper armour layer is less secured as the toe rocks can slip over

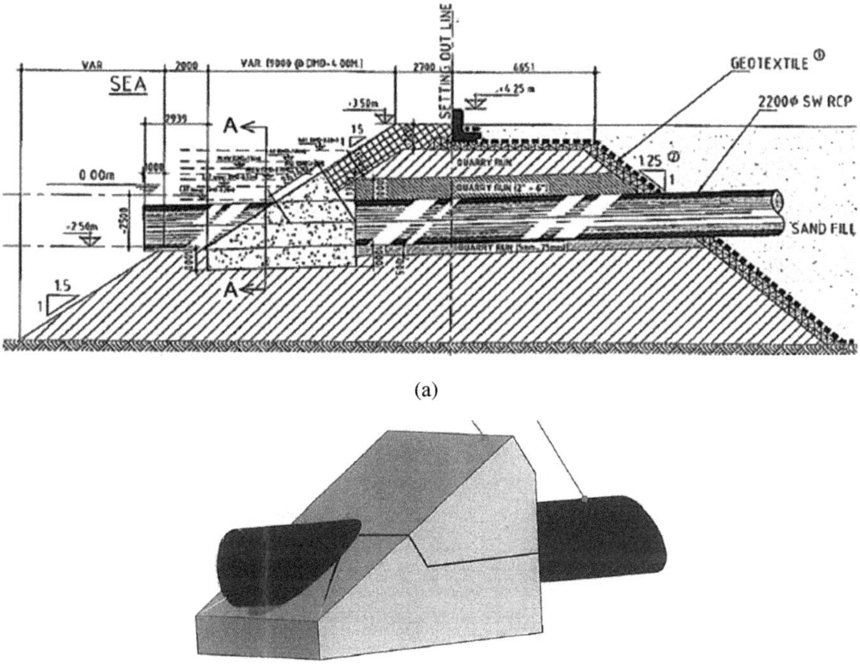

(a)

(b)

Figure 19.6. An outlet structure with a large diameter outfall pipe (2200 ϕ): (a) Longitudinal section; (b) Isometric illustration of the Outlet structure.

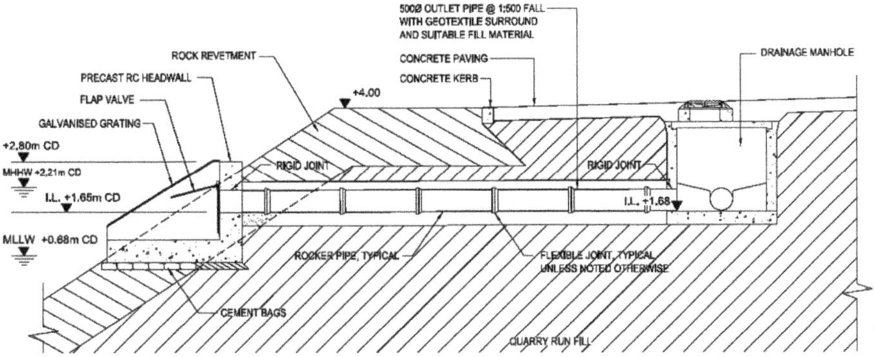

(a)

Figure 19.7. A protruding outlet structure with a small diameter pipe (500 ϕ): (a) Longitudinal section; (b) Plan view.

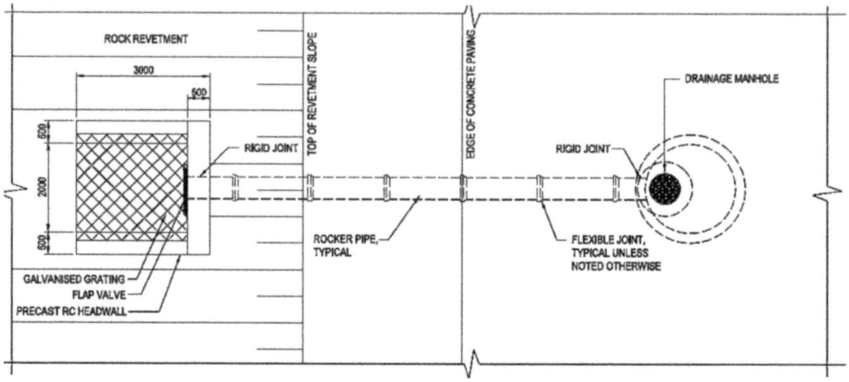

(b)

Figure 19.7. (*Continued*)

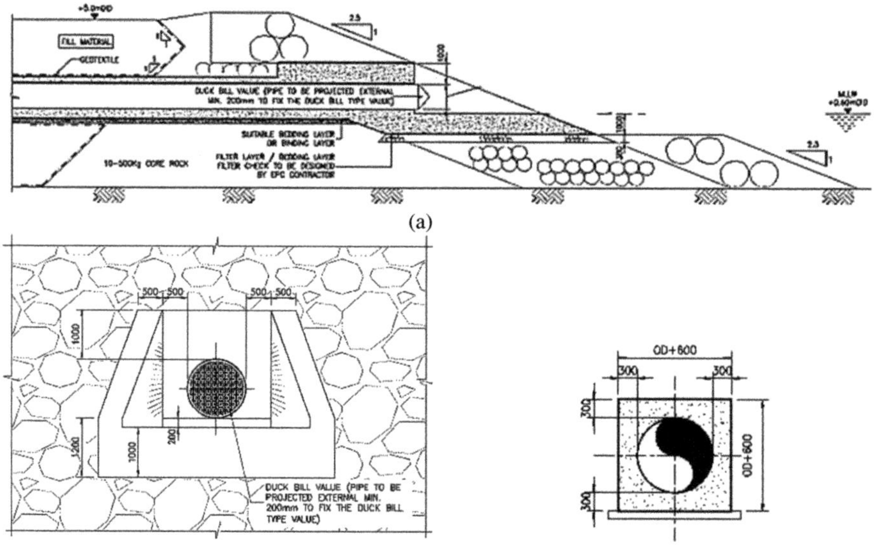

(a) (b) (c)

Figure 19.8. A flushed outlet structure with a single outfall pipeline: (a) Longitudinal section; (b) Outlet structure — front view; (c) Outfall pipe concrete encasement.

the headwall. In this example, a duckbill type of a non-return valve is provided.

Using the multi-pipeline outfall system, as described in Section 19.2, is also possible with the rock armour. Figure 19.9 shows an example of

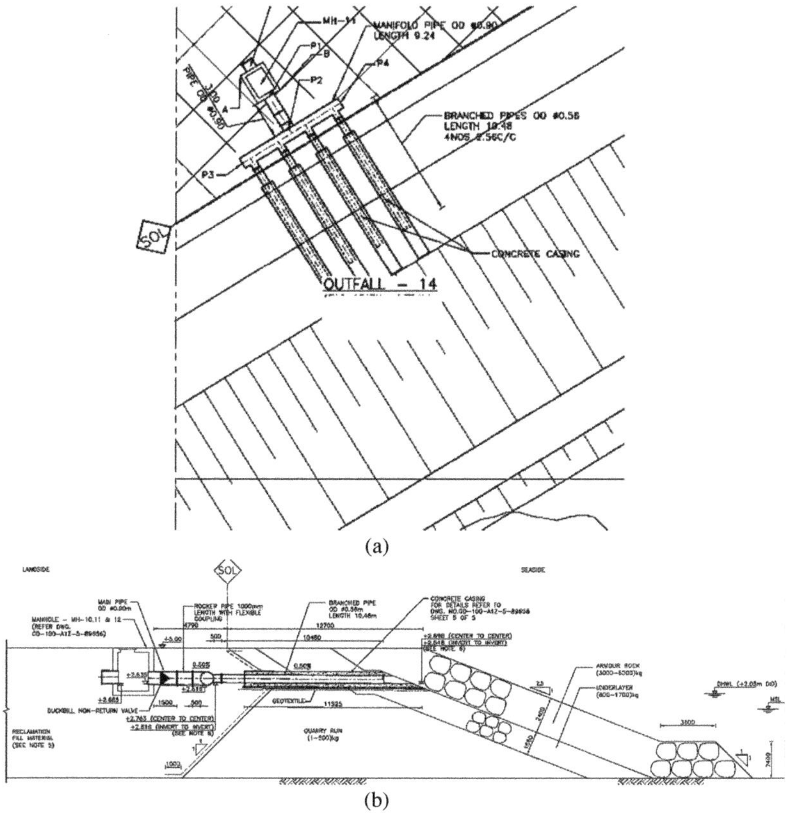

Figure 19.9. Multi-pipeline outfall system in a revetment with rock armour: (a) Plan view; (b) Longitudinal section.

such an outfall system, which is actually from the same project as the example presented in Figure 19.1. In this example, the rock armour size is 3,000–6,000 kg, and the underlayer is 800–1,700 kg (which is larger than what is required, and this is due to temporary protection requirements during construction, otherwise a rock class of 300–1,000 kg could have been used). The pipe hole size and the spacing between the outfall pipes are the same as in Figure 19.1, and considered satisfactory. It is, however, important to note that the porosity of a randomly placed two-layer rock armour is about 35% (see Table 9.6), which is sufficient for gravity drainage flow, but it is not for pumped flows. Furthermore, the construction

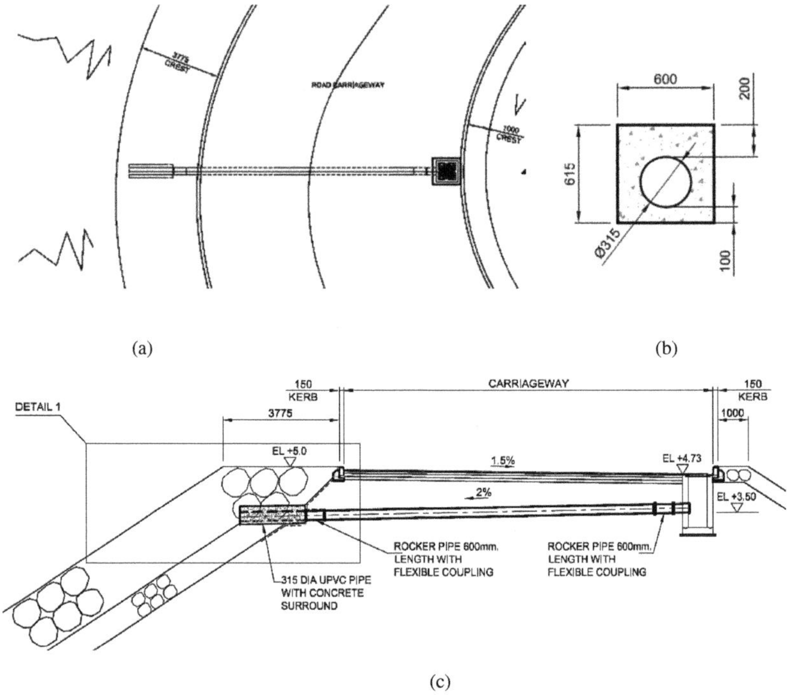

Figure 19.10. Single outfall pipeline with rock armour: (a) Plan view; (b) Pipe concrete encasement; (c) Longitudinal section.

team must be instructed to be very careful while placing the armour rocks on top of the outfall pipes so that no rocks block any of the pipes.

A single outfall pipeline example is shown in Figure 19.10 with the pipe not penetrating through the rock armour layer. A plan view is shown in (a), a section in the concrete casing in (b) and a section in the revetment in (c).

19.4. Subsea Pipelines and Cables

Pipelines and cables may connect land industrial or communication centres to off-shore destinations. These pipelines and cables are frequently required to cross a revetment, being the borderline between land and sea. The crossing normally takes place under the revetment structure so that the integrity of the revetment structure is not compromised.

Different methods and ways could be used to protect these pipelines and cables depending on many factors. However, the protection method could also be dictated by having the pipelines and the cables already on the seabed — or trenched before the construction of the revetment, or they are to be installed during or after the revetment construction. Different examples are presented in this section to illustrate some of the basic important issues.

Two cases shall be presented for the subsea pipelines, the first where the pipelines already exist on the seabed before the construction of the revetment and the reclaimed area behind, as illustrated in Figures 19.11–19.13. The other case is to have the pipelines installed through an existing revetment, as shown in Figure 19.14.

Figure 19.11 presents an example of two 14" effluent pipelines already existing on the seabed before the construction of the revetment. These pipelines shall be buried under a new reclaimed area and a rubble mound revetment protecting the water edge of the reclamation. The first point to check is the pipe strength if it is to withstand the new added weight and load, then the pipe protection against possible damage during construction. A plane view for the revetment crossing is shown in Figure 19.11(a), the revetment design section with no pipelines in (b), a longitudinal section through the pipe protection in (c), which defines the extent of the various protection arrangements, three cross-sections for the various protection arrangements in (d) and the details of the concrete hat protection in (e). Originally the pipelines were designed to be supported by sleepers with certain free span intervals, these free spans cannot withstand the new loads from the reclamation sand or the rubble mound rock material. Therefore, in all the cross-sections in (d) the pipeline is shown to be continuously supported by polypropylene bags, filled by cement/ sand mixture, which are to be pushed in place by divers.

The pipe protection extends through three distinct zones, namely the reclamation sand, quarry run (i.e. small rocks) and main armour and toe (i.e. large CAU/rocks). For the protection in the reclamation sand zone, Figure 19.11(d)_C-S 3-3, the pipe is shown to be covered with polypropylene bags filled with sand as a mild protection against the possible existence of small stones in the reclamation sand. Then a multilayer protection is used in the small rocks zone, i.e. the quarry run and underlayer rocks, as shown in (d)_C-S 2-2, where the pipe is wrapped with sandbags, geotextile, 20 mm crushed rocks and 50-150 mm crushed rock. Although the construction team should be instructed to handle the dumping

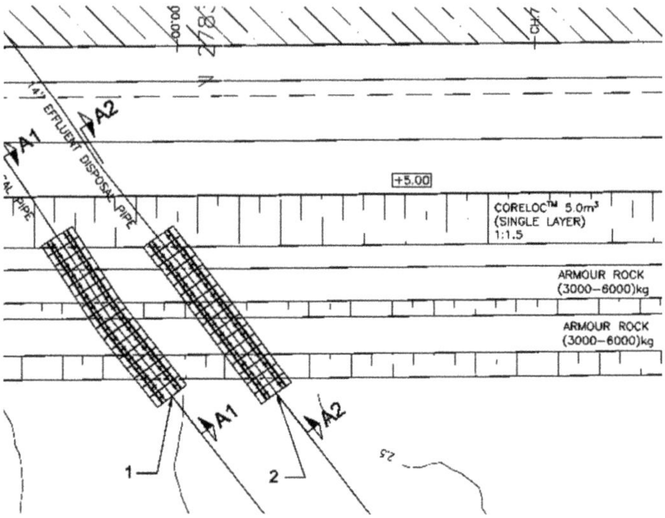

(a)

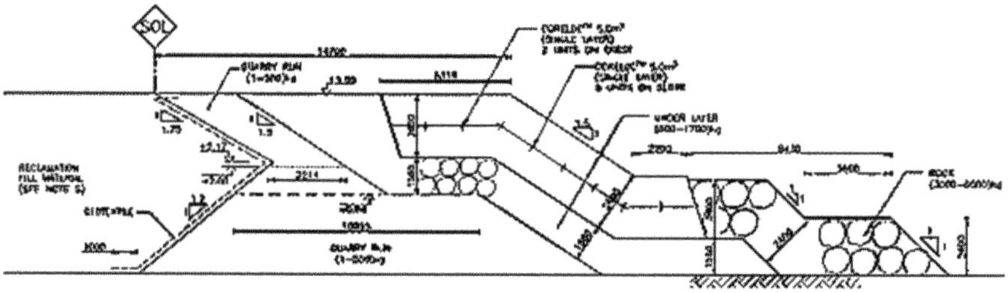

(b)

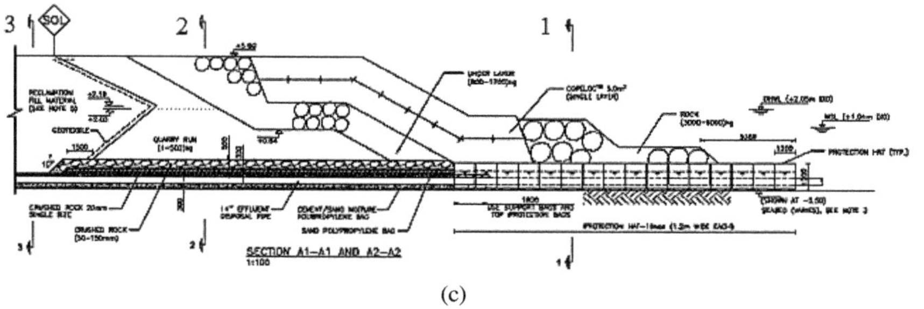

(c)

Figure 19.11. Pipe protection with concrete hats — existing pipeline (pre-revetment construction): (a) Plan; (b) Revetment section — no pipelines; (c) Longitudinal section A1-A1/A2-A2; (d) Cross-sections 1-1, 2-2 and 3-3; (e) Concrete hat pipe protection configuration.

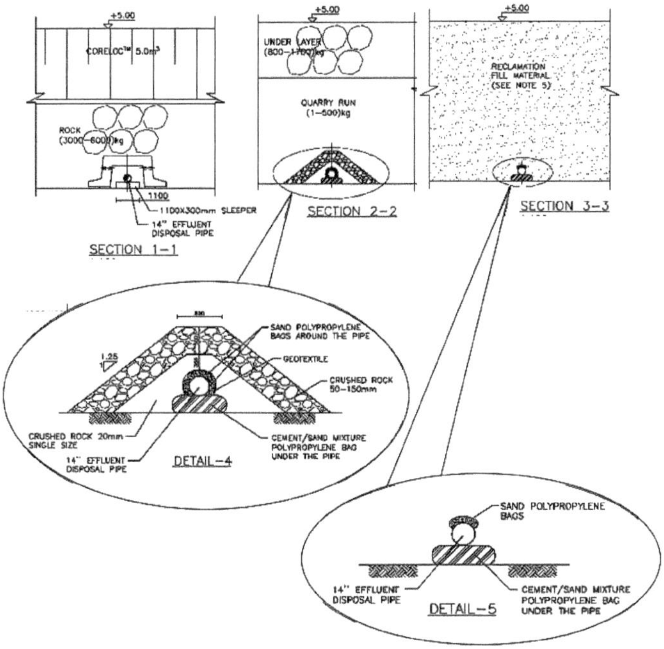

(d)

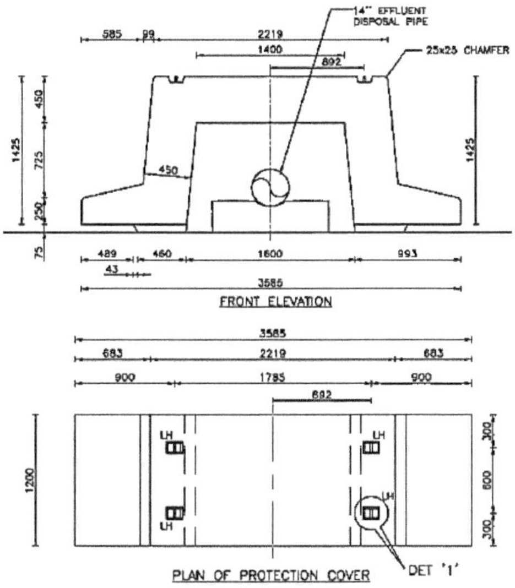

(e)

Figure 19.11. (*Continued*)

operation of the rock material very carefully over the pipes and to lower the release point of the rocks as low as possible, caution is still necessary against possible high-energy impact. The last zone of protection is when the pipe is under the larger CAU/rocks units of the main armour and the toe, as indicated in (d)_C-S 1-1, as in this zone a more robust pipe protection arrangement will be required. Small rocks cannot be used as these cannot support the larger armour units and will also be exposed to waves and currents, hence a concrete hat is used to protect the pipe, providing a solid foundation to the toe and facilitating smooth transition from the original revetment section to the pipeline section. Note that this type of a concrete hat, or similar, could be used throughout the buried length of the pipeline if the pipe cannot withstand the overburden weight.

An alternative design to the example in Figure 19.11 is given in Figure 19.12, the change is only in cross-section 1-1 of the protection, i.e. the zone under the armour and toe units, whereas C-S 2-2 and 3-3 remain as shown in Figure 19.11(d). A grout bag (5.0 m wide x 0.30 m thick) is used instead of the concrete hat. In order to prevent possible washing out of the pipe protection small rocks, the grout bag will be extended outwards to wrap the pipe after stopping the rock protection, as illustrated in Figure 19.12(d). The extra 3.0 m of the grout bag protection beyond the last toe rock is to safeguard against possible rolling out of a dislodged toe rock.

A different protection method is presented in Figure 19.13 with RC slab, known as “Impact Protection Slab”. The slab is to protect a buried pipe from possible fallen large rocks, construction accidental impact or surcharge weight. This method could be used for existing pipelines under a new revetment, or beside a construction site, or even for a new pipeline to be placed near an existing coastal structure. Note that the RC slab is separated from the pipe trench fill by a layer of soft fill, i.e uncompacted fill, to absorb possible impact loads.

Figure 19.14 presents the design for installing three large diameter HDPE outfall pipelines from an inland industrial compound through an existing revetment to an offshore discharge location. A trench needs to be opened in the revetment, and to be deep enough to allow restoring the revetment section without compromising its structural integrity. A plan view for the pipes is shown in (a), and a longitudinal section along the pipelines in (b). It also shows the revetment section details which comprise 1–1,000 kg quarry run core and an armour layer of 300–1,000 kg. The open trench in the revetment is shown in (c), whereas the backfilled

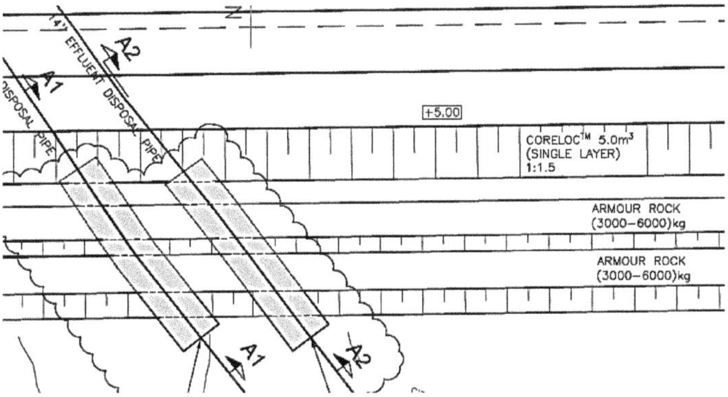

(a)

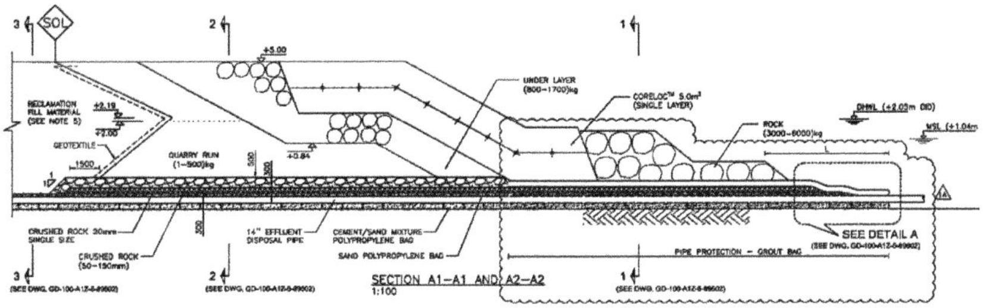

(b)

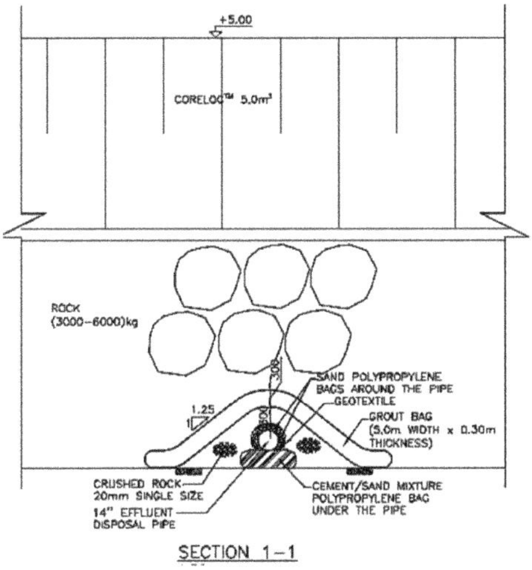

(c)

Figure 19.12. Pipe protection with grout bags — existing pipelines (pre-revetment construction): (a) Plan; (b) Longitudinal section; (c) Cross-section 1-1; (d) Pipe exit details.

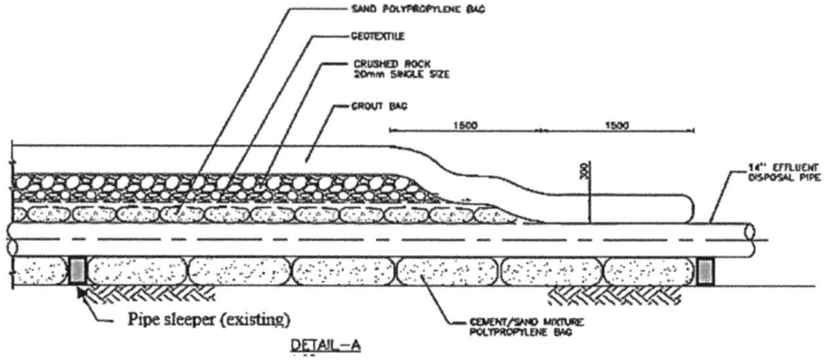

(d)

Figure 19.12. (*Continued*)

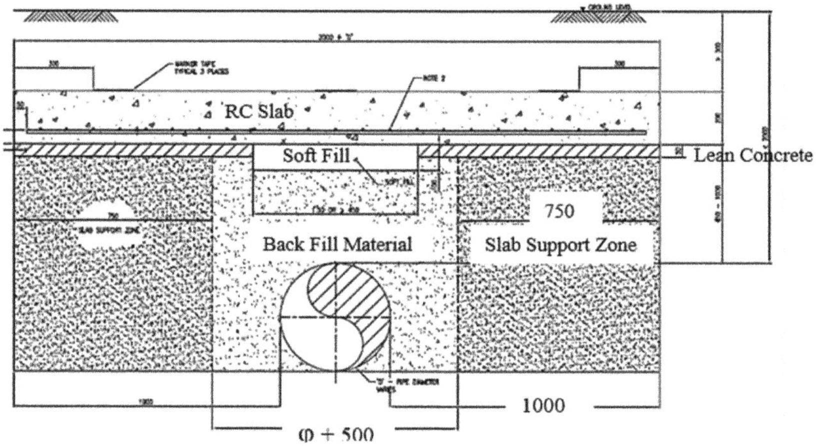

Figure 19.13. Pipe protection with a concrete slab (UKOPA, 2015).

trench and the restored revetment section is shown in (d). The pipe collars shall be placed on a 0.15 m bedding layer of 20–75 mm stones, which shall also be used to bury the pipes up to the top of the collars to ensure no large rocks will come in contact with the pipes. Then the original 1–1,000 kg quarry run shall be used, in varying thicknesses, to bring the slope to its original gradient before placing the armour layer. The pipes

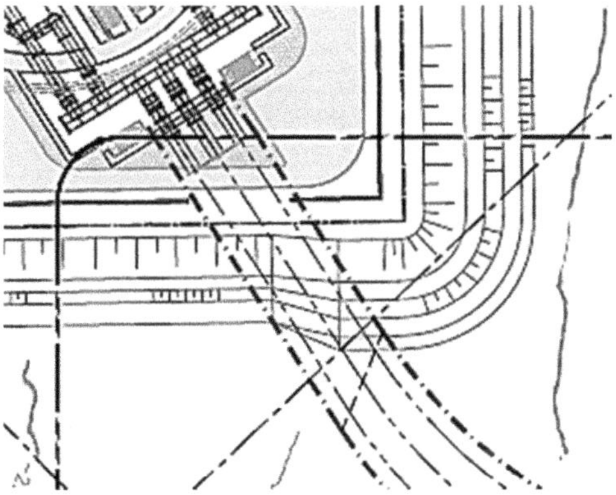

(a)

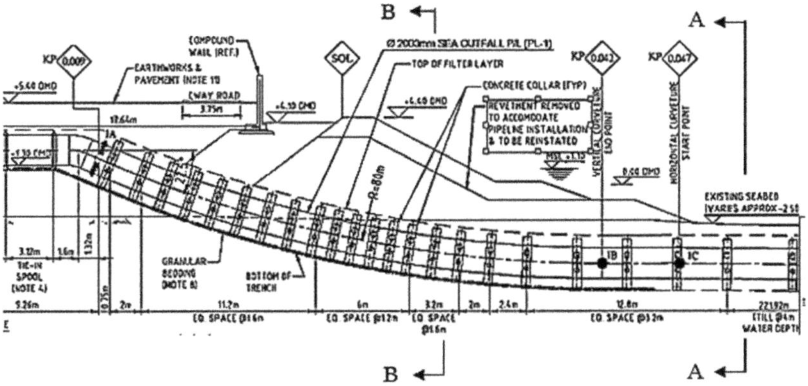

(b)

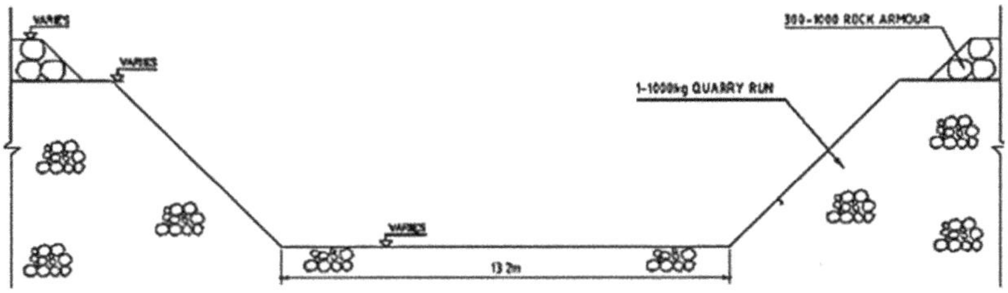

(c)

Figure 19.14. Pipe protection with granular filter — new pipelines (post-revetment construction): (a) Plan; (b) Longitudinal section; (c) Cross-section B-B (stage-1: open trench in the revetment); (d) Cross-section B-B (stage-2: backfilled trench, restored revetment); (e) Cross-section A-A (offshore of the revetment section).

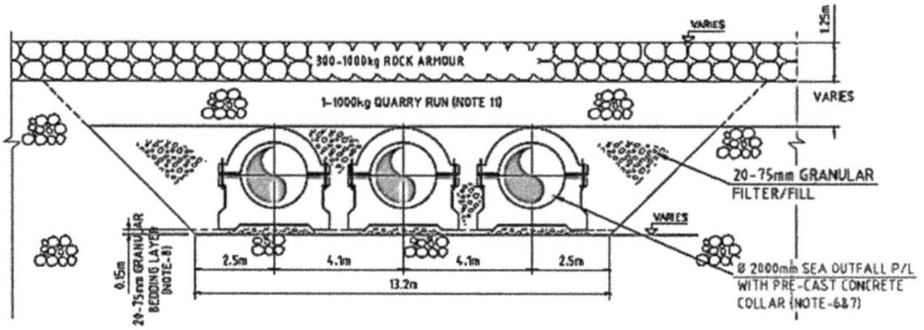

(d)

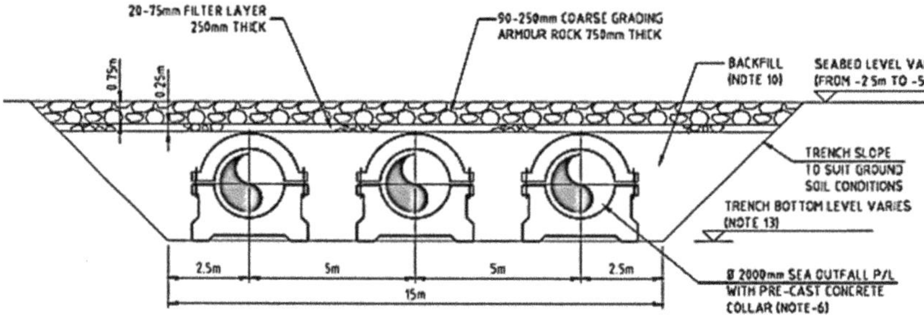

(e)

Figure 19.14. (*Continued*)

will continue after that, being trenched in the seabed with the necessary protection, as shown in (e), and backfilled with the excavated seabed material, if suitable to be used again as a backfill, with the proper surface protection layers.

Two cases shall be presented for the subsea cables, the first where the cables already exist on the seabed before the construction of the revetment and the reclaimed area behind, as shown in Figure 19.15. The other case is to have the cables installed after the construction of the revetment including placing the necessary cable ducts as part of the revetment construction, as shown in Figure 19.16.

Subsea cables are normally made with a reinforced multi-layer protection and placed directly on the seabed, designed to withstand high overburden loads. Therefore, when a new revetment is to be constructed on top of an existing cable, as in Figure 19.15, there will be no problem

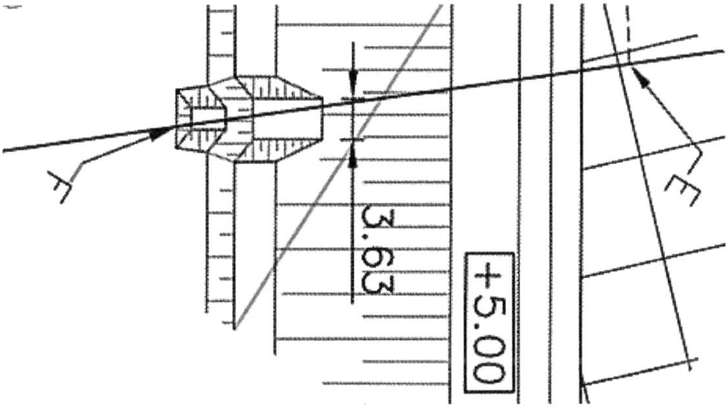

(a)

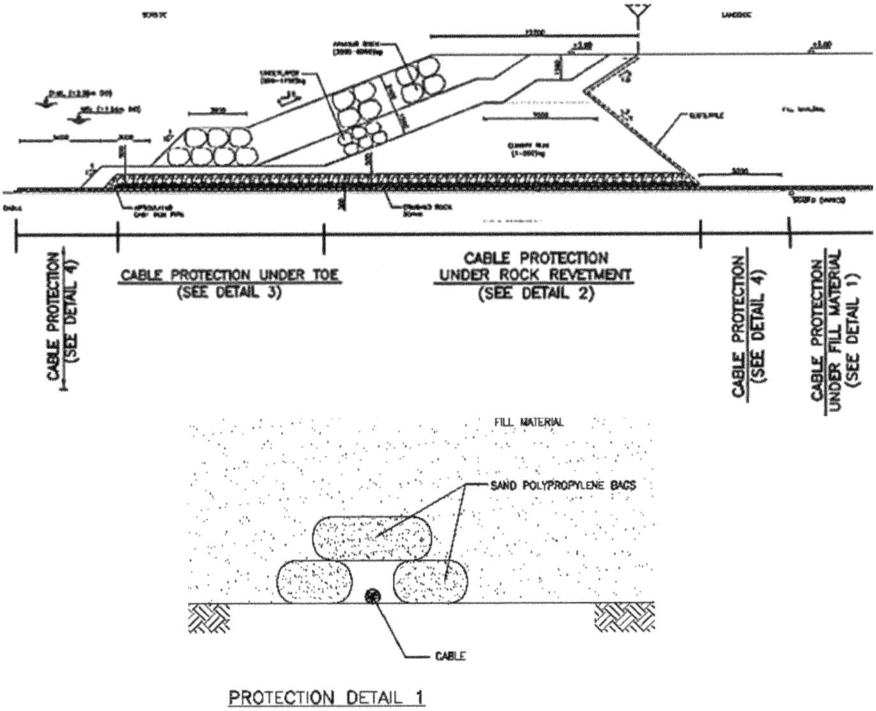

(b)

Figure 19.15. Cable protection — existing subsea cables (pre-revetment construction): (a) Plan; (b) Longitudinal section; (c) Protection details 1, 2, 3 and 4.

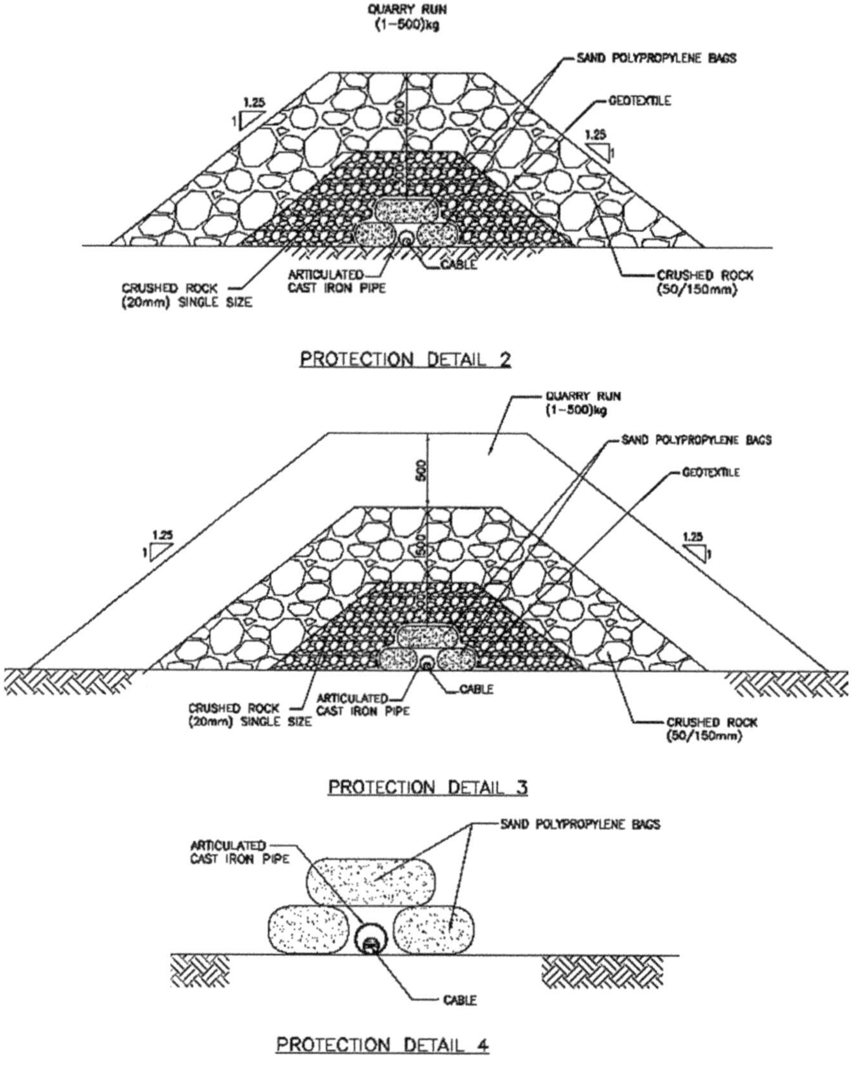

(c)

Figure 19.15. (*Continued*)

with the new extra weight, however, certain protection measures are to be provided, especially under the revetment. The plan view of the cable crossing is shown in (a), and a longitudinal section is shown in (b) identifying the different zones along the cable with the details of the required

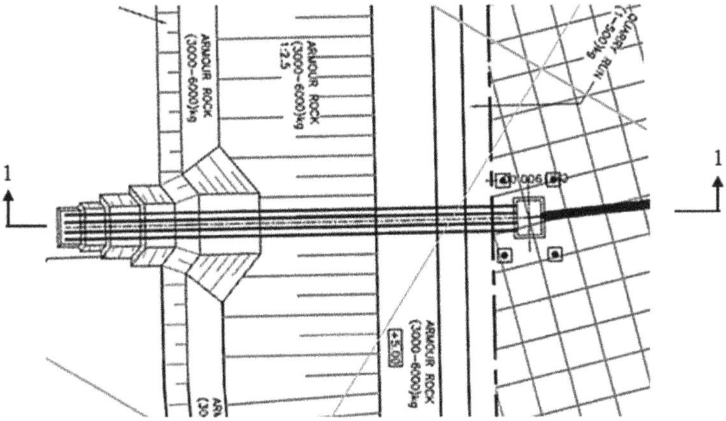

(a)

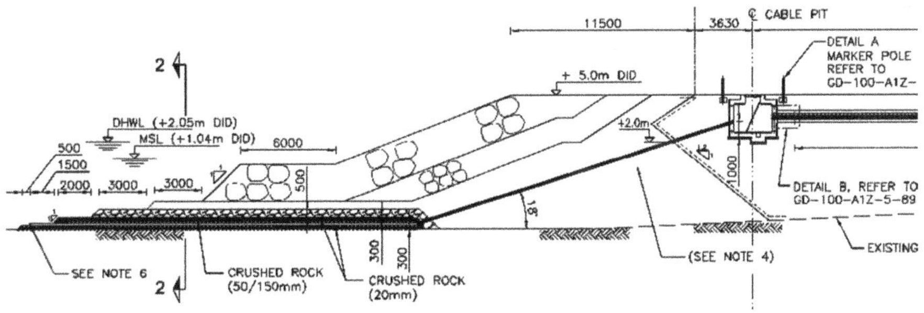

(b)

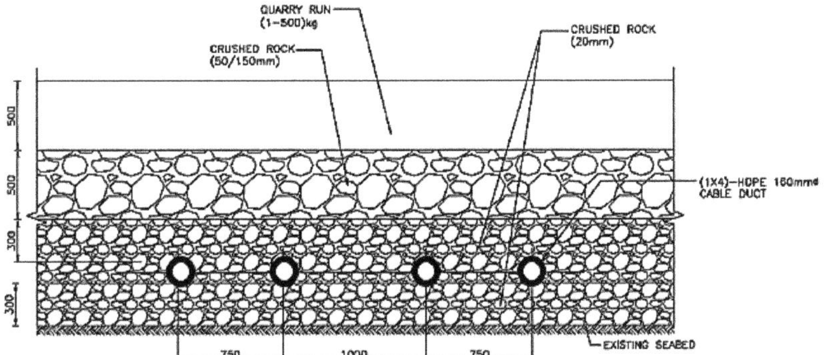

(c)

Figure 19.16. Cable protection — new cable ducks (part of the revetment construction): (a) Plan; (b) Longitudinal section 1-1; (c) Section 2-2, where the cable ducts exist.

protection in each zone as shown in (c). Detail-1 is for the sandbags protection in the reclamation zone, which is simple but necessary to separate the cable from possible contact with small rocks that might exist in the sand. Detail-2 shows the protection inside the revetment rock bund, which consists of an articulated cast iron encasement protector, sandbags wrapped by a geotextile, 200 mm layer of 20 mm crushed rocks and 500 mm layer of 50–150 mm crushed rocks. Detail-3 is for the protection under the toe, which is practically the same as in Detail-2 after adding a 500 mm layer of quarry run under the large rocks. Detail-4 covers the zones extending about 5.0 m before and after the rock bund, and this protection is the same as in Detail-1 after adding the cable encasement protector.

The protection of the cables will be easier if they are to be placed after the construction of the revetment, as in Figure 19.16, where appropriate cable ducts are installed during the construction to facilitate the installation of the cables through the revetment. A general plan view is shown in (a), and a longitudinal section in (b), where the horizontal and vertical alignment of 4 × 160 mm HDPE cable ducts are illustrated connecting an inland cable pit to the seabed outside the revetment. In (c), cross-section 2–2 at the toe of the revetment illustrates the protection arrangement at the apron of the toe and how the ducts terminate at approximately 8.0 m from the toe.

19.5. Shoreline Outfalls

Outfalls at the shoreline are commonly used if the effluent is relatively clean and will not cause any environmental impact at the shoreline. Basically, there are two categories for shoreline outfalls, pipes and channels. However, both should have an appropriate scour protection on the seabed, and to ensure that the outfall will not induce any morphological change to the shoreline equilibrium.

Most of the large pipe outfalls do not discharge directly into the sea but into a stilling chamber that dissipates a significant part of the flow energy before releasing it to the sea. This means that the seabed scour protection will not need to be of large rocks or extend a long distance. However, the length of the scour protection should be long enough such that the outfall discharge current reduces to where it cannot scour the seabed, which is normally below 0.3–0.4 m/s for sand beds. It is, however, highly recommended to provide extra rock material at the scour protection edges, in the form of extra thickness or additional length, this is for launching a falling apron in case of bed scour.

Channel outfalls, unlike the pipe outfalls, are not usually provided with a stilling basin but discharge directly to the sea as the current velocity is expected to be rather low. However, the exit current velocity from the channel to the sea could be reduced, if required, by widening the channel at the exit line or by having an offshore groyne, perpendicular to the current direction, say at 25–50 m from the exit, to stop and break the discharge jet. Either of these two solutions will significantly reduce the current velocity and consequently the area for the scour protection. Furthermore, in most cases the outflow discharge is primarily fresh water, and probably warmer than the receiving sea water. This means there will be a stratified flow situation due to the difference in densities between the outflow and the saline sea water, with the outflow dispersing in a plume-like flow on the surface. The characteristics of this type of flow are project-specific as many factors must be considered, e.g. sea currents, waves and water depth, therefore the designer must be cautious in considering the flow stratification in the design of the scour protection.

Two sets of examples shall be presented in the following, the first is for pipe outfalls and the second is for channel outfalls.

An example of a relatively small outfall is shown in Figure 19.17 with a pipe of φ 1.0 m and a maximum discharge of 2.83 m^3/s, the pipe invert level is relatively high as it is higher than the MHW. A plan view is shown in (a), and different cross-sections in (b), (c) and (d). The outfall consists of two parts, the wing wall with its apron where a significant part of the outflow energy will be dissipated, either by energy dissipation blocks, or in this case using large rocks to cause turbulence. The second part is the plunge pool, which acts as a stilling basin and provides a gentle transition to the natural seabed, and of course as a scour protection. The rock size in this part shall withstand the outflow current, sea current and the breaking waves at the shoreline. The inland area surrounding the wing wall shall also be protected against erosion.

A much larger outfall pipeline than the one in the previous example is shown in Figure 19.18, with a pipe of ϕ 2.60 m and a maximum discharge of 13.6 m^3/s, the pipe invert level is at –3.80 mCD. The location plan is shown in (a), and the outfall arrangement in (b). Sheet piles with struts are used to support the pipe, whereas rocks are used to form the wing walls

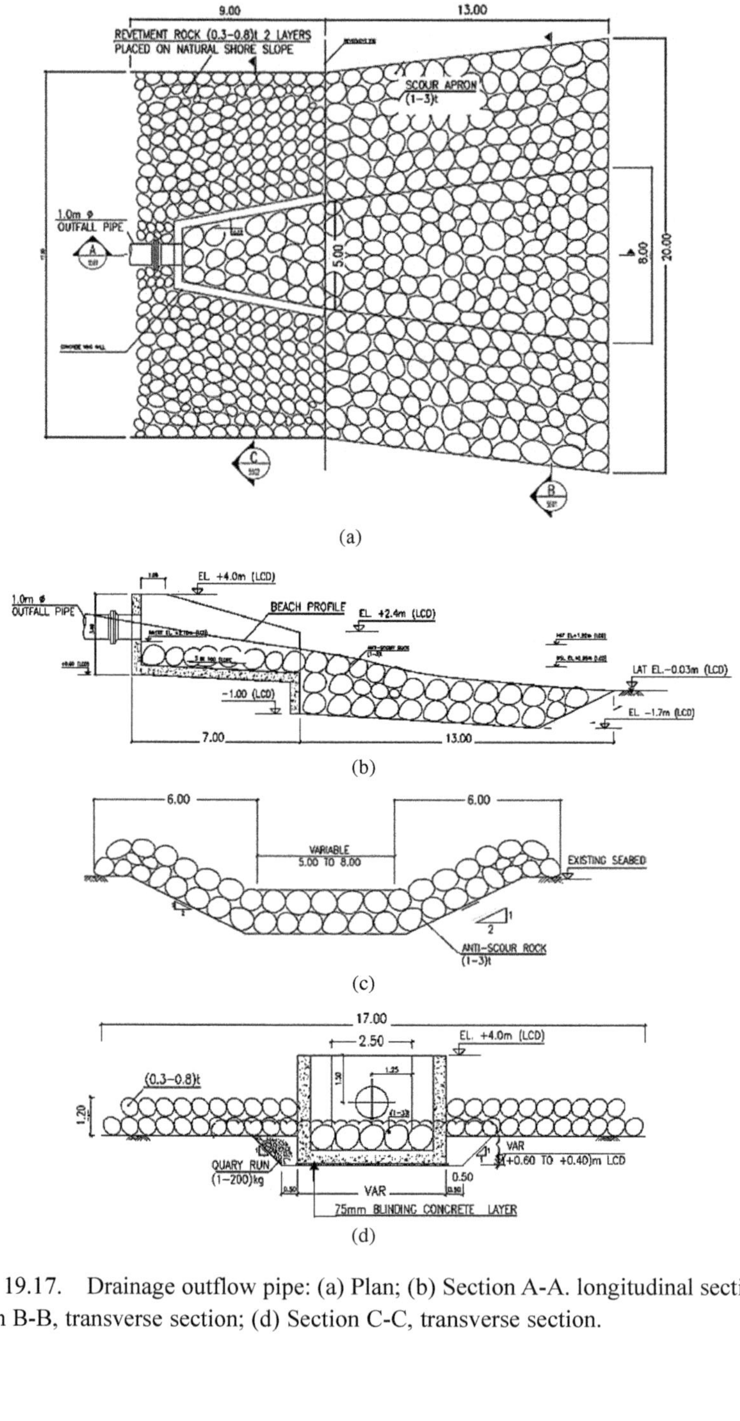

Figure 19.17. Drainage outflow pipe: (a) Plan; (b) Section A-A. longitudinal section; (c) Section B-B, transverse section; (d) Section C-C, transverse section.

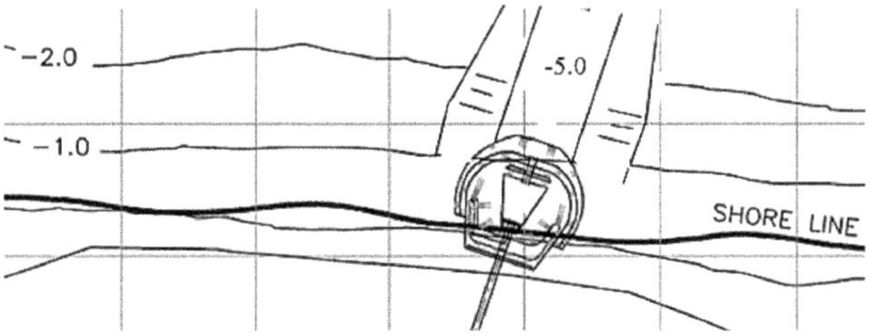

(a)

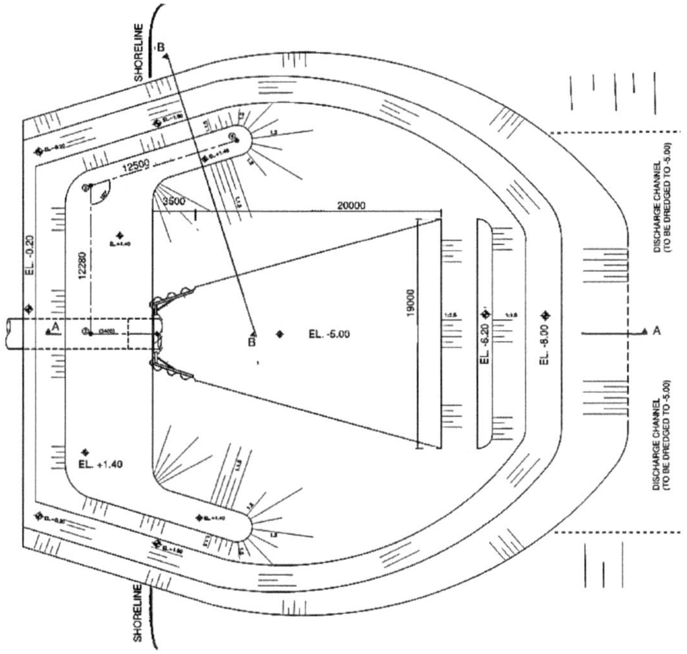

(b)

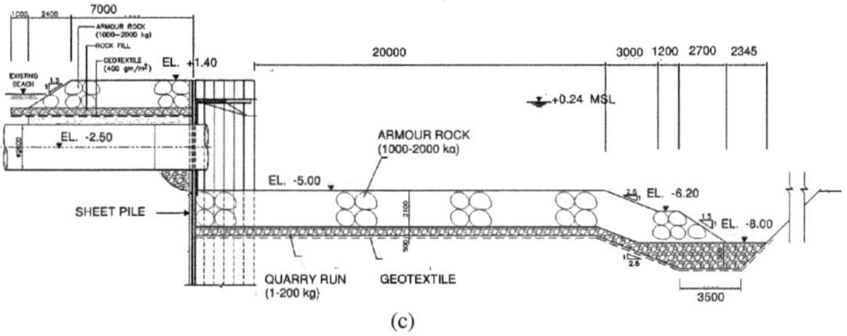

(c)

Figure 19.18. Large pipe outfall structure, scour protection and dredged channel: (a) Location plan; (b) Plan; (c) Section A-A, longitudinal section through the scour protection; (d) Section B-B, transverse section.

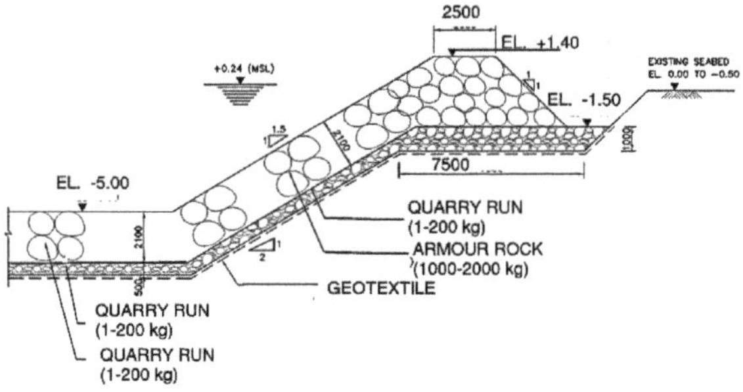

(d)

Figure 19.18. (*Continued*)

and the scour protection. Longitudinal and transverse sections are shown in (c) and (d), respectively. The toe of the scour protection is well-secured against possible under-scouring by having it embedded and keyed in the seabed. As the invert level of the pipe is low, and much lower the surrounding seabed, hence, a channel must be dredged at the same level as the scour protection until it reaches the same seabed contour, i.e. –5.0 mCD. After a careful assessment, this outfall design has not been approved due to its interference with the shoreline regime. Dredging a channel in a sand seabed located on an active longshore sand drift shoreline is not recommended and should be avoided. Accordingly, the design has been completely revised to the outfall shown in Figure 19.19, where the main objective is to avoid having a dredged channel.

An outfall chamber/stilling basin is needed to raise the flow exit level and it will also reduce the outflow energy. An exit level above +2.0 mCD is required to avoid any dredging to the seabed. The surface level of the scour protection shall be at approximately the same level as the original seabed to minimise any disturbance to the shoreline equilibrium levels. A general location plan is shown in Figure 19.19(a), and a detailed plan in (b). Longitudinal sections through the scour protection and the pipe inlet are shown in (c) and (d), respectively, and a transverse section in (e). Note that the scour protection extends to the landside of the outfall chambers to prevent any erosion. A 10.0 m wide strip of the underlayer surrounds the main armour layer, adding an extra safety by extending the scour protection and also providing enough material to act as a falling apron in case of any scouring taking place.

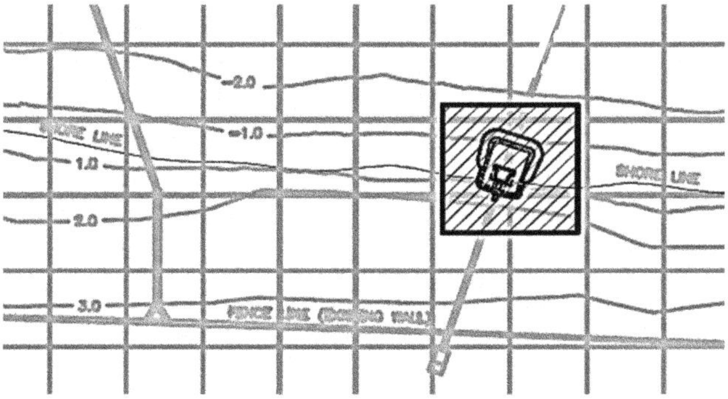

(a)

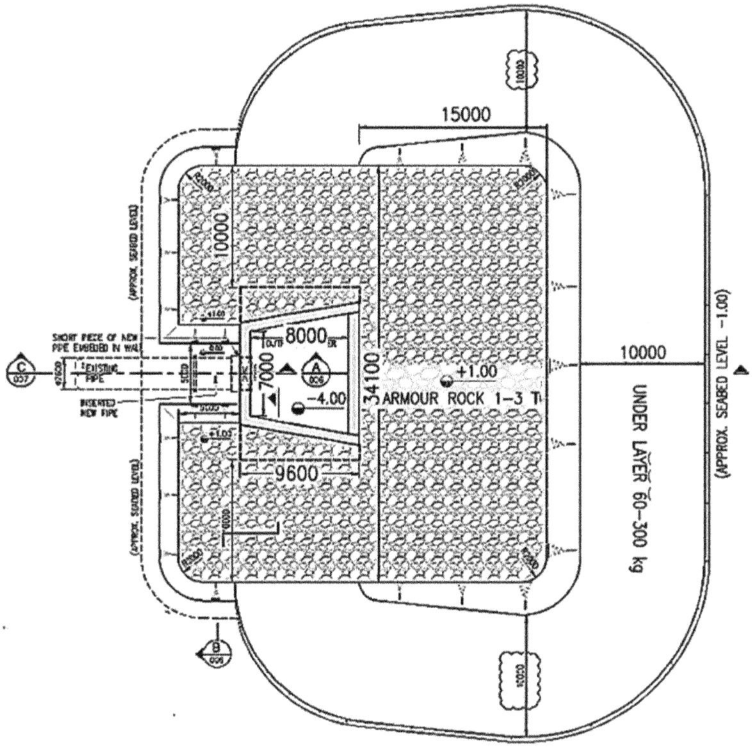

(b)

Figure 19.19. Large pipe outfall — outlet structure and scour protection: (a) Location plan; (b) Plan; (c) Section A-A, longitudinal section through the scour protection; (d) Section B-B, transverse section; (e) Section C-C, longitudinal section through the pipe inlet.

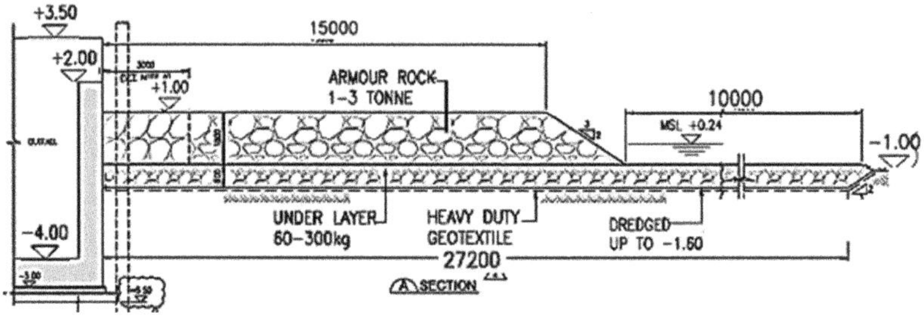

(c)

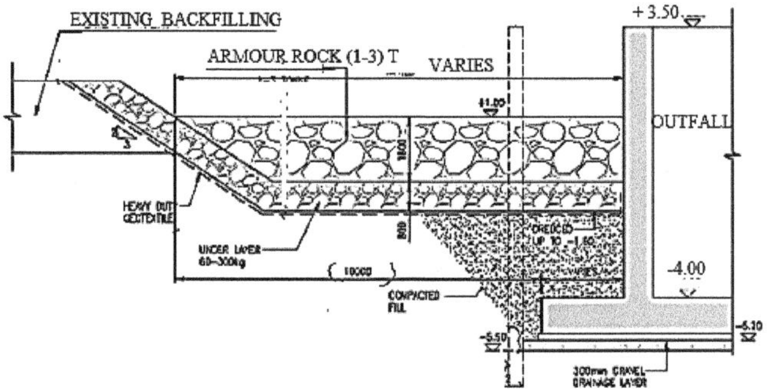

(d)

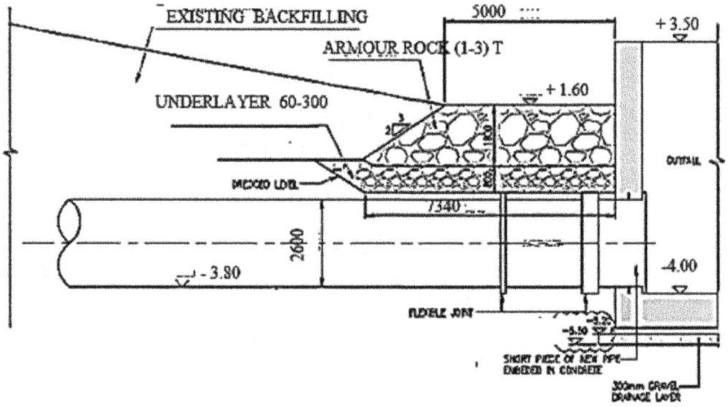

(e)

Figure 19.19. (*Continued*)

A relatively large outfall with three pipelines of φ 3.0 m each is shown in Figure 19.20. A general plan is shown in (a), the outfall structure plan in (b) and a longitudinal section in (c), with all levels relative to the Project datum. Although the pipe invert levels are low at –7.75 m, the

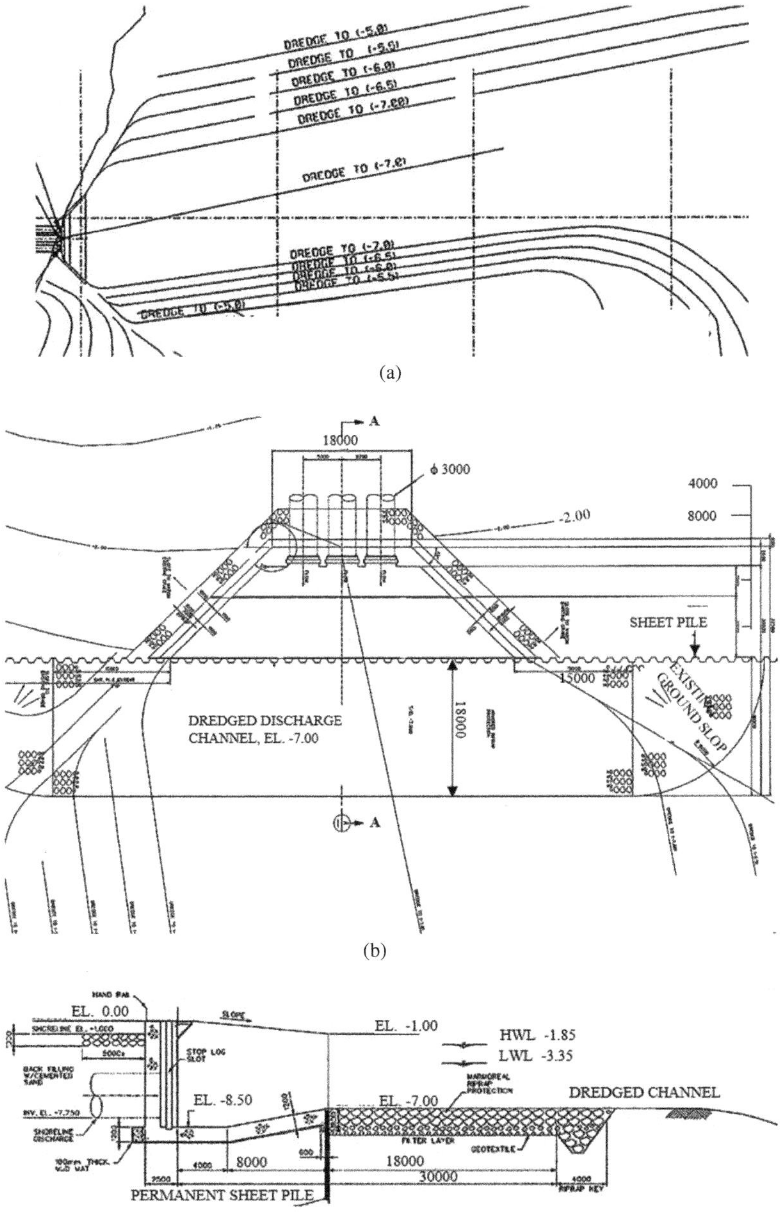

Figure 19.20. Three large pipelines outfall structure, scour protection and a dredged channel: (a) General plan; (b) Outfall structure plan; (c) Section A-A, longitudinal section.

outfall structure exit level was kept low at –7.00 m, which is well below the adjacent seabed level at –2.00 m. A scour protection is to be provided for a length of 30.0 m followed by a dredged channel at a level of –7.00 m in the sandy seabed until reaching the –7.00 m contour. A dredged channel at this project location could be tolerated as there is very little littoral drift and whatever little loose sand is deposited in the channel will be washed out by the outflow current.

An outfall chamber combined with a stepped concrete apron is shown in Figure 19.21 with a pipe of φ 3.5 m and an invert level at –7.50 m, all levels are relative to the Project datum. The outflow exit level from the chamber is at –1.5 m, to a relatively large scour protection area. The outfall chamber and the shoreline are delineated by a sheet-pile wall to protect the inland area and the seabed in front of the chamber is dredged so that the surface level of the scour protection will be approximately 1.0 m below the adjacent seabed. This slight depression in the seabed in front of the outfall could be tolerated in this project location, similar to the above example, as there is very little littoral drift. Although the outflow current velocity is not large, as the scour protection top layer is only 50–300 kg, the scour protection extends a distance of 75 m from the outflow chamber to reach the –5.0 m contour. After using such an intricate outfall structure, the extent of the scour protection and the need for a dredged channel could be optimised.

Two examples are presented to illustrate the channel shoreline outfalls. The first is shown in Figure 19.22 for a relatively narrow double channel outfall approximately 13.0 m wide. The channel has a straight concrete slab apron, 37 m long, thereafter it will gradually widen for a length of 18 m to a width of about 25.0 m. A rock scour protection will follow the concrete slab apron for a length of 33 m to reach a final width of 39 m. The longitudinal slope of the concrete slab apron is 1:50 and the slope for the rock scour protection is 1:28, approximately following the natural seabed slope.

A wider outfall channel is shown in Figure 19.23, and it is not actually at the shoreline but at a revetment that is slightly offshore from the natural shoreline. The channel is 26.0 m wide with a design discharge of 110 m^3/s. A rock scour protection is provided for a length of 37.0 m and width of 56.0 m, with a surface level slightly above the surrounding natural seabed. No extra rock material has been provided at the edges of the scour protection to act as a falling apron in case of bed scouring.

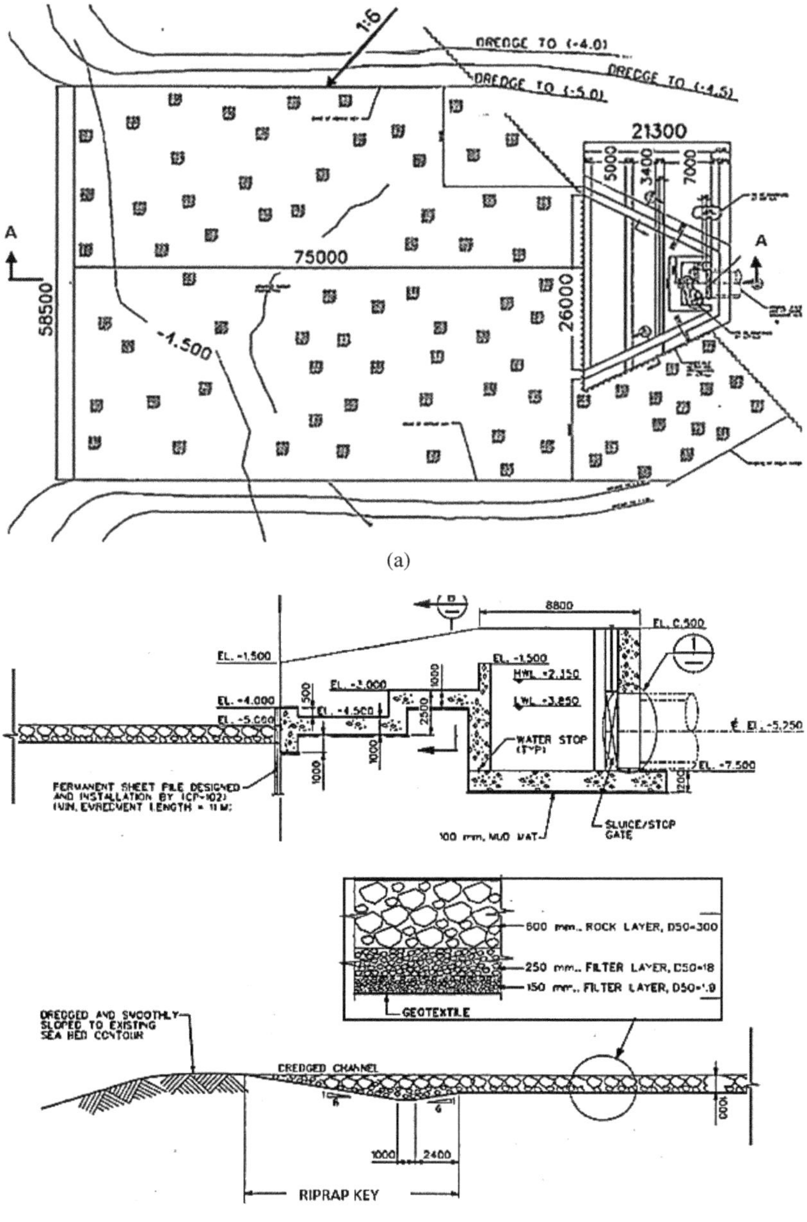

Figure 19.21. Large pipe outfall with extensive scour protection and dredged channel: (a) General plan; (b) Section A-A, longitudinal section (landward and seaward stretches).

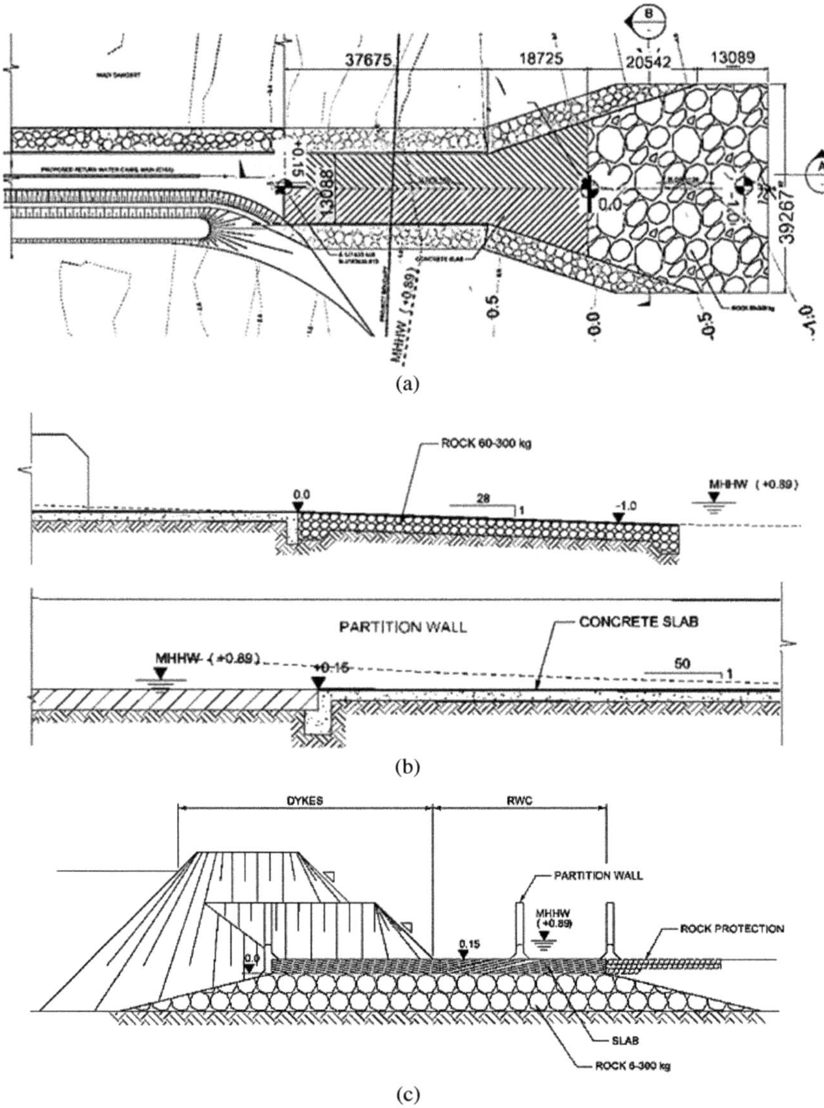

Figure 19.22. Outfall channel with concrete slab and rock scour protection: (a) Plan; (b) Section A-A, longitudinal section; (c) Section B-B, transverse section.

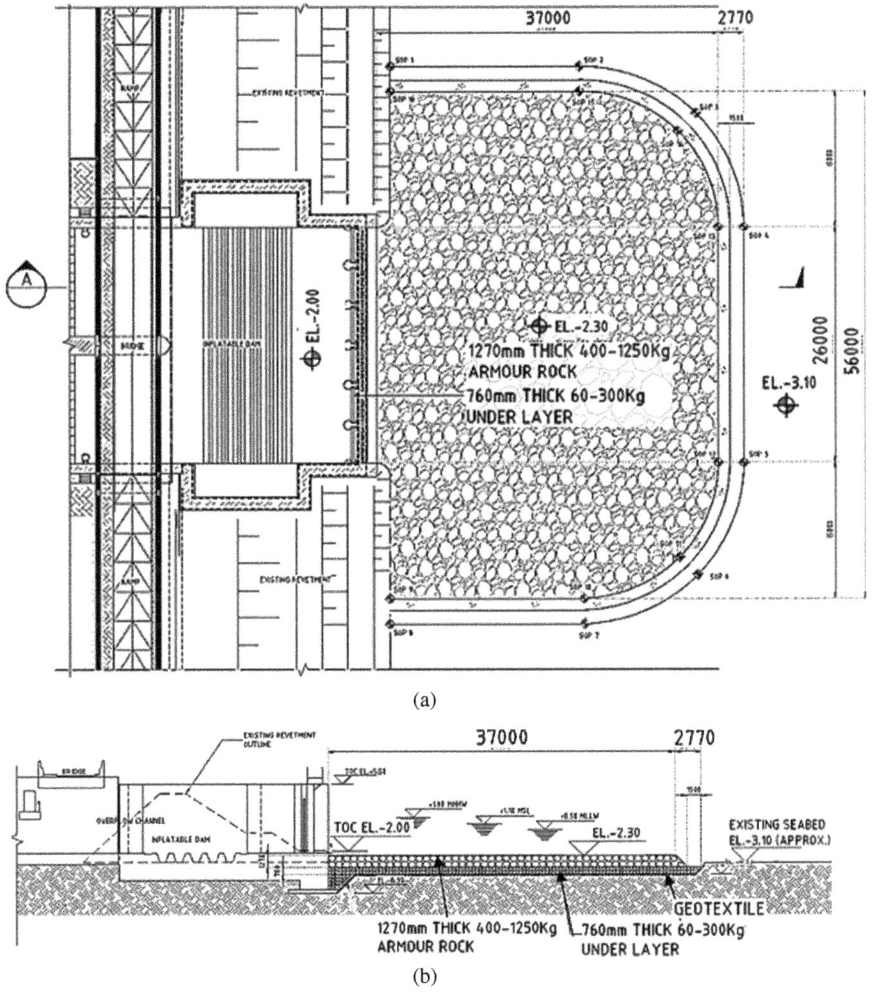

Figure 19.23. Outfall channel with rock scour protection: (a) Plan; (b) Section A-A, longitudinal section.

19.6. Outfalls and Intake Channels

Thermal power plants that are located on shorelines have very large requirements for sea water for cooling. The sea water intake system to the plant is either through subsea pipelines from an offshore submerged intake structure or through an intake channel with an onshore pump house. The outfall system, similar to the intake, could either be through subsea pipelines and offshore submerged diffusers or at the shoreline, directly or through a channel. The selection between the surface channel intake/outfall system and the subsea pipelines with offshore intake/outfall system is project-specific. Each system has its advantages and disadvantages; however, the selection will be dependent on many factors, mainly:

- Bathymetry and topography
- Oceanographic data (wave condition, tidal levels, currents, storm surges)
- Impact on shoreline
- Geotechnical data
- Outfall thermal dispersion characteristics
- EIA assessment
- Cost

The outfall and intake channels typically consist of two breakwaters/groynes with a channel in between reaching some distance beyond the shoreline. The outfall channel is normally designed as a narrow channel with bed scour protection due to the high discharge velocity. Whereas the intake channel is normally designed as a wide channel to keep the current velocities low and reduce the sediment load reaching the pumps. The channel could be dredged to increase its cross-section and reduce the current velocity even more, however there could also be a requirement for sediment trap before the pumps. No bed scour protection will be needed to protect the channel bed as the current velocities are normally kept below the bed material scour threshold. Few examples are presented herein to illustrate different design concepts and arrangements for the outfall/intake channels.

The project shown in Figure 19.24 includes both, an outfall and an intake channel, with the nearshore in this project having a shallow surf zone. Therefore, the two channels are to be extended to reach the deep water. The outfall channel, as shown in (a), has an elongated S-shape to reach the deep waters and to prevent wave intrusion inside the channel. It is a narrow channel, 17.0 m wide, with rock material for the bed scour protection, with heavier rocks near the exit being exposed to sea waves, as shown in (b) and (c). The intake channel is actually a large, enclosed basin that is completely protected from any wave action. The water is fed into the basin by a number of pipe siphons as shown in (d) and in (e).

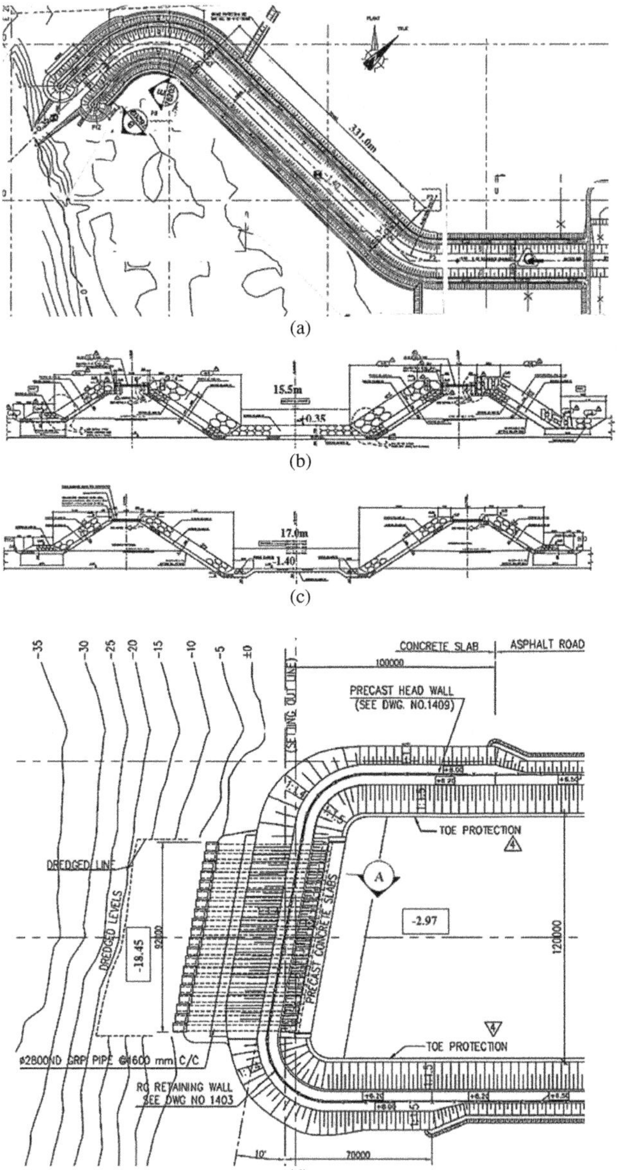

Figure 19.24. S-shape Outfall channel and an intake channel with siphons over a shallow reef: (a) Plan — outfall channel; (b) Section B-B; (c) Section E-E; (d) Plan — intake channel with siphons; (e) Section A-A; (f) Aerial image of the outfall and intake channels (Google Earth).

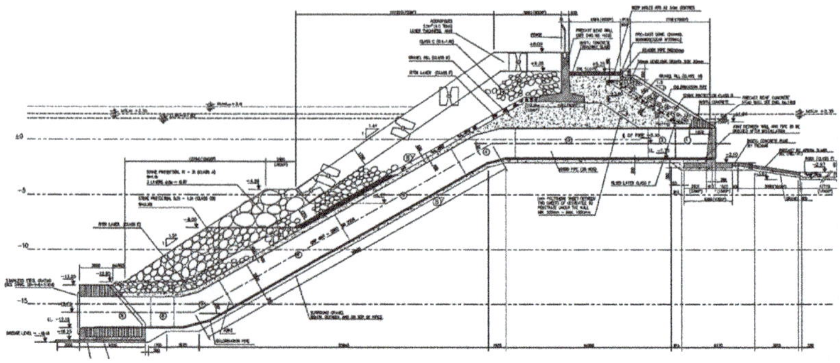

(e)

(f)

Figure 19.24. (*Continued*)

These siphons will help reducing/eliminating the intrusion of suspended sediments, debris and floating oil spills. An aerial image of the project is shown in (f) with breaking waves at the edge of the shallow shelf.

A different arrangement is shown in the example of Figure 19.25 for a channel intake with a straight alignment and without any shelter to the opening. This type of configuration could obviously be used in places where the sea is very calm, or the direction of the channel does not face the principal wave direction. The channel is about 1,400.0 m long with the bed level at the entrance at the same level as the surrounding seabed,

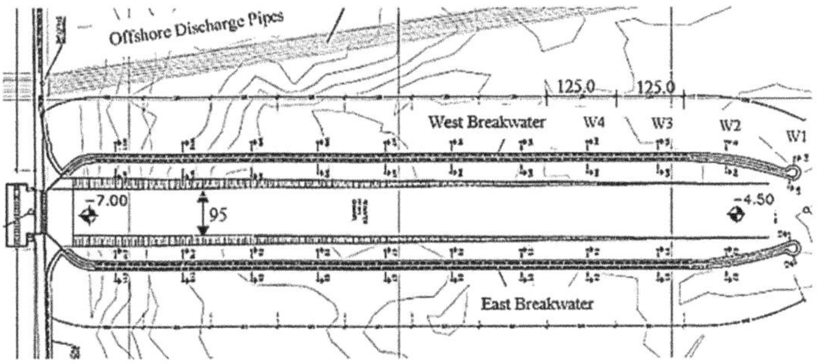

(a)

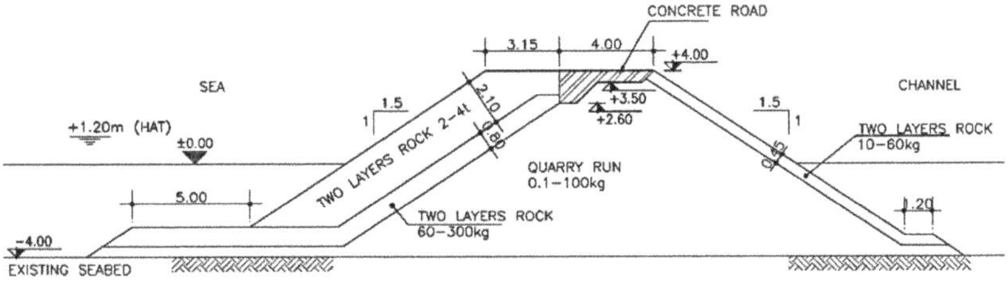

(b)

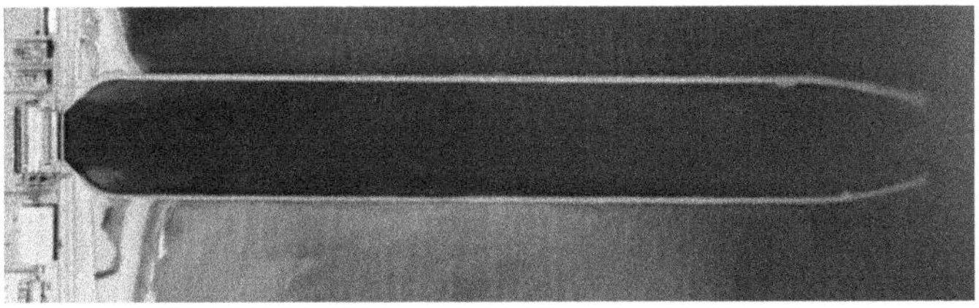

(c)

Figure 19.25. A straight intake channel — no shelter: (a) Plan — intake channel; (b) Section W4-W4 West Breakwater; (c) Aerial image of the intake channel (Google Earth).

though the channel bed is dredged to deeper level at the pumphouse. This extra depth will help reduce the current velocity further and also will act as a stilling basin. Maintenance dredging will be carried out at set intervals to maintain the channel operational levels. A typical cross-section of the West Breakwater is shown in (b), and an aerial image of the project in (c).

A similar type of an intake channel with a straight alignment is shown in Figure 19.26, though with the channel entrance fully sheltered from sea waves by an offshore breakwater. It is important to have relatively calm

Figure 19.26. A straight intake channel — with shelter.

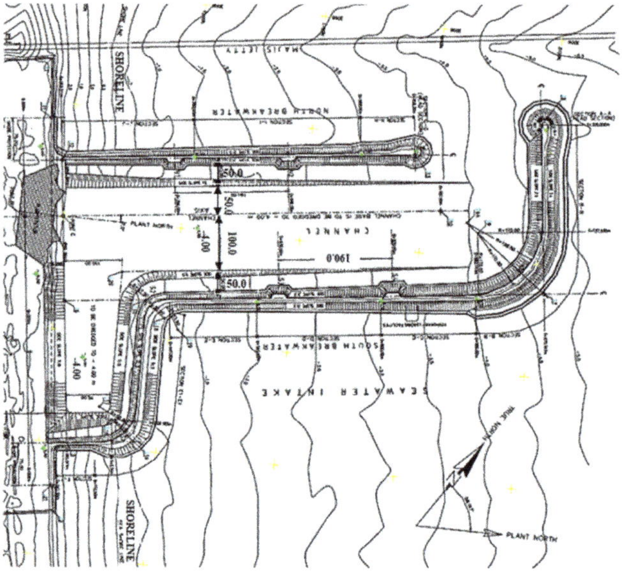

(a)

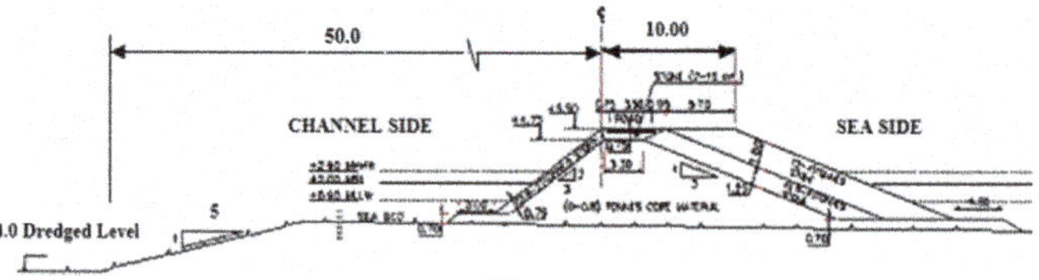

(b)

Figure 19.27. L-shape intake channel (Derun *et al.*, 2005): (a) Plan — intake channel; (b) Section E-E, South Breakwater; (c) Ariel image of the intake channel (Google Earth).

(c)

Figure 19.27. (*Continued*)

water conditions inside the channel to avoid any disruption to the pump house operating conditions, and so as not to increase the sediment load inside the channel.

L-shaped intake channel is shown in Figure 19.27, with a length of approximately 800.0 m and a width of 150.0 m. The channel is sheltered from sea wave action by an L-shape entrance, as shown in (a). The channel bed is at –4.0 m at the entrance, which is the natural seabed level, with dredging of the nearshore stretch of the channel undertaken to maintain the bed level. A typical cross-section of the South Breakwater is shown in (b) and an aerial image in (c).

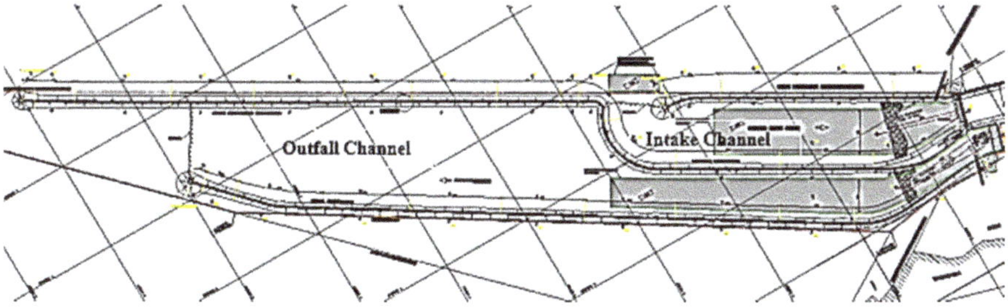

Figure 19.28. Combined intake and outfall channels.

Figure 19.28 is given as an example to demonstrate the wide possibilities for selecting the configuration of the channels to achieve a working and optimised layout. In this project, the two channels extend from the shoreline side by side, which made it possible to use three breakwaters to form the two channels.

Chapter 20

Marine Pipeline Protection

20.1. Outfalls and Intake Channels

An offshore pipeline system could consist of an intake structure and a pipeline, and/or an outfall pipeline with a diffuser assembly. Several techniques could be used to install marine pipelines (also known as submarine, subsea or offshore pipelines), these are tunnelling, trenching and surface routing.

Tunnelling techniques have been perfected during the past decade with the microtunnelling machines can now directionally drill tunnels up to 2.0 m in diameter. Tunnelling generally is the most expensive offshore pipe installation technique, although it is the least obtrusive. It is used to connect the land installation with the marine part of the pipeline avoiding interfering with any structures already existing on the coastline.

Marine pipelines are normally trenched in the landfall and nearshore stretches. Conventional marine excavation or dredging is possible to depths approaching 15.0 m. Dredging is possible in deeper waters, although the costs become prohibitive. The trenched pipelines could be armoured with rocks, concrete mattresses or simply covered by spoils from the dredging activity.

The least expensive configuration is simply laying the pipelines on the seabed. The pipelines must be anchored to mitigate current and wave-induced forces. Pipe materials that are less dense than water require ballast systems to neutralize the buoyancy forces.

Based on the above, it becomes clear that proper selection of the pipeline route plays a major role in optimising the design of the pipeline and how it will be installed. The key points to be considered could be summarised as follows:

- Seabed geotechnical conditions, specially at the shore approach (hard/soft seabed)
- Seabed topography (avoidance of rough areas, i.e. rapid sea level changes)
- Landfall site characteristics
- Existing pipelines (avoidance of pipe crossing)
- Spoil ground and borrow pits' locations
- Anchorage areas, fishing zones and shipping routes (avoidance of such zone due to risk of anchor or dropped-object damage)
- Hydrodynamic factors (avoidance, or shortening the route in high tidal current locations and shallow waters, especially with wave breaking)
- Safety policy and risk assessment

20.2. Landfall

A marine pipeline reaches the landfall location by way of a shore approach. The shore approach is normally in shallow waters and therefore the pipeline will be exposed to wave action forces and to longshore currents, hence most shore approaches are trenched and do require proper protection. The trench, with the necessary protection, will continue through the shoreline forming the pipeline landfall.

Alternative to pipeline trenching is the use of HDD (Horizontal Directional Drilling) where interference with the beach and the surf zone could be completely avoided. This construction methodology makes it possible to select shore crossing locations that otherwise would be prohibitively expensive and impractical because of existing obstructions and environmental impact.

20.3. Nearshore/Shore Approach Stretch

When the seabed depth is greater than half the wavelength, the wave-induced velocities become negligibly small. Therefore, the stretch between this limit and the shoreline shall be considered as the nearshore zone.

Marine pipelines are frequently protected in this stretch, either by being trenched and buried with backfill material and possibly including armour protection, or by being placed on the seabed and covered by some fill material, rubble mound and armour protection. The backfill material may need to be an engineered soil material, and in case of the need for pre-lay intervention, this material replaces the natural seabed as a foundation layer for the pipeline.

The stability of marine pipelines on the seabed under hydrodynamic loads (waves and currents) are normally checked for the critical installation stage where the pipe is empty and under 1-year Return Period conditions. The life-time stability is checked with a 100-year storm event with the pipe full.

20.4. Offshore Stretch

The pipeline in this stretch is effectively not exposed to significant hydraulic loadings, and as such it is frequently placed on the seabed, not trenched or covered by any type of protection. It is, however, required to have the pipeline trenched, buried and protected if there are risks of dropped object loads, fishing activities with trawl loads (as these are dragged on the seabed) and anchors (as these could be dropped either in an emergency situation or by accident). The required burial depth to avoid possible anchor damage depends on the type of seabed material and the expected weight of the anchor (Braestrup *et al.*, 2005).

20.5. Design Examples

Different examples of intake and outfall pipelines are shown in this section with different protection concepts along the three primary zones, i.e. landfall, nearshore and offshore. The presented examples consist of the pipe profile, extracted from the alignment sheet in segments for clarity, and representative cross-sections along the pipeline with some other details to illustrate principal areas of interest.

Figure 20.1 gives a basic example of trenched outfall pipelines in all zones. A longitudinal profile is shown in (a) illustrating the different layers of the trench fill and the protection along the pipeline. The nearshore stretch is protected by a rock armour layer, whereas such relatively heavy

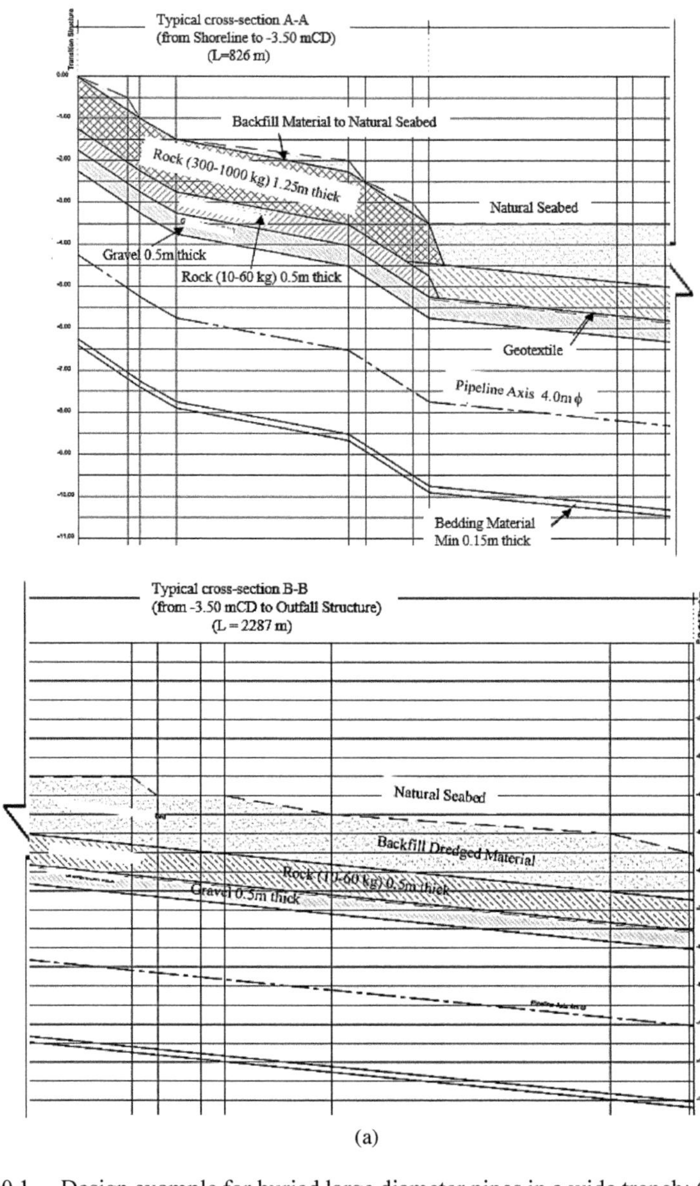

(a)

Figure 20.1. Design example for buried large diameter pipes in a wide trench: (a) Simple alignment sheet (longitudinal profile); (b) Typical cross-section A-A; (c) Typical cross-section B-B.

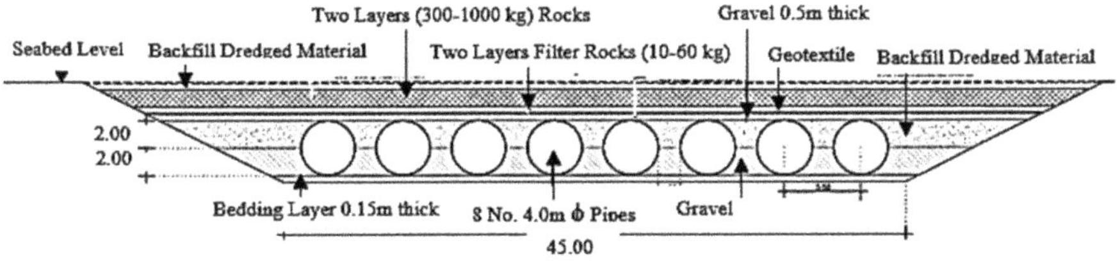

(b)

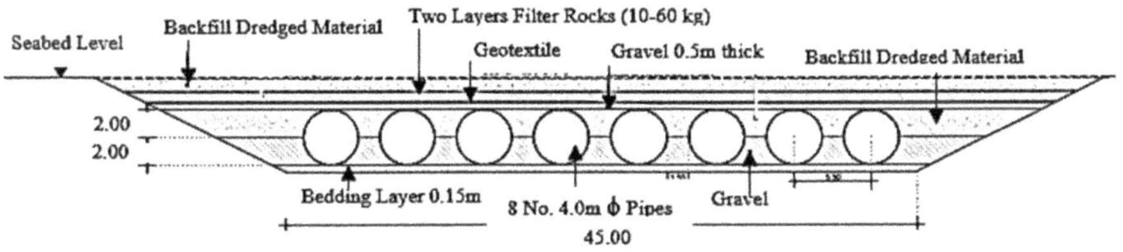

(c)

Figure 20.1. (*Continued*)

rock protection is not required in the offshore stretch. The different layers of bedding, support and protection are shown in (b) and (c).

A second example using rock armour protection is given in Figure 20.2. A general plan is shown in (a) with an existing and a new set of intake pipelines. The new four intake pipelines slightly vary in length, with the shortest pipeline of 1,962.0 m and the longest, 2,037.0 m. A more detailed plan for the landfall and part of the shore approach for the new pipelines is shown in (b), with the corresponding pipelines profile in (c). A detailed plan for the end of the offshore stretch is shown in (d), which includes the intake structures A & B. The profile for pipeline B at the intake structure is shown in (e). The pipelines are protected throughout their entire length, being trenched in the shallow nearshore part, as shown in (f), then the trench depth will gradually be reduced, as shown in (g), until they are completely above the seabed for the rest of the offshore stretch, as shown in (h).

More elaborate details are shown in Figure 20.3 for three long outfall pipelines. The pipes are of 1,800 mm dia. and the longest of the pipelines is 2,570 m long. A general layout plan is shown in (a), and the overall alignment sheet, in (b). For clarity, the alignment sheet is divided into four segments with an enlarged view of the profile, as shown in (c, d, e and f) indicating the relevant cross-sections along the pipelines. The landfall

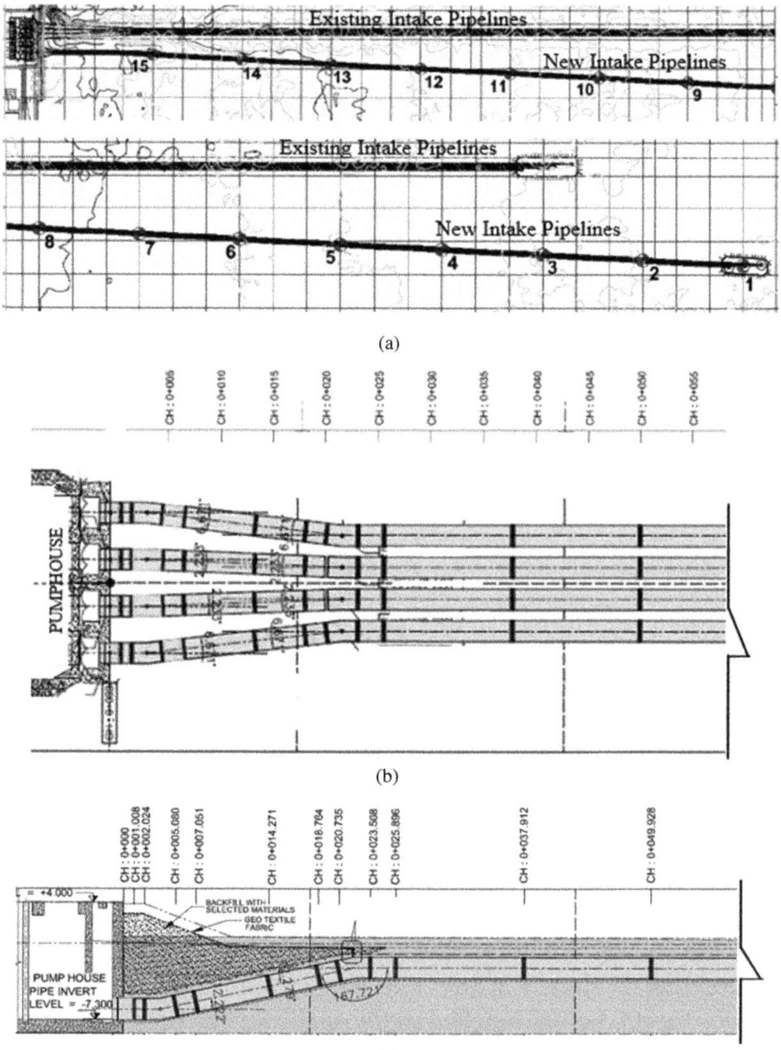

Figure 20.2. Intake pipelines with rock protection.

(The number and spacing between the pipes are indicative): (a) New intake pipelines — general plan; (b) New intake pipelines — nearshore plan; (c) New intake pipeline B — nearshore profile; (d) New intake pipelines — offshore plane (CH.1935-CH.1990) including Intake Risers A & B; (e) New intake pipeline B — offshore profile (CH.1939-CH.1987.5) including Intake Riser; (f) Typical cross-section for the nearshore stretch; (g) Typical cross-section for the transition from nearshore to offshore; (h) Typical cross-section for the offshore stretch.

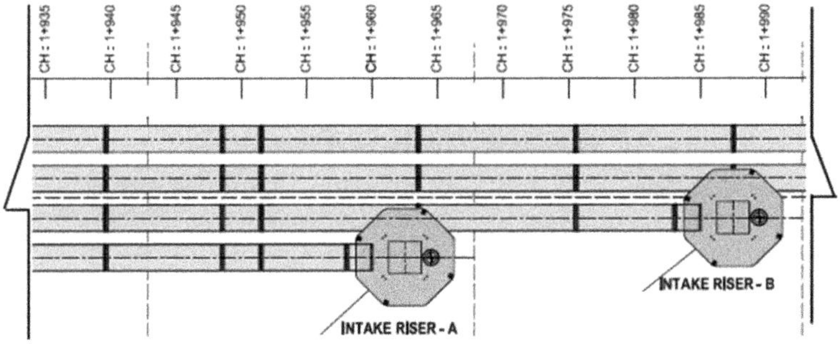

(d)

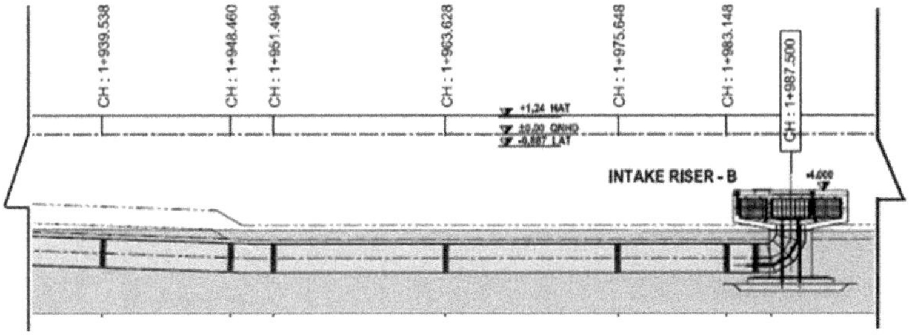

(e)

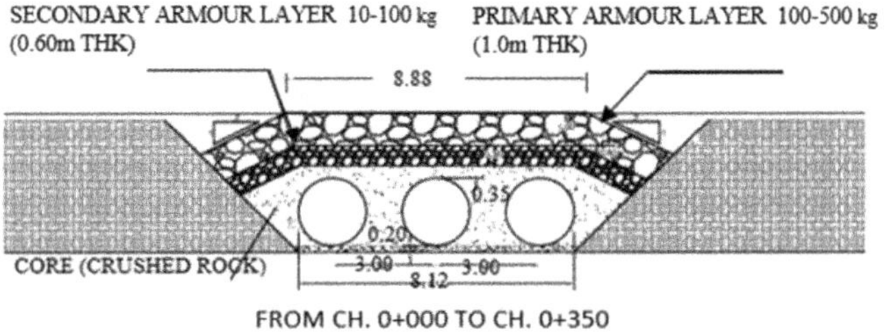

(f)

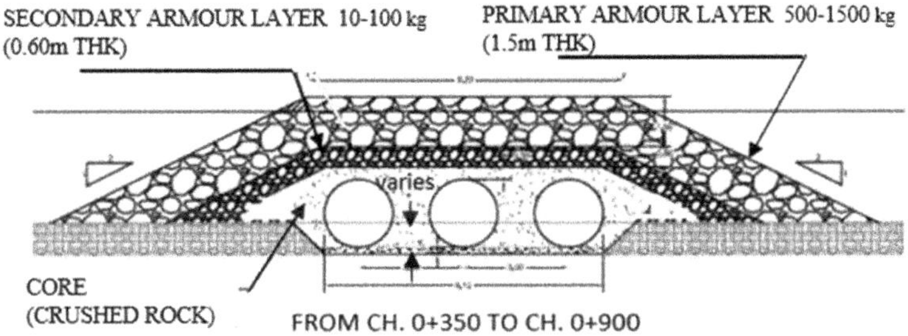

(g)

Figure 20.2. (*Continued*)

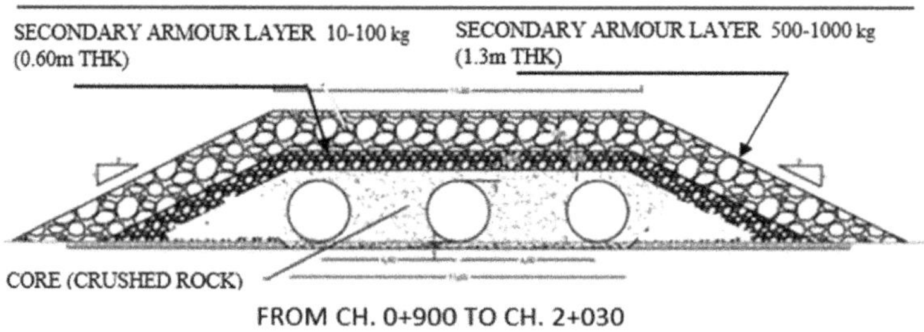

(h)

Figure 20.2. (*Continued*)

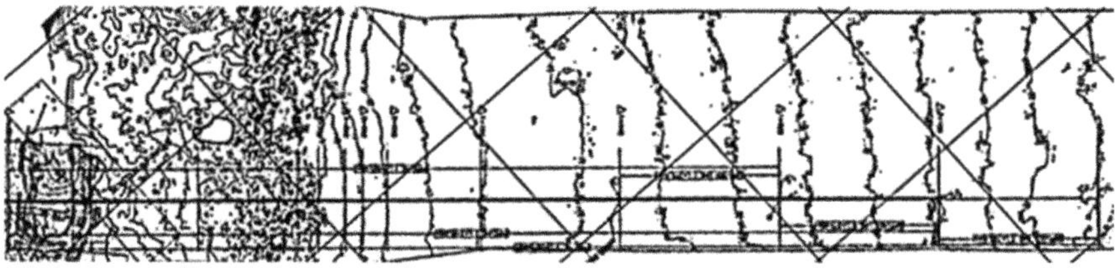

(a)

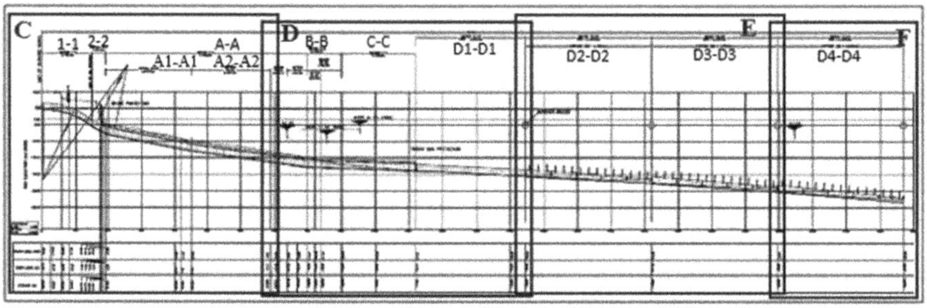

(b)

Figure 20.3. Outfall pipelines with concrete mattress protection: (a) Pipelines plan; (b) Full alignment sheet; (c) Enlarged view of the onshore, landfall and near-shore parts; (d) Enlarged view of the near-shore and transition to the off-shore; (e) Enlarged view of the off-shore part; (f) Enlarged view of the pipelines end; (g) Typical Pipelines plan illustrating the collars' arrangement; (h) Cross-Section A1-A1; (i) Section A4-A4; (j) Section B1-B1; (k) Section C-C; (l) Section D-D; (m) Detail-1, shore protection at the landfall; (n) Detail-2, end of trenching-plan; (o) Detail-2, cross-section 1-1 in plan; (p) Detail-2, cross-section 2-2 in plan.

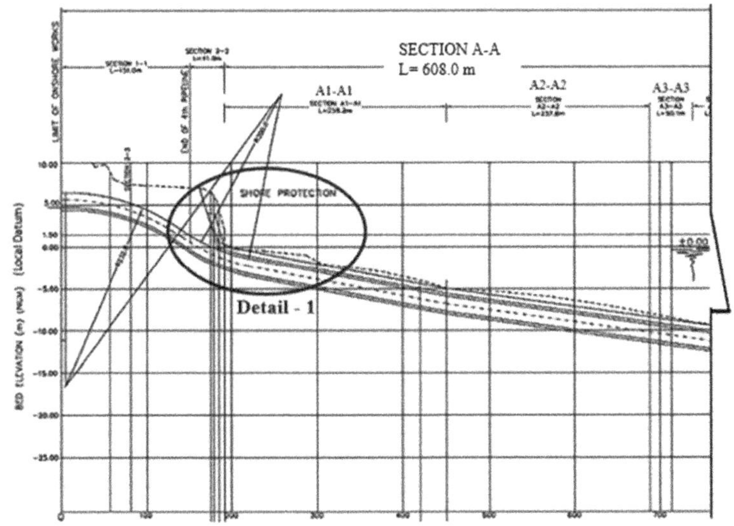

(c)

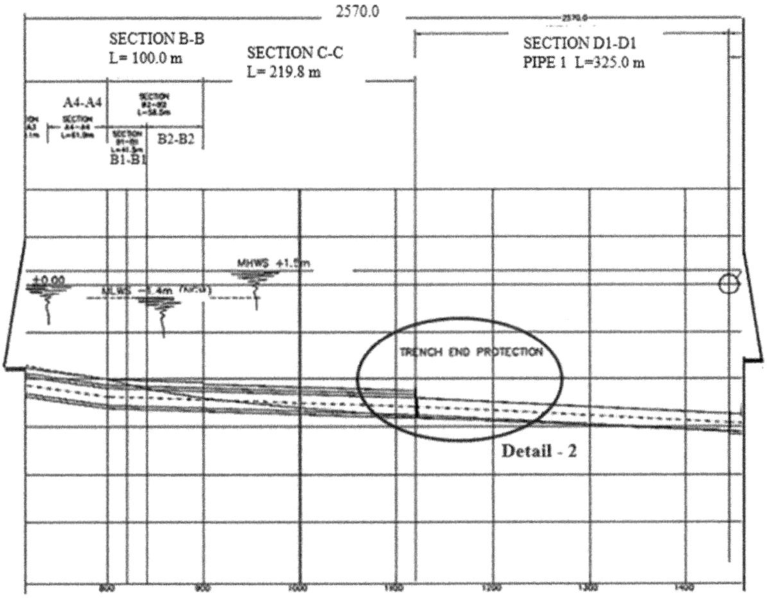

(d)

Figure 20.3. (*Continued*)

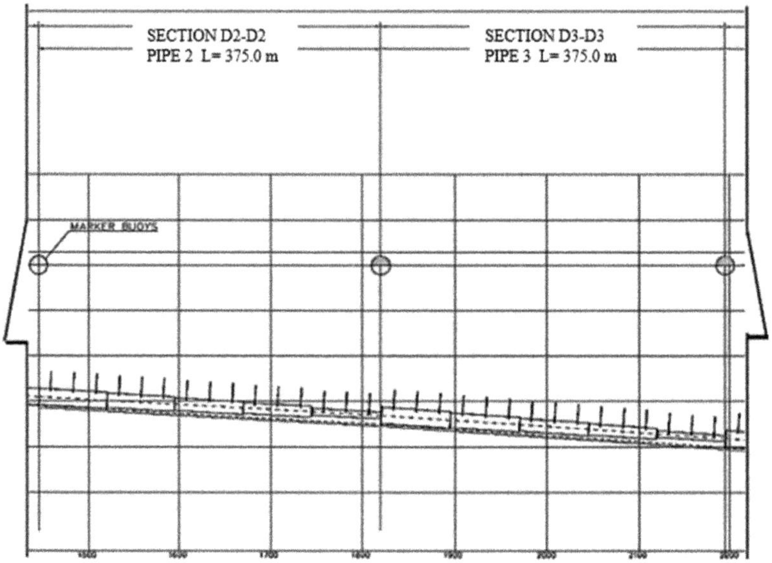

(e)

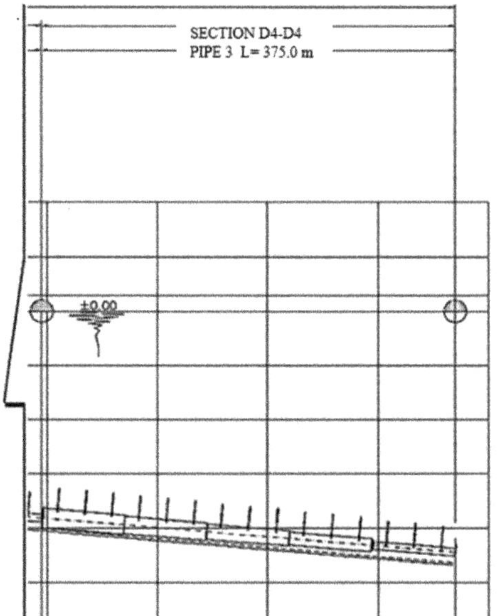

(f)

Figure 20.3. (*Continued*)

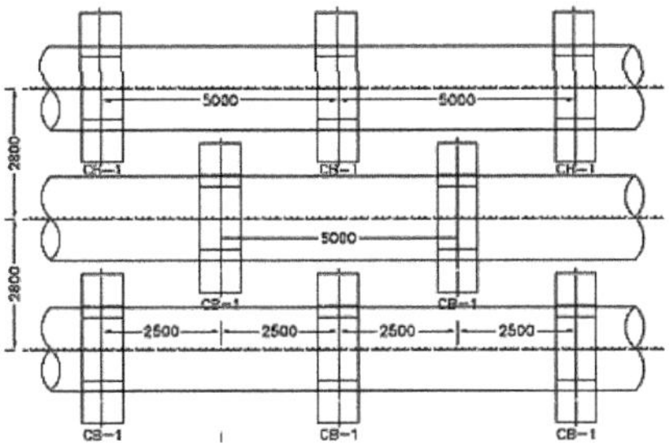

(g)

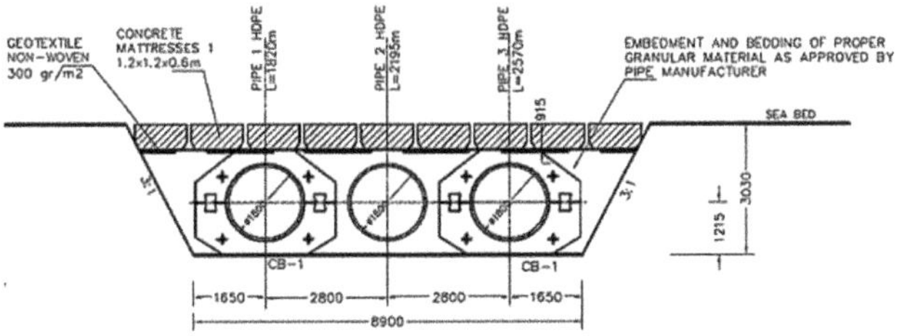

(h)

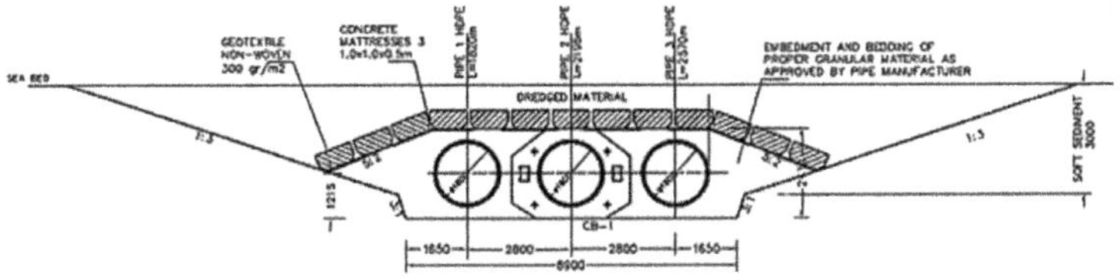

(i)

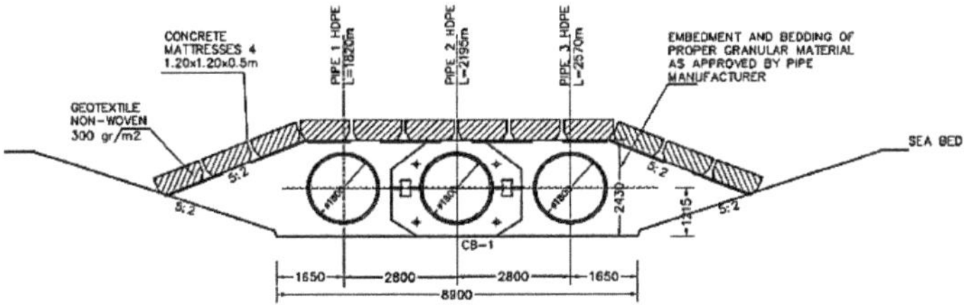

(j)

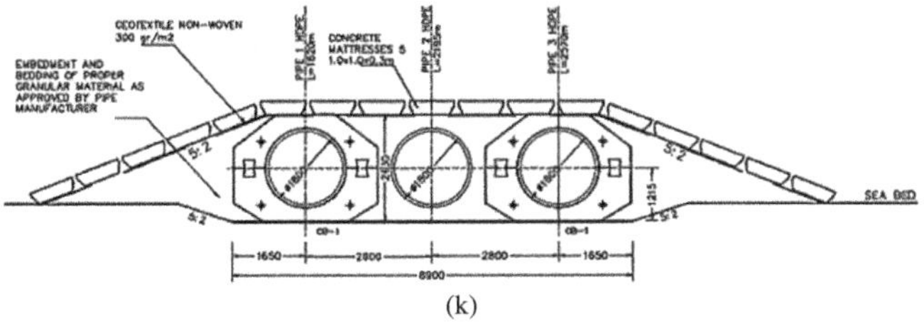

(k)

Figure 20.3. (*Continued*)

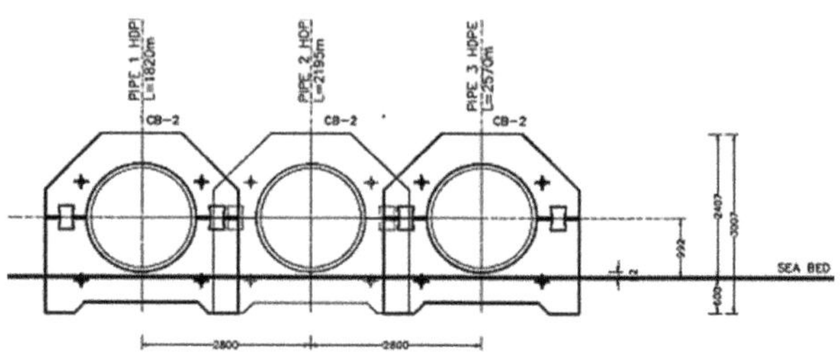

(l)

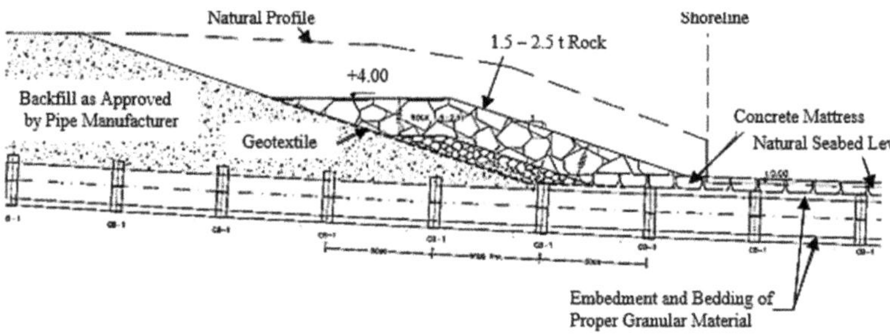

(m)

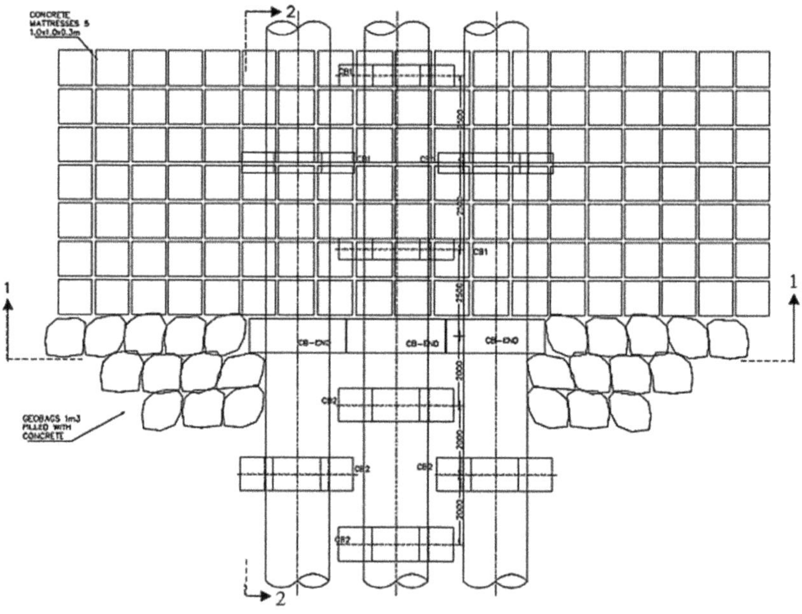

(n)

Figure 20.3. (*Continued*)

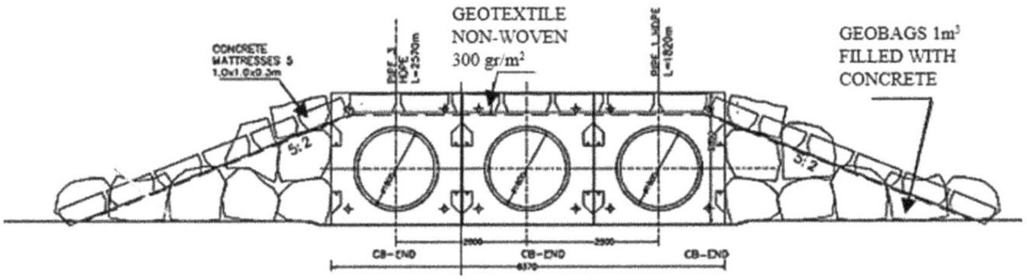

(o)

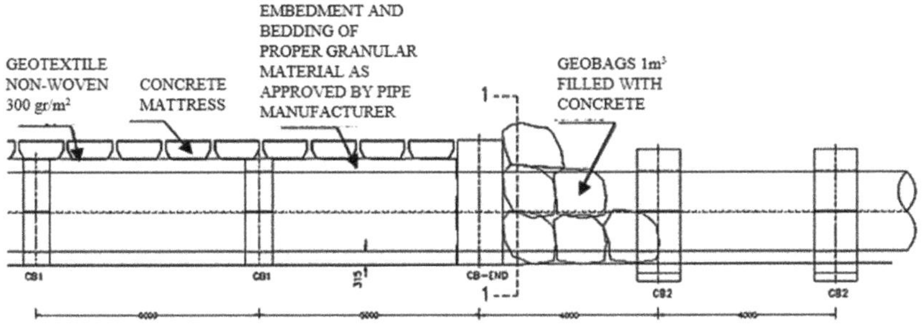

(p)

Figure 20.3. (*Continued*)

location is indicated in (c) and a detailed longitudinal profile for the shore protection is shown in (m). The pipelines are fitted with concrete collars and these are to be arranged to be staggered after placement in order to optimise the trench width, as shown in (g). The pipelines in the nearshore stretch (Sections A–A, B–B and C–C) are well-protected by having them trenched, in varying depths, and covered by a concrete mattress. The pipelines are fully trenched and buried in Section A–A for a distance of 608.0 m, i.e. the first part of the stretch, as shown in (h) and (i). The depth of the trench will gradually be reduced in Section B–B, which covers a distance of 100.0 m, as shown in (j), then the pipelines will no longer be trenched in Section C–C, but still remain under the protection of the concrete mattress for a distance of 219.8 m, as shown in (k). Such a transition from being fully trenched pipelines to being placed over the seabed could be attributed to a change in the seabed material and difficulty of dredging, or due to some kind of fishing or other human activities near the shoreline that necessitates the trenching. In this case, however, the reduction in the

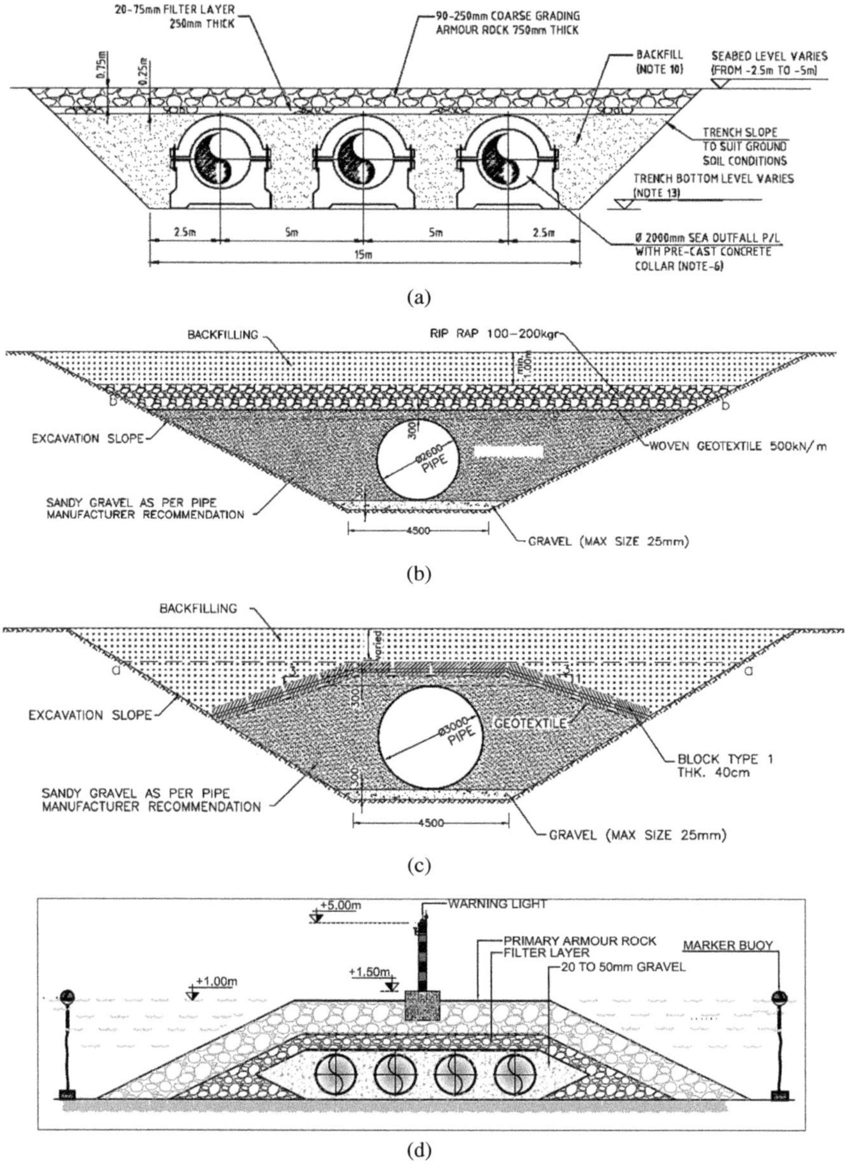

Figure 20.4. Selected illustrative cross-sections: (a) Wide trench; (b) Narrow trench for a single pipeline with buried rock protection; (c) Narrow trench for a single pipeline with buried concrete mattress protection; (d) Marker buoys and warning light arrangement in shallow waters.

burial depth is to have a smooth transition from the trenched pipelines in the nearshore stretch to the unprotected untrenched pipelines in the offshore stretch. The pipelines, and the diffusers, will continue for 1,500.0 m in the offshore stretch on the seabed, as shown in (l). The transition from the protected to the unprotected pipelines is shown in Detail 2. The location of this transition is shown in (d), and the detailed design is shown in (n, o, and p). Note that a different type of concrete has been used to form a separation wall between the protected and the unprotected stretches. The two triangles on the sides of the pipelines are sealed by 1.0 m^3 geobags filled with concrete.

Further illustrative cross-sections are given in Figure 20.4 to demonstrate various trench filling and protection arrangements and possible marking arrangements for buried pipelines.

Chapter 21

Seabed — Outcrops and Depressions

21.1. Introduction

As part of the required input to the design is a detailed bathymetric survey, and if required a geophysical survey. These surveys are usually carried out by single-beam (for large offshore areas) and multibeam echo sounding, side scan sonar, sub-bottom profiling and magnetometer in conjunction with Differential Global Positioning System (DGPS). The surveys must also include the detection of possible obstructions and debris that might be present in the project area and may require removal prior to commencement of the construction works.

A full geotechnical investigation is also to be carried out with as many boreholes as deemed necessary, and this is a project-specific requirement. The obtained information will determine the composition of the seabed and if any special treatment is required to strengthen and enhance the seabed.

This chapter is not intended to discuss the details of any survey techniques or the geotechnical investigation, it will however present two frequently encountered simple issues related to the seabed levels and composition, which need to be resolved during the design. These are localised seabed surface anomalies and the existence of a thin layer of loose sediments.

21.2. Simple Probing Survey

If a seabed is composed of cap rock as a top layer, it will almost invariably be covered by a thin layer of sand/silt-sand deposits, especially in lower spots. The thickness of this layer needs to be properly determined, especially along the toe line of the new structure. The layer thickness will influence the toe design, if it is thin (<0.30–0.40 m), the toe could be considered as resting directly on the bed rock, otherwise dredging/cleaning of the sediment deposits along the toe strip might be considered or an adequate scour protection measure is to be applied.

A simple probing survey by a diver with a steel bar will give a detailed information about the sediments' layer thickness along the toe line. Accordingly, the toe design will be completed, with localised adjustments if needed. Few practical examples are given in the following to illustrate the survey methodology and the outcome.

The first example is of a reclaimed rectangular land extension on a cap rock seabed, as shown in Figure 21.1, with the seabed covered by a thin layer of sediments. The prepared plan for the diver probing survey is shown in the figure, which includes the SOL of the revetment and a 9.0 m wide strip covering the preliminary toe location. The outer edge of the

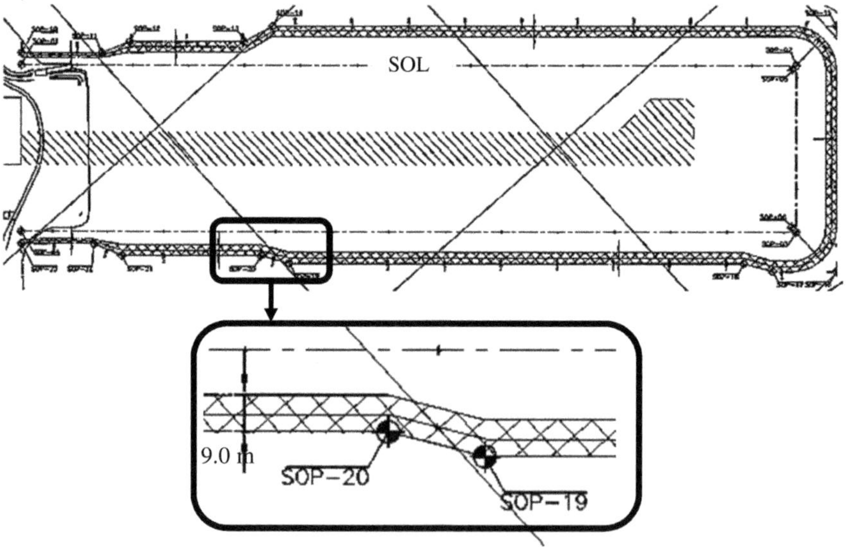

Figure 21.1. 9.0 m wide toe strip with 50.0 m intervals probing points.

strip is marked by a series of SOPs (Setting Out Points). Three probing measurements across the strip shall be taken every 50.0 m. Note that the 50.0 m interval is rather long, it was however satisfactory for this particular project as the seabed is very flat and uniform with almost a consistent thickness of deposits.

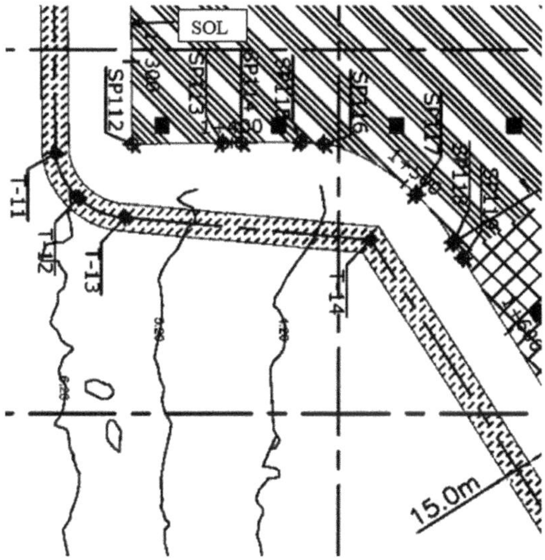

(a)

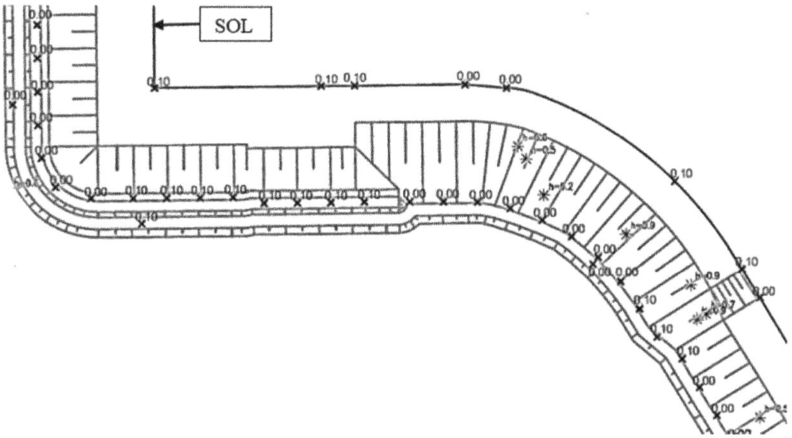

(b)

Figure 21.2. 15.0 m toe strip with 10.0 m interval probing: (a) Plan for the probing survey strip; (b) Survey results, in meters.

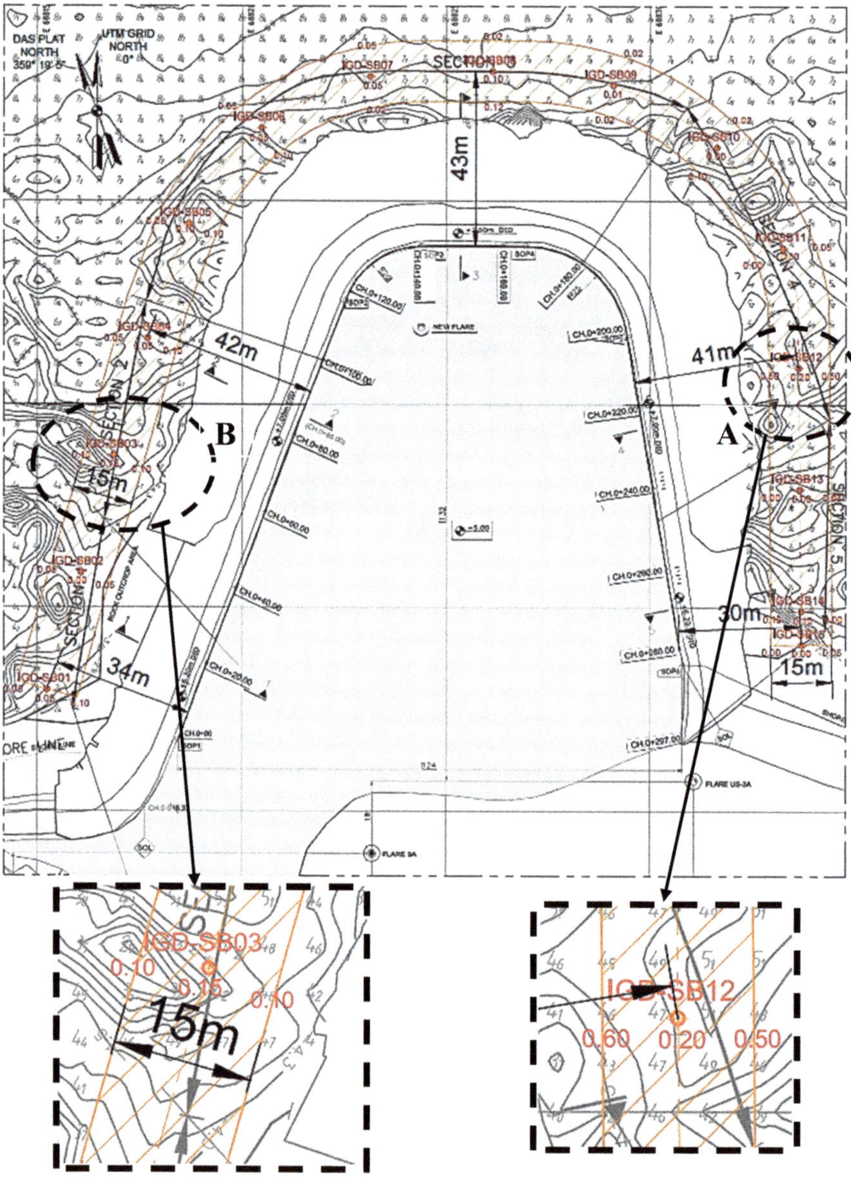

Figure 21.3. 15.0 m toe strip with 30.0 m interval probing.

In Figure 21.2, the prepared plan for the diver survey is shown in (a) which includes the revetment SOL and a 15.0 m wide toe strip to be surveyed. The alignment of the toe strip is well-defined by SOPs along the strip centreline. It is worth noting that the selection of a wide strip in this example is recommended as the toe design, although still in its preliminary stage, was envisaged to be a wide toe. Hence, the 15.0 m strip width was selected rather than the 9.0 m of the previous example. The survey results are shown in (b), which is the average of three readings across the strip at 10.0 m intervals. Few probing measurements were taken at the revetment SOL as an extra information.

A third example is shown in Figure 21.3, which is for the same project discussed in Section 4.4.2. A 15.0 m toe strip was defined for the diver survey, as shown in the figure, with three measurements across the strip at 30.0 m intervals. All measurements indicate no, to little, deposits with a thickness of $\leq$ 0.10 m, apart from two locations, A and B. The area marked "A" on the east side of the project has deposits with thickness reaching up to 0.60 m. The other area, marked as "B", on the west side of the project has an outcrop protruding above the WL (water level) and deposits with thickness reaching up to 0.15 m. Both areas shall be discussed in Section 20.3.

21.3. Outcrops and Depressions

Caprock seabed surface is not always flat, but there could be some anomalies, namely depressions and outcrops (bulges), and both can vary in size and shape. These two seabed features, if they cannot be avoided, will have an impact on the design of the coastal structures. Both of these anomalies actually existed in the example shown in Figure 21.3, where the encircled area "A" on the east side is for a depression and the area "B" on the west side is for an outcrop.

An enlargement of area "A" for the depression survey (bathymetric and probing) is given in Figure 21.3, as it is covered with sand up to a depth of 0.60 m. The relevant design section for this area is shown in Figure 21.4, indicating that the sand is to be removed and the depression filled with rock material. The rock fill could be quarry run, preferably with 1–100 kg, or larger rocks (300–1,000 kg) as in the shown example if it is exposed to some waves and currents action.

An enlargement of area "B" of the outcrop bathymetric survey is also shown in Figure 21.3, as it extends about 15.0 m along the revetment and

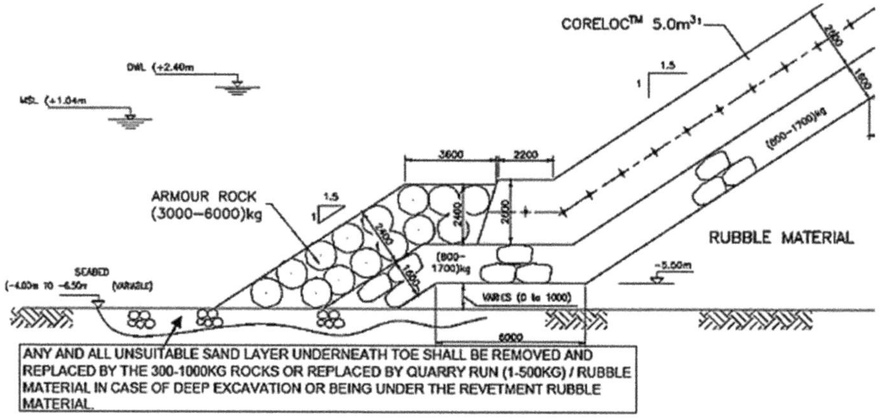

Figure 21.4. A seabed depression in the toe strip and the relevant design section.

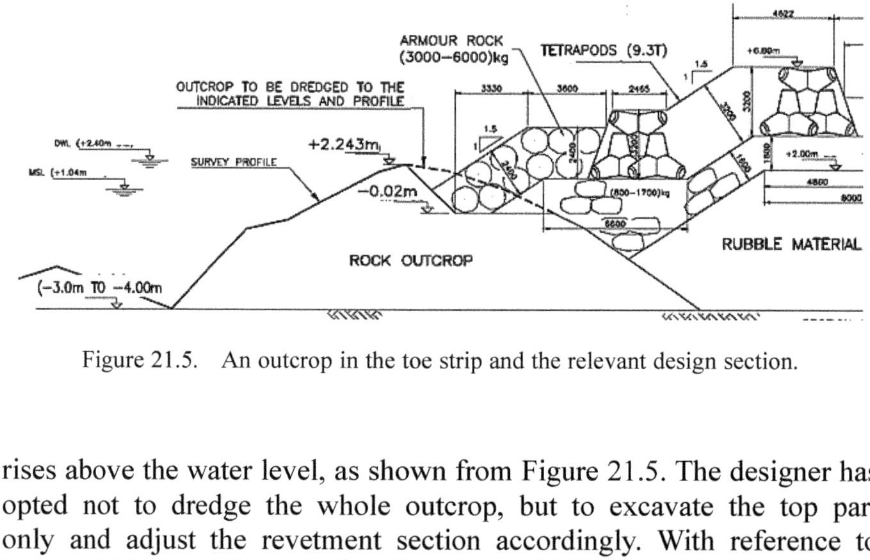

Figure 21.5. An outcrop in the toe strip and the relevant design section.

rises above the water level, as shown from Figure 21.5. The designer has opted not to dredge the whole outcrop, but to excavate the top part only and adjust the revetment section accordingly. With reference to Figure 4.15, it can be seen that Tetrapods of 9.3 t have been used in the area of the outcrop as the armour units (which are available on site as salvaged units from an older revetment), thereafter Core-locs of 5.0 m^3. It is apparent that any single layer CAU with a placement pattern cannot be used in the outcrop area with such a sharp change in levels, whereas a double-layer random placement CAU could be used, see Figure 10.16. However, if the outcrop is to be completely dredged, then the Core-loc armour units could be used, but the decision not to dredge in this case was influenced by the availability of the Tetrapod units on site and the preference of the construction team.

Another example of an outcrop which might affect the design of a revetment is shown in Figure 21.6 for a nearshore outcrop. This outcrop is about 50.0 m in length along the revetment and rises above the water level. It is not in the toe strip zone but just offshore of the revetment, which makes it possible that it might act like a lens in focusing the wave energy on the revetment, i.e. will be exposed to wave conditions

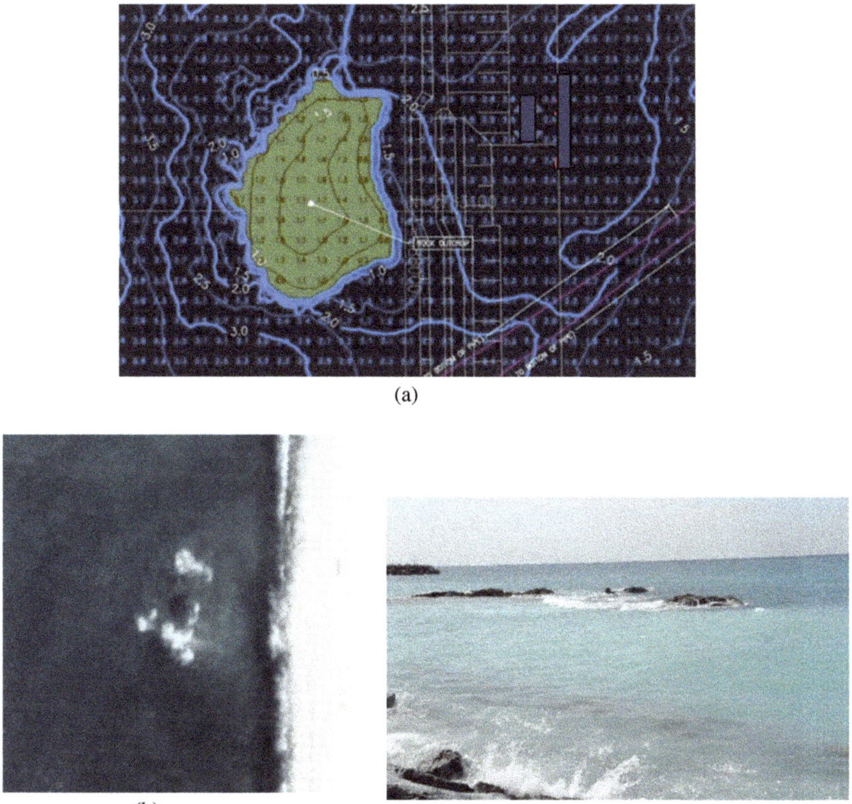

(a)

(b)

(c)

Figure 21.6. A nearshore outcrop: (a) Bathymetric and topographic survey of a nearshore outcrop; (b) Satellite image of the outcrop (Google Earth); (c) The outcrop as seen from the revetment side.

higher than the design conditions. Therefore, this type of a situation must be assessed and evaluated very carefully as the length of the outcrop, its distance from the revetment and the wave conditions (height, period and direction) all play a part in wave conditions behind the outcrop.

Chapter 22

Beach Control Structures

22.1. Introduction

Natural beaches are always subject to erosion and accretion, however due to operational, economic and safety reasons often some measures are to be taken to protect and stabilize the beaches. The design philosophy and design methodology of coastal structures is to be developed after considering the following basic issues:

- Function of the structure
- Structural integrity
- Physical environment
- Amenity requirements
- Construction methods
- Operation and maintenance

Coastal defences/structures are commonly classified as hard and soft engineering works. However, it is more representative and realistic to include an "intermediate" class for arrangements which include both hard and soft structures working together. Main examples for these three classes could include "breakwaters and revetments" for the hard type, "nourished beaches and artificial beaches/reefs" for the soft type and "groynes, offshore/detached breakwaters and perched beaches" for the intermediate type. This intermediate type of structures will be the subject of this chapter.

22.2. Groynes and Headlands

The groyne provides a physical barrier to longshore sediment transport along the beach by deflecting nearshore wave breaking generated currents. Consequently, the orientation of the beach will be altered to be aligned with the prevailing wave direction. Sediments are deposited on the updrift side of the groyne relative to the main direction of the sediment transport, while erosion occurs in the lee side. Therefore, there will be a difference in the beach level between two sides of the groyne. Groynes may extend across part or all of the intertidal zone with the typical longitudinal profile consisting of inshore level section (the groyne root), sloping middle section parallel to the beach slope and a horizontal seaward section (the groyne head).

Groynes are normally grouped to form a groyne field if the groynes are nearly similar, or a groyne system if there are different types of groynes. A groyne system could incorporate terminal/primary groynes/breakwaters with intermediate secondary groynes.

Different materials could be used to construct the groynes, or it could be a composite construction, however, only rubble-mound material shall be presented in this section. Groynes may be partly permeable to currents and sediments (in suspension), and this may reduce the efficiency of the groynes, but this could be beneficial as it minimises the downdrift erosion while still stabilising the updrift beach.

Although groynes are mainly straight and almost perpendicular to the shoreline, they could however have various plan shapes (see CIRIA, 2020, C793), and for different reasons. Design examples of different shapes and configurations of groynes and headlands are presented in this section.

A typical groyne is shown in Figure 22.1 where it starts at level +2.0 mCD at the root, continues at this level for 10.0 m, and then it slopes down towards the head of the groyne at 1:10, similar to the beach slope. The top level of the groyne at the construction time is 0.5 m above the sand. This height difference is considered sufficient provided that the beach will be constructed by aligning the sand as close as possible to the expected equilibrium orientation. Strong storms could however produce sand updrift in the excess of the 0.5 m, which will lead to sand spilling over the groyne.

A long artificial beach in relatively shallow and calm sea is subdivided by a groyne field, a longitudinal section and cross-sections of the

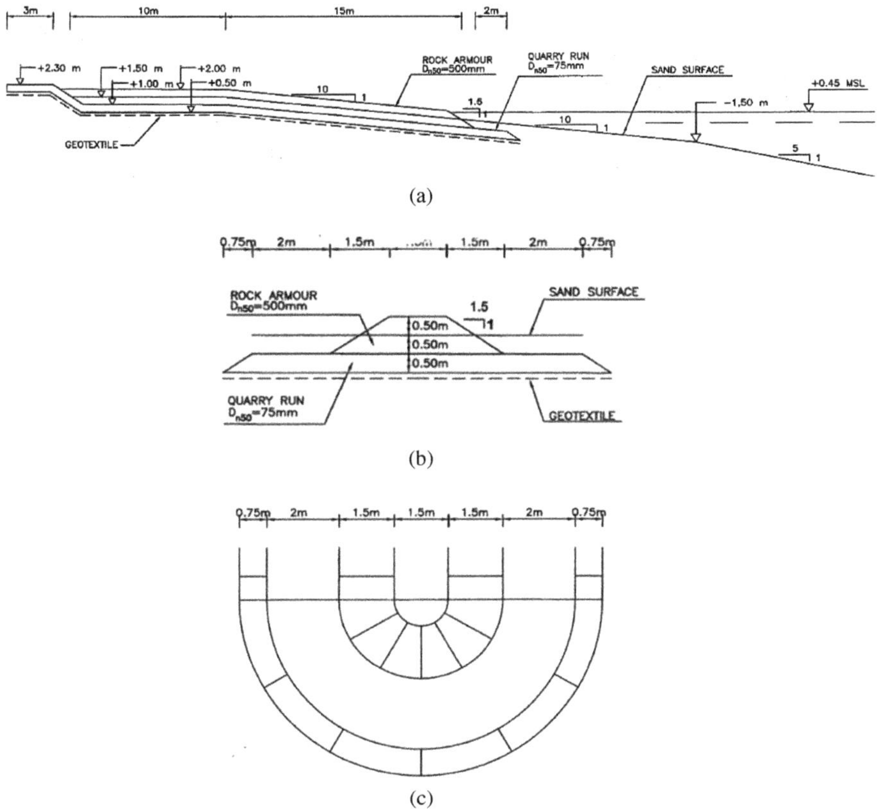

Figure 22.1. Typical small groyne: (a) Longitudinal section; (b) Cross-section; (c) Plan of groyne head.

groyne are shown in Figure 22.2. The groyne does not reach the beach crest and is only on the slope, however, it does extend slightly above the HWL.

A more elaborate groyne shape is presented in Figure 22.3 with an oblique orientation to the shoreline and with an F-shaped head. The overall layout of the groyne field is shown in (a), the groyne plan in (b) and cross-sections in (c, d, e and f) for the deepest area of each. A geotextile has been used on the beach side of sections (a and b) to prevent loss of sand through these sections. Whereas no geotextile is used in sections (c and d) as the sand layer is relatively thin and deep with the possibility

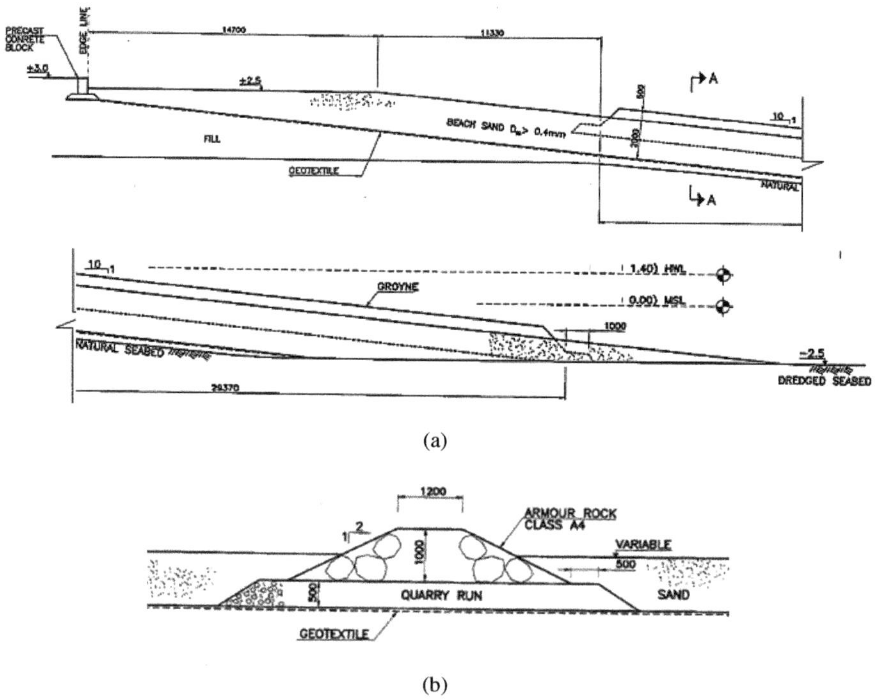

Figure 22.2. A small groyne on artificial beach: (a) mm Longitudinal section; (b) Cross-section.

of sand loss through the groyne being minimal. The top level of the groyne descends from +1.8 mCD down to −1.5 mCD.

Part of an Island shore protection is shown in Figure 22.4, where a combination of a groyne (Southwest Groyne) and a groyne-headland (Northwest Groyne) are used. These are relatively large structures that are actually designed more like breakwaters than groynes, however, their function is mainly to protect the Island beaches. A geotextile has been used in sections (d and g), where the groynes are in contact with the beach sand.

An artificial pocket beach is shown in Figure 22.5 where two groynes, or in this case they could also be called breakwaters, are protecting the beach from being washed out. A general layout is shown in (a) and a detailed plan of groyne-1 is shown in (b). The beach is expected to assume a flatter slope in the wave active zone between the beach crest,

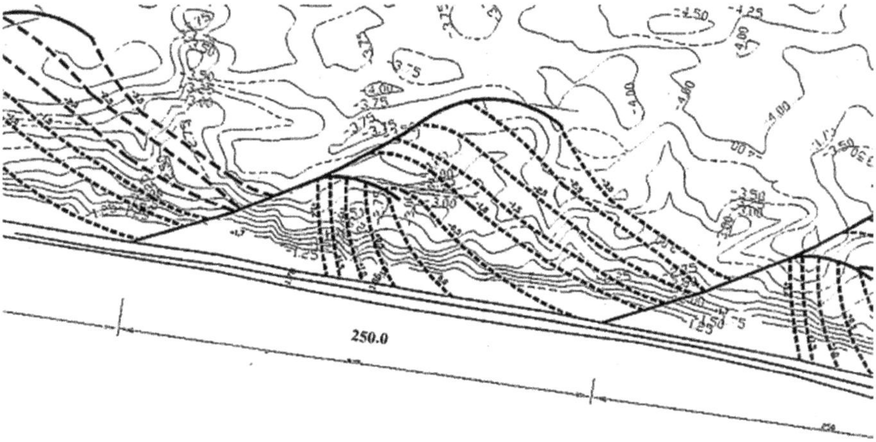

(a)

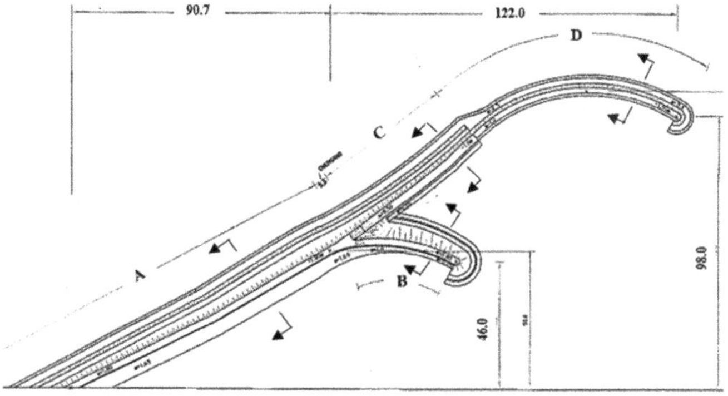

(b)

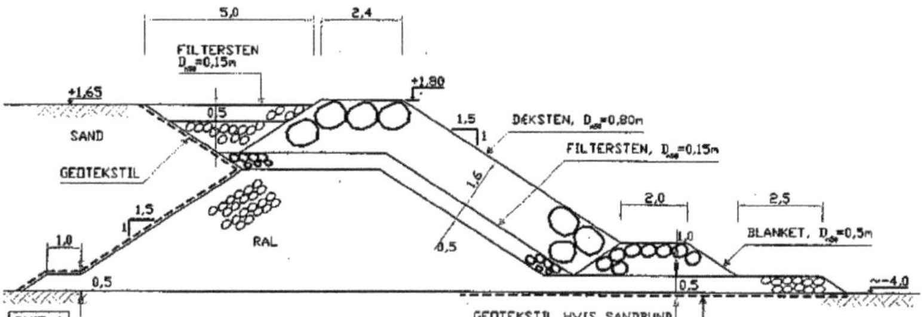

(c)

Figure 22.3. Oblique F-shape groynes: (a) General layout; (b) Groyne plan; (c) Cross-section A; (d) Cross-section B; (e) Cross-section C; (f) Cross-section D.

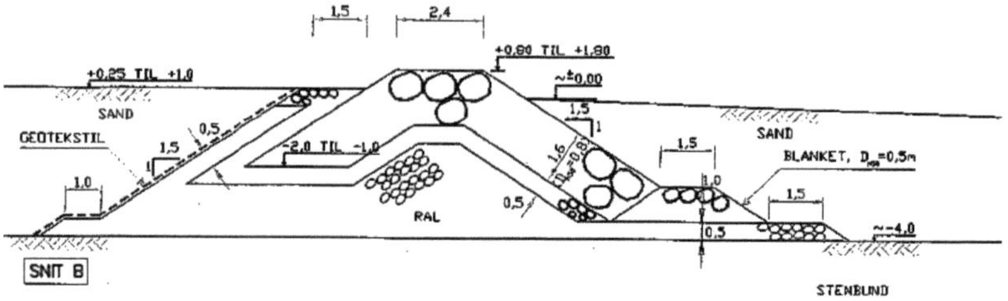

(d)

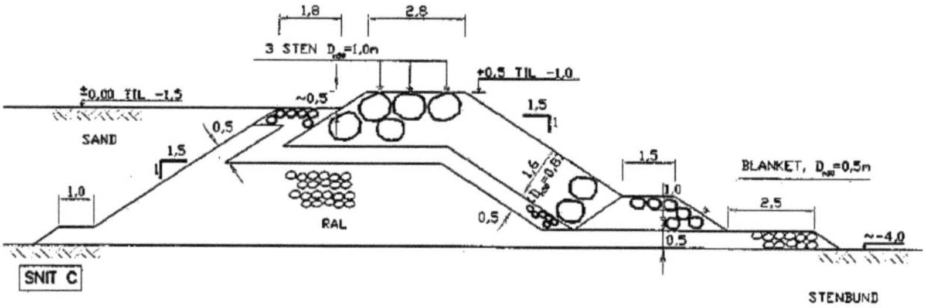

(e)

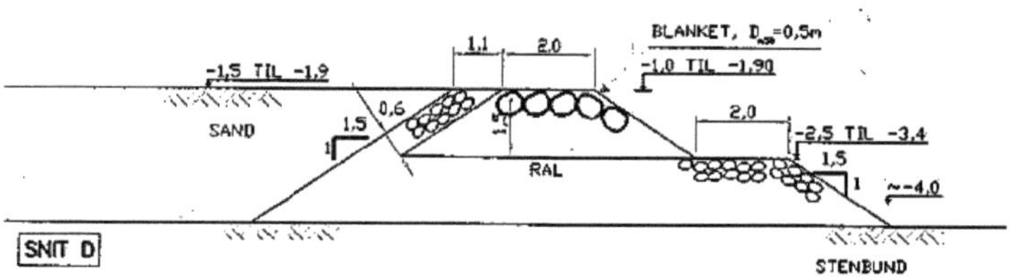

(f)

Figure 22.3. (*Continued*)

at +1.0 mCD MSL and –2.0 mCD, thereafter the slope will be steeper. A longitudinal section of the groyne is shown in (c), which also indicates the expected beach slopes, and a cross-section at groyne root is shown in (d). The required extent of the geotextile to contain the beach sand has been indicated in the longitudinal section and the cross-section. The geotextile will be used from the top down to level –3.20 mCD, and from the root of the groyne up to the point where the beach sand profile reaches the –3.20 mCD level. This level was chosen after considering two criteria, first: providing a cover layer of the quarry run to the geotextile with a thickness of at least 1.5 m, and secondly: the infiltrated sand from the beach, which is assumed to attain a 1:7 slope, will not intersect with the

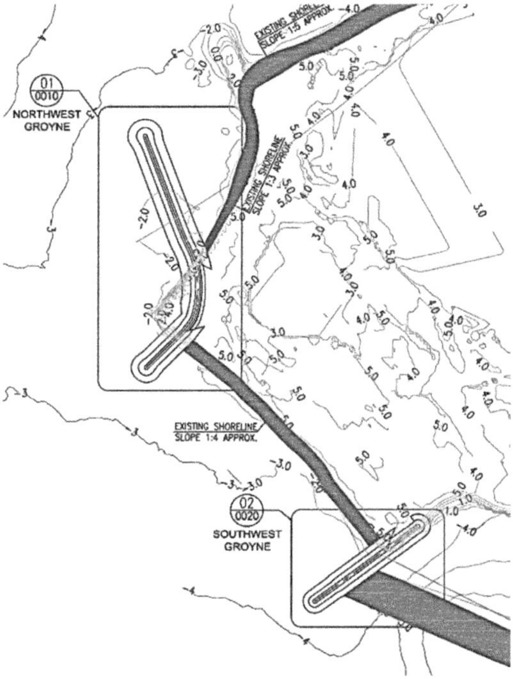

(a)

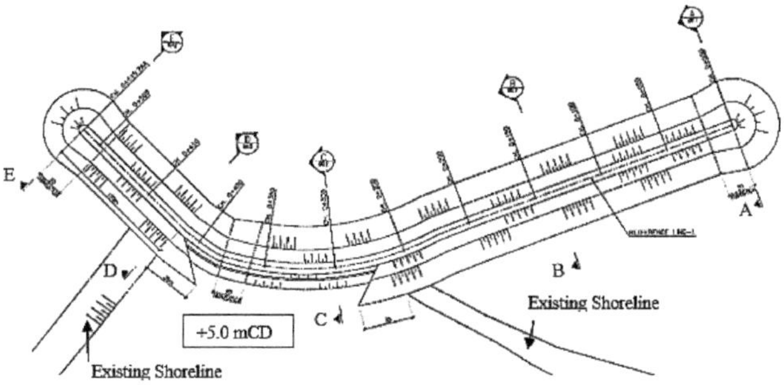

(b)

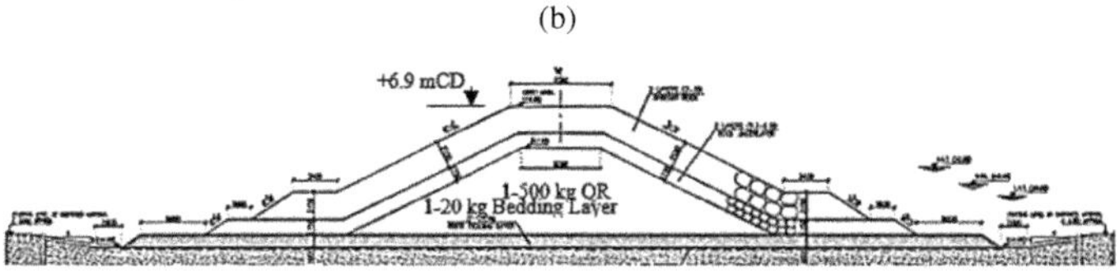

(c)

Figure 22.4. Example configuration of groynes and headlands combination: (a) General layout; (b) Plan of the Northwest Groyne; (c) Cross-section A–A; (d) Cross-section C–C; (e) Plan of the Southwest Groyne; (f) Cross-section F; (g) Cross-section G.

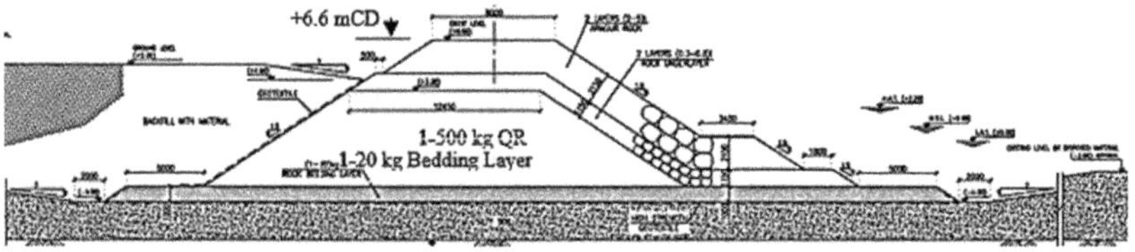

(d)

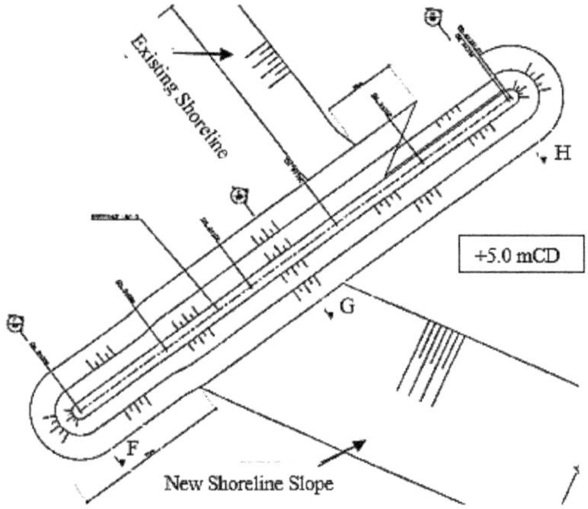

(e)

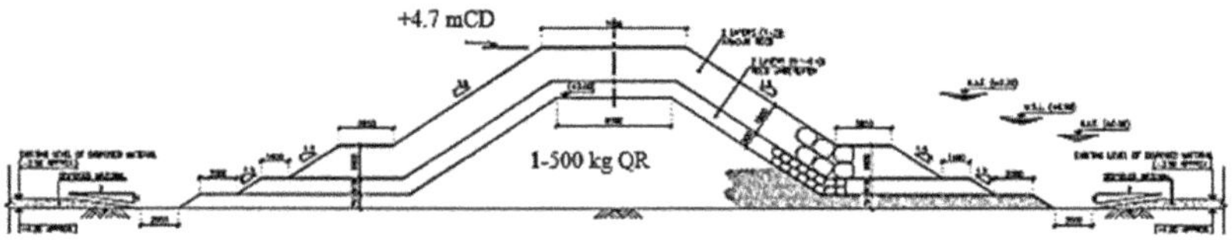

(f)

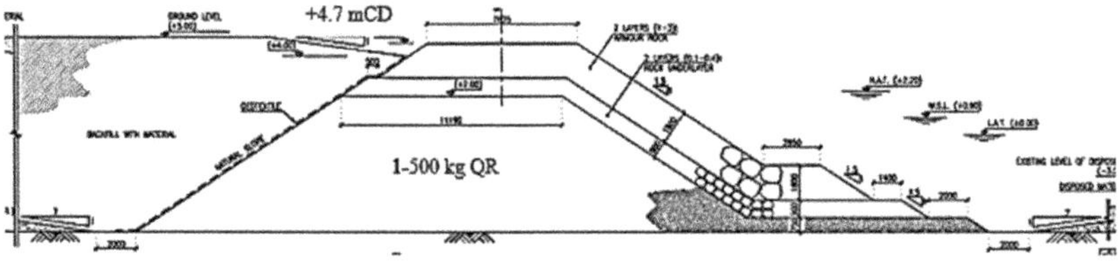

(g)

Figure 22.4. (*Continued*)

groyne side slope (see also Figure 13.4 for further clarification). It is however to note that although the design of beaches relies heavily on the numerical modelling results in establishing the profile and the shape of the beach, it is recommended not to completely rely on these results and

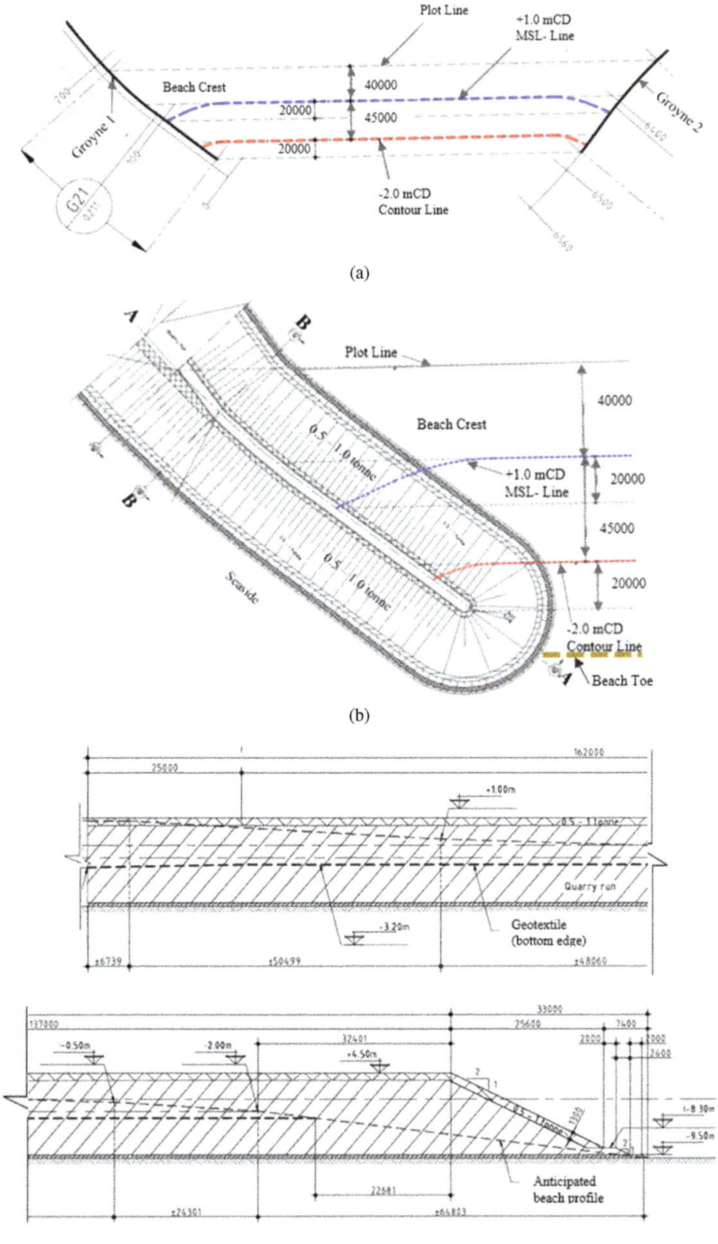

Figure 22.5. Groyne details of an artificial pocket beach: (a) General layout; (b) Groyne-1 plan; (c) Longitudinal section A-A; (d) Cross-section B-B.

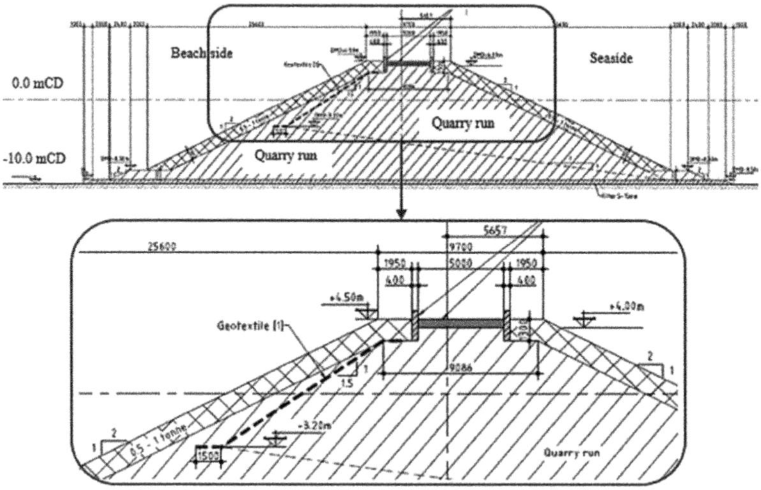

(d)

Figure 22.5. (*Continued*)

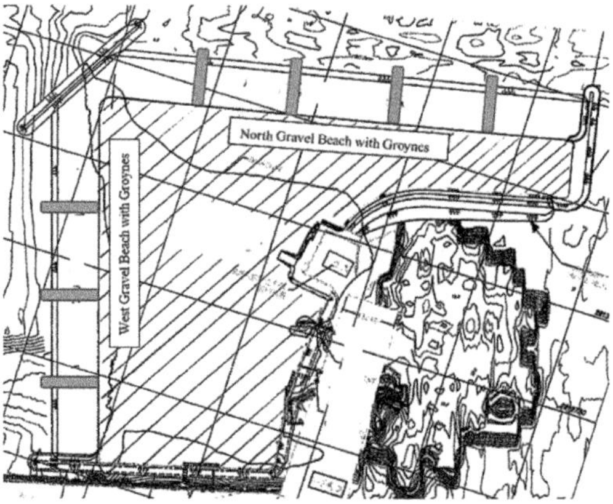

(a)

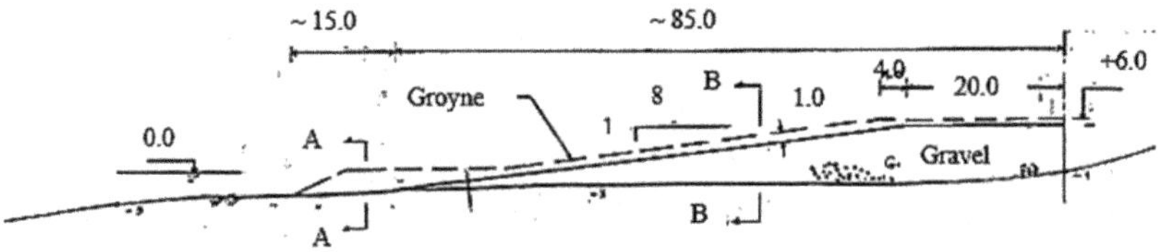

(b)

Figure 22.6. Groyne system in a gravel beach: (a) General Layout; (b) Longitudinal section in the beach groyne; (c) Cross-section A-A; (d) Cross-section B-B.

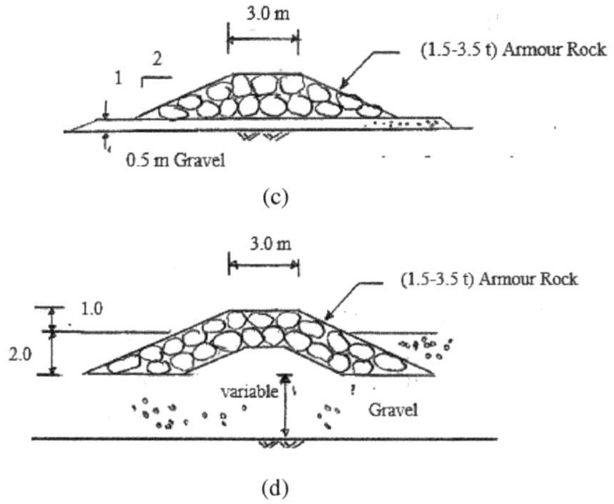

Figure 22.6. (*Continued*)

design with some margin of safety. The beach toe in this example is made to be at the extreme edge of the groyne head. It would have been a better design if the groyne is made little longer, say by 3 to 4 m to safeguard against possible sand loss.

Groynes are not only used with sand beaches but are also used with gravel or shingle beaches, as illustrated in Figure 22.6. A general layout is shown in (a), where the extreme edges of the two gravel beaches are contained by primary groynes/breakwaters and the corner by a groyne-headland/breakwater. Secondary typical groynes are used in the North and West gravel beaches in between the primary groynes to stop littoral drift of the gravel. A longitudinal section in the groyne is shown in (b), and two cross-sections in (c and d). The top level of the groyne is at 1.0 m above the gravel beach.

22.3. Detached/Nearshore Breakwaters

This type of a shore parallel structure is described as nearshore, offshore or a detached breakwater, which provides shoreline protection by intercepting incoming waves and creating stable pocket beaches in between. The crest level normally ranges between 1.0 m above the MSL (i.e. emerged crest) down to MLWS (i.e. submerged crest). The crest could be narrow or wide depending on wave characteristics, required degree of protection, crest level and few other factors. The configuration of the detached breakwater system could be of a single or segmented breakwaters with gaps in between. The offshore distance of these breakwaters will play an important part in the functional performance of these breakwaters. The shoreline will form a series of salients behind the breakwaters, their shape and extent depend on the gap length between the breakwaters and the distance from the shoreline. If the salient reaches the detached breakwater, then it is called a tombolo.

The detached breakwaters could be a traditional rubble mound structure or a simple pile of large rocks. It is however normal in most designs that the same armour block size is used on both slopes and the crest. For larger structures, the construction will be carried out by marine equipment, whereas smaller structures could sometimes be constructed by land-based equipment from the shoreline. The designer, together with the construction team, needs to agree on an effective and relatively easy to construct design. Few examples of illustrative designs are presented in the following to demonstrate some of the design concepts.

In Figure 22.7, a barely submerged, narrow crest detached breakwater is shown. The armour is of a single layer of heavy rocks (8.0–14.0 t), whereas the core is made of fairly large rocks (1.0–3.0 t), which makes the structure very porous. However, being porous could be tolerated for this type of structure as its main function is to reduce the wave height and deflect the currents.

A different design concept for a porous breakwater is shown in Figure 22.8, where the same rocks are used as armour and core. The coarse core is not placed directly on the bedding layer but separated by a layer with smaller rocks. Two different side slopes are used in this breakwater, 1:2 and 1:3, with the latter towards the seaside.

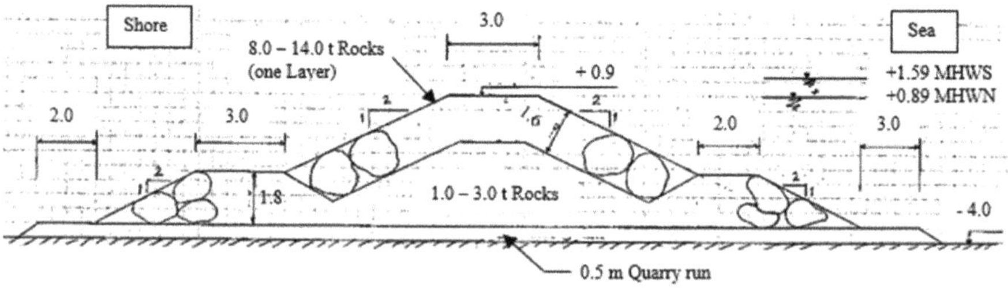

Figure 22.7. A detached breakwater with heavy single-layer rock armour.

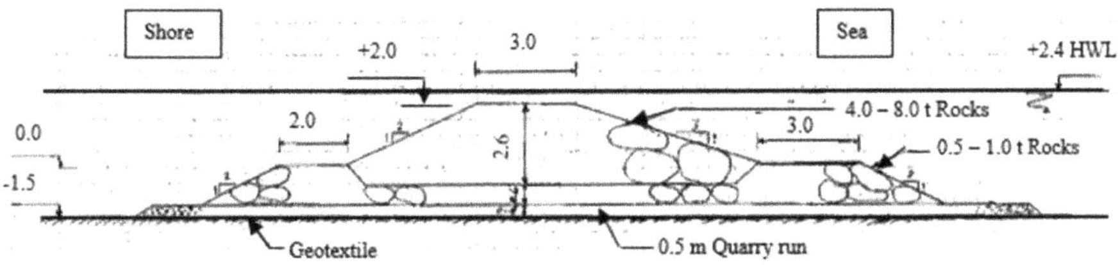

Figure 22.8. A detached breakwater made mainly of armour rocks.

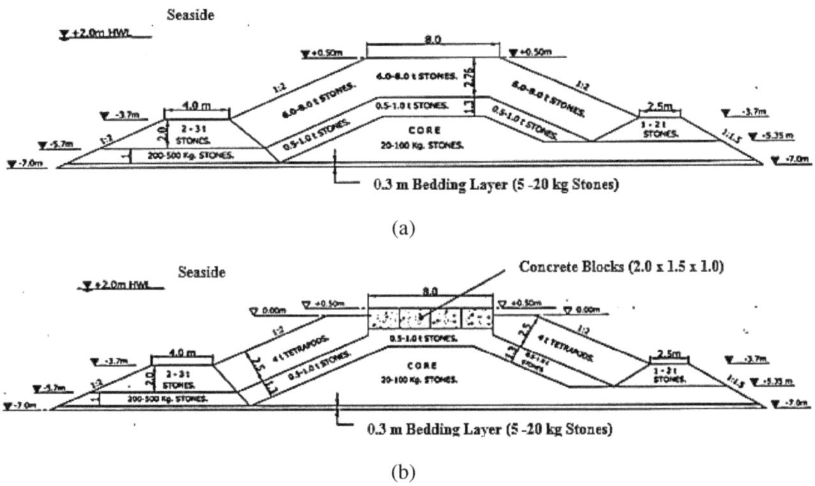

Figure 22.9. Two alternatives for a wide crest, submerged detached breakwater (Mahalingaiaha *et al.*, 2015): (a) Traditional rubble mound structure; (b) Rubble mound with CAU on the slopes and concrete blocks on the crest.

Two alternative designs are shown in Figure 22.9 for a wide crest detached breakwater. The two alternatives are designed as traditional rubble mound structures with armour layer, underlayer and a core, the difference between the two is the armour. In (a), 6.0–8.0 t rock armour is used for the two slopes and the crest, whereas in (b) 4.0 t, Tetrapods are used on the slopes and 7.5 t concrete blocks on the crest.

In Figure 22.10, V-shape detached breakwaters are shown. The breakwaters are actually designed to be connected to the shoreline by a tombola as shown in (a). A cross-section in the breakwater is shown in (b), the cross-section is after the tombola has been fully developed and protected against wave overtopping by a layer of shingle. The breakwater is to be

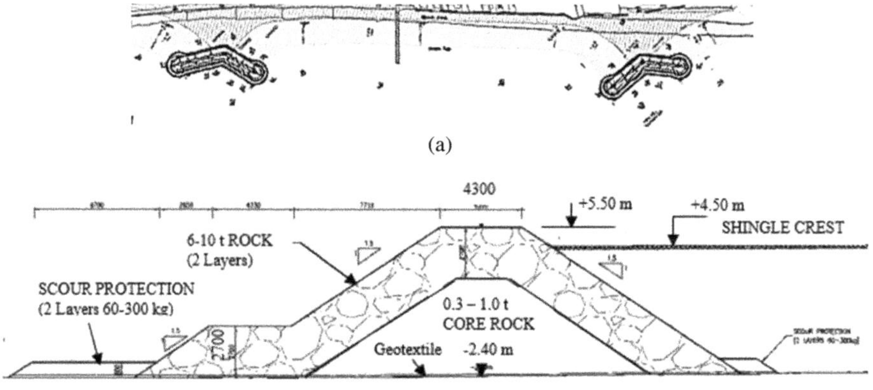

Figure 22.10. Detached breakwaters with tombola (Wareing *et al.*, 2013): (a) Plan; (b) Cross-section in the detached breakwater.

designed as fully detached so that the seaside and lee side slopes are stable under design condition as the tombola will take time to fully develop and reach the breakwater. It is, however, important to note that there should have been a geotextile or a filter layer on the lee side to prevent loss of sand (see Figures 22.4(d, g)).

22.4. Artificial Beaches

During storms, a wide beach can perform as an effective energy absorber with wave energy dissipated across the surf zone and the wide beach rather than impacting the upland structures. However, creating new beaches is primarily not for the purpose of coastal protection, but as an amenity feature and for recreational activities. Designing artificial beaches aim to effectively creating a beach which will increase the level of user comfort and coastal aesthetics in areas where there is no natural beach formation or in coastal areas of lower quality.

Apart from the beach terminal structures, a submerged breakwater/sill is quite often used to retain the beach in a perched position. This will lead to a significant reduction of sand filling volume, and it could offer some control over possible offshore sand losses.

The beach shall be designed and constructed, in plan and profile, to be as stable as possible. The orientation of a stable beach is to be perpendicular to the prevailing waves, with the necessary terminal and/or protection structures, a close to zero net littoral transport could be achieved with no longshore sand loss. Whereas to have a cross-shore stabile beach requires good quality sand with the right grain size and gradation, and to be constructed close to the expected equilibrium slope.

The recommended beach sand characteristics could be summarised as follows (Mangor, 2007):

- $0.25 < d_{50} < 0.50$ mm
- Well sorted, $u < 2.0$ ($u = d_{60}/d_{10}$)
- No fines, silt content < 2%
- No large stones, gravel and shell content < 3%
- No organic matter

It is, however, important to note that only the top layer, and not all the sand fill material used to construct the beach, needs to satisfy the above criteria. It is recommended to use a minimum 1.0 m thickness for this top layer of the beach sand for sheltered and mildly exposed beaches. The layer thickness is to be increased for beaches exposed to moderate or high waves, however, a thickness more than 2.0 m is not normally used.

Typical sand and gravel beaches in shallow locations are shown in Figure 22.11. The existing seabed level is at –2.5 m as shown in (a), and the dredged level for the boat harbour basin and the approach channel is –4.0 m, with the beaches crest level at +2.5 m. The sand beach, as shown in (c), is for recreational use, whereas the gravel beach, as shown in (b), is for shore protection under the gangway and floating dock, similar to Figure 18.15. The sand beach and the boat harbour basin are contained between two terminal groynes. No perched beach arrangement, i.e. no sill, is required in such a shallow location.

A different type of an artificial sand beach is shown in Figure 22.12, nearshore detached breakwaters are used instead of the groynes to protect the beach and preserve its profile and plan in a stable condition. This arrangement for an artificial beach is better used in shallow locations, as is the case with seabed level at –1.8 m, as shown in (a). A cross-section through the emerged detached breakwater and the mid-beach front is shown in (b) to define the required shortest beach crest. Note that the beach toe rests on the breakwater without a filter layer or geotextile as this is not required because the tip of the sand slope does not protrude outside the breakwater (as indicated by the broken line with a slope of 1:7),

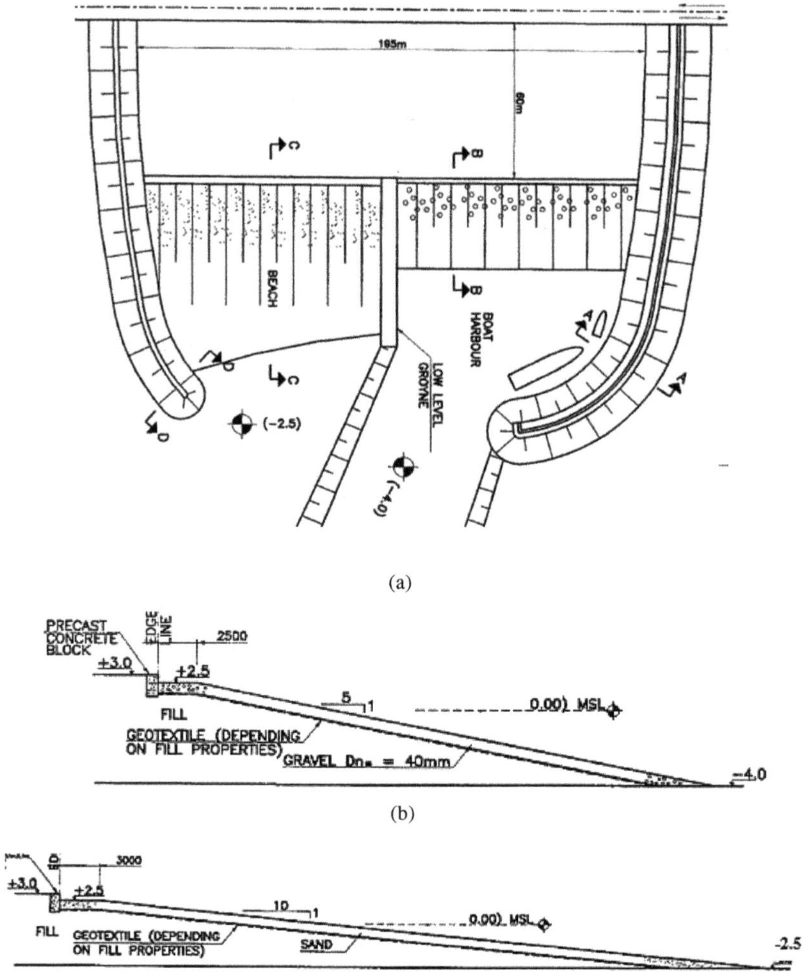

Figure 22.11. A combined artificial gravel and sand beach: (a) Plan; (b) Gravel beach — cross-section B–B; (c) Sand beach — cross-section C–C.

as such no sand will be washed away. However, using a filter layer/ geotextile is always recommended to prevent the minimal sand loss sucked out through the porous rock material.

As the beach height, from the seabed to the beach crest, gets larger, it means more sand will be required, then a perched beach concept will be

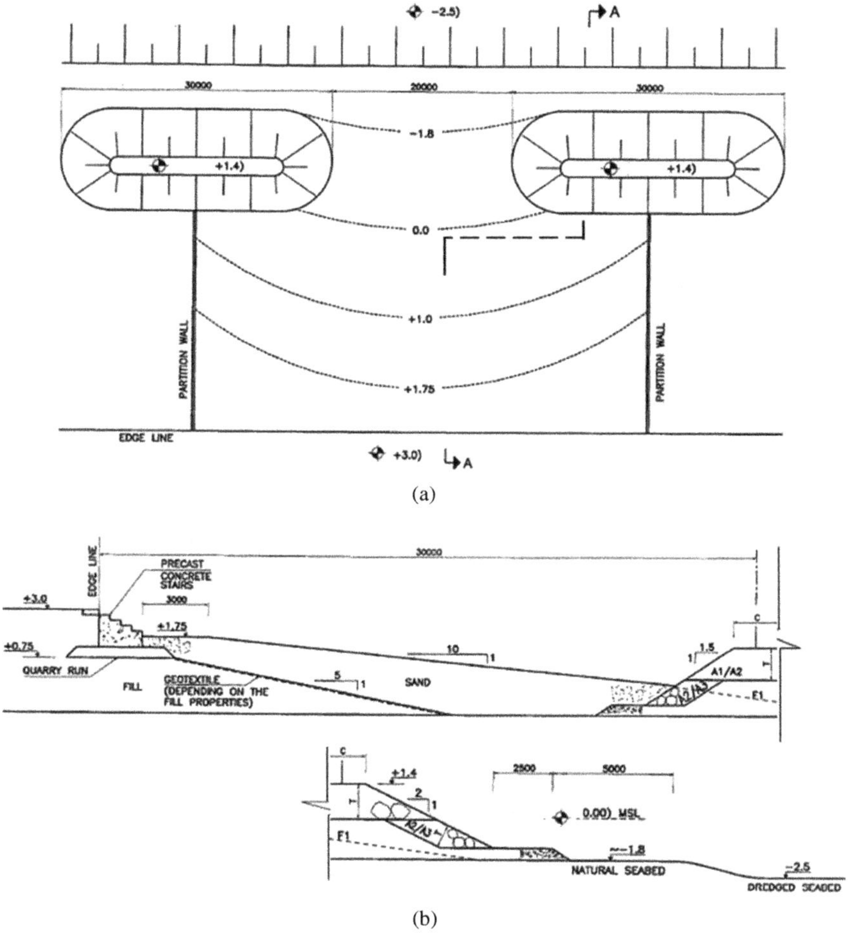

Figure 22.12. Artificial sand beach behind nearshore emerged breakwaters: (a) Plan; (b) Cross-section A-A.

the recommended approach. A sill shall be used to retain the beach, the height and the crest width/configuration of the sill are two important factors to be selected. Both of these factors contribute to the reduction of the impinging wave heights and improve the stability of the beach behind. There is no recommended value for either of the two (i.e. height and width of the sill), but some general guidelines are available and as per the designer's preference. The sill crest level normally ranges from just below the MLWS (so it will not be seen by the swimmers at low water) to just below MSL (to be more effective in protecting the beach during storms). The sill crest width is normally to be wider as the sill gets lower to retain some of its function in reducing the impinging wave height. However, the low crest is quite often combined with an apron on the shore side covering and protecting the beach toe from the increased risk of being washed out (Musumeci *et al.*, 2012).

A typical example for a perched beach with rubble mound sill is shown in Figure 22.13. The sill crest level is just below the LAT, and a geotextile is used at the interface between the sill and the beach fill material. At the top of the interface an additional granular filter is used to secure the 1.0 m top layer of the beach sand, ensuring no loss of sand through the rubble mound, and to provide a short apron.

A combined perched beach with a rubble mound sill is shown in Figure 22.14, where beach sand is used in the lower part and a gravel slope covered by sand beach layer in the upper part. The fill material which forms the embankment under the beach crest is covered by a 0.70 m layer of gravel with a 1:5 slope, separated by a geotextile. Beach sand is used to make up the rest of the beach and covers the gravel slope by a 0.50 m thick layer. This arrangement will secure the embankment with its 1:5 slope not to be any steeper and provide a stable slope during the construction stage. The thin layer of beach sand is however likely to be displaced due to wave action and will require regular maintenance and replenishment.

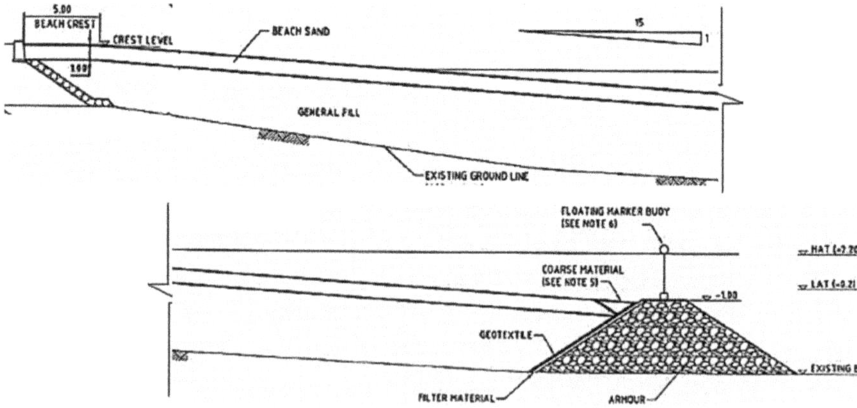

Figure 22.13. Typical perched beach with rubble mound sill.

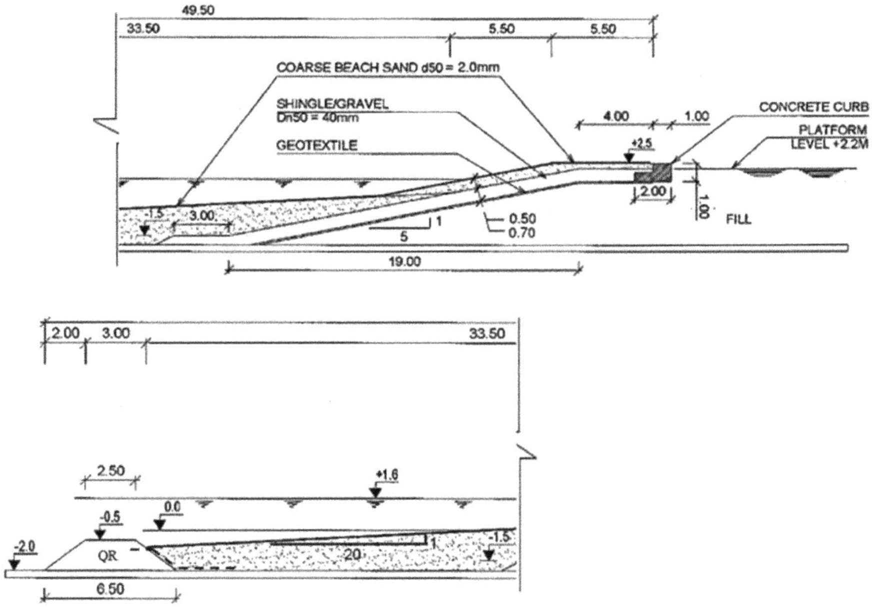

Figure 22.14. Perched beach with a small rubble mound sill and a gravel underlayer.

The same concept of using a combined perched beach is used in Figure 22.15, however, the embankment under the beach crest is of quarry run and the sill is a concrete block.

A detailed example for a perched beach in a sheltered location is given in Figure 22.16, the plan is shown in (a) and a cross-section at the middle part of the beach in (b). The beach is confined between two groynes joined across the beach by a sill of quarry run. Sand fill is used for the bulk of the material, whereas the top layer of 1.0 m thickness is beach sand. A geotextile is used at the interface between the sand fill and the sill. A geotextile is also used to isolate the sand fill from the sill.

Another detailed example is given in Figure 22.17 for a perched beach exposed to moderate waves. The plan is shown in (a) and a cross-section at the middle part of the beach in (b) shows the sill. For clarity, the cross-section is shown in an enlarged scale in (c). Granular filter is used in this example at the interface between the sand fill and the sill, protected by a layer of quarry run.

The third detailed example is given in Figure 22.18 for a perched beach at an exposed location with high waves. The plan is shown in

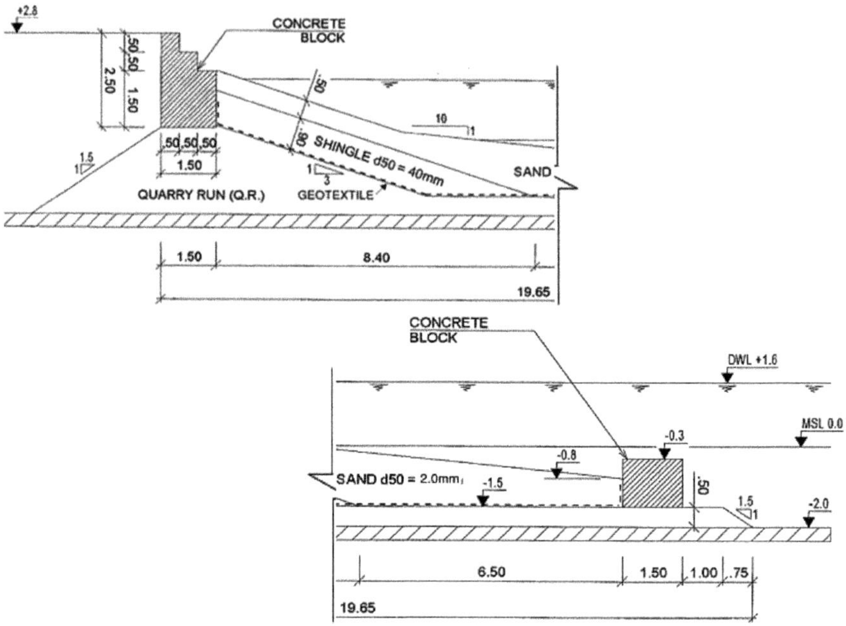

Figure 22.15. Perched beach with a concrete block sill and a gravel underlayer.

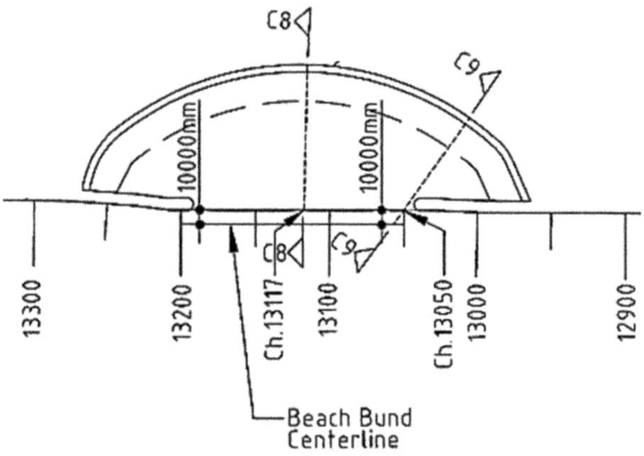

(a)

Figure 22.16. Perched beach example in a sheltered location exposed to small waves: (a) Plan; (b) Cross-section C8-C8.

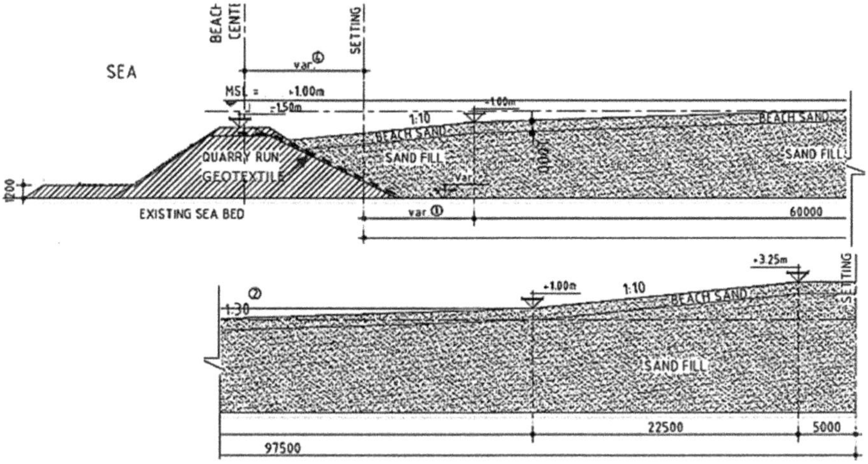

(b)

Figure 22.16. (*Continued*)

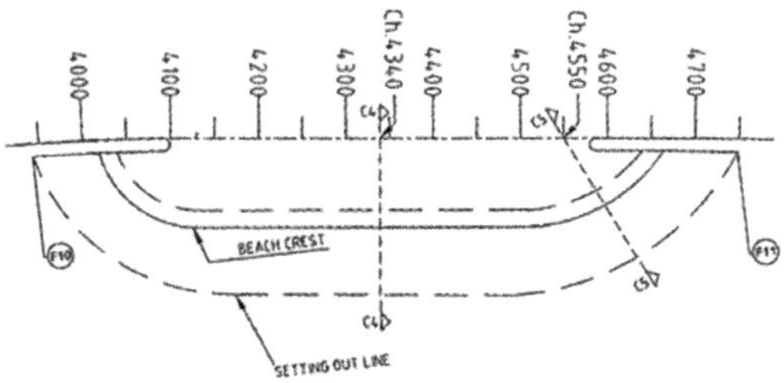

(a)

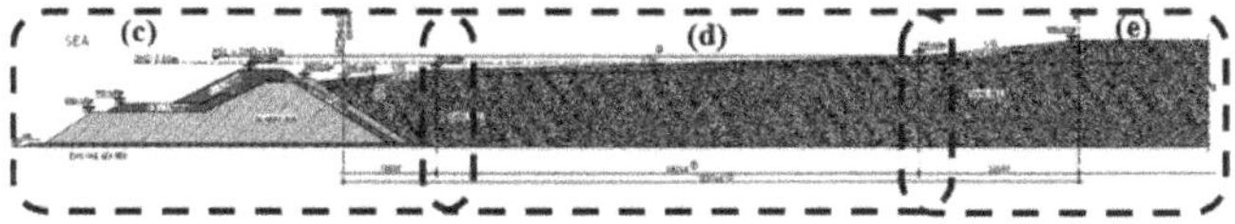

(b)

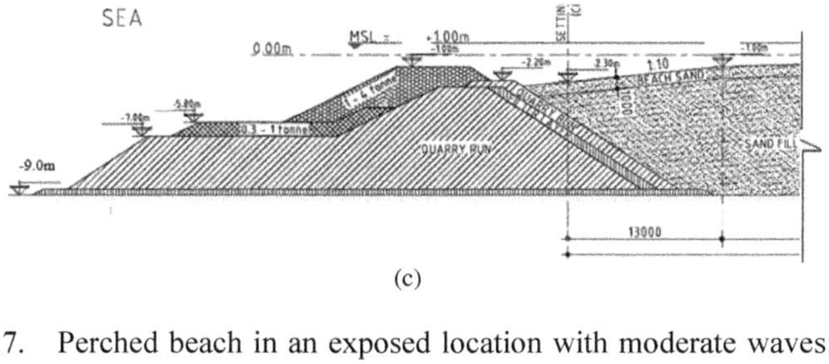

(c)

Figure 22.17. Perched beach in an exposed location with moderate waves: (a) General layout; (b) Cross-section C4-C4 through sill and beach; (c) Enlarged view of the seaside part of section C4-C4; (d) Enlarged view of the middle part of section C4-42; (e) Enlarged view of the beach crest in cross-section C4-C4.

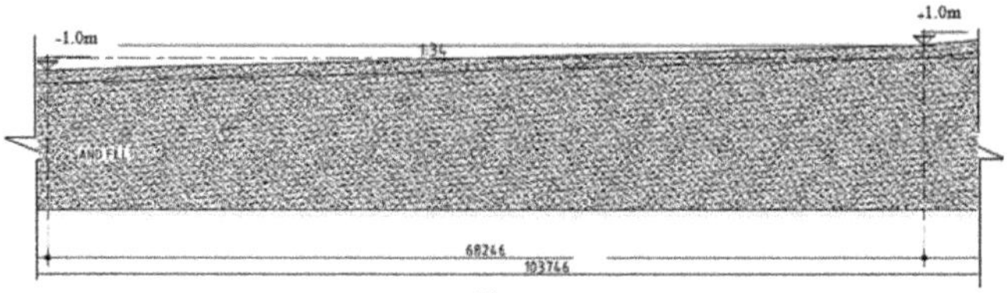

(d)

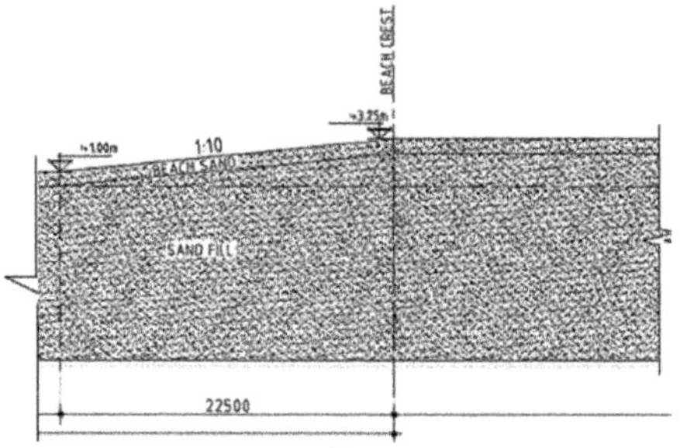

(e)

Figure 22.17. (*Continued*)

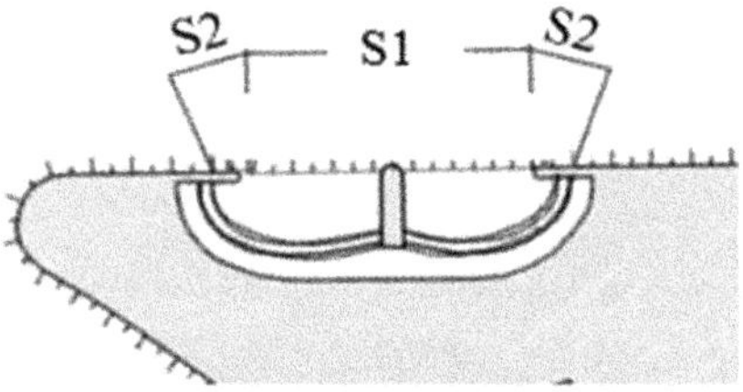

(a)

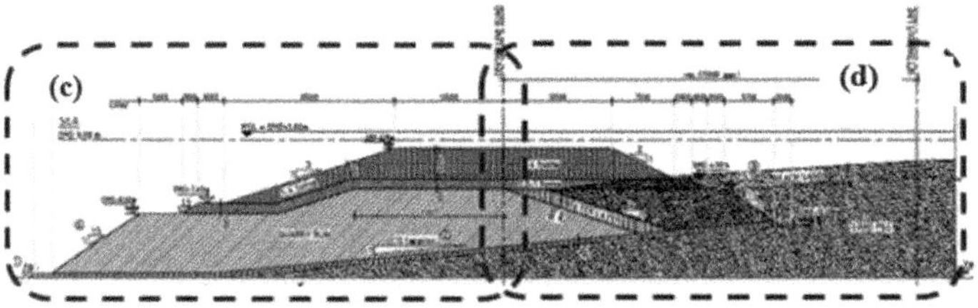

(b)

Figure 22.18. Perched beach in an exposed location with high waves: (a) General layout; (b) S1-sill section; (c) Enlarged view of the seaside part of the S1-sill cross-section; (d) Enlarged view of the shore side part of the S1-sill cross-section; (e) S2-beach groyne cross-section; (f) Enlarged view of the seaside part of the S2-beach groyne cross-section; (g) Enlarged view of the shore side part of the S2-beach groyne cross-section.

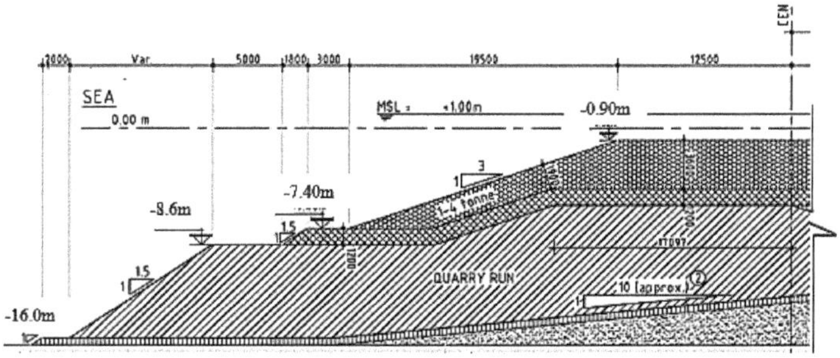

(c)

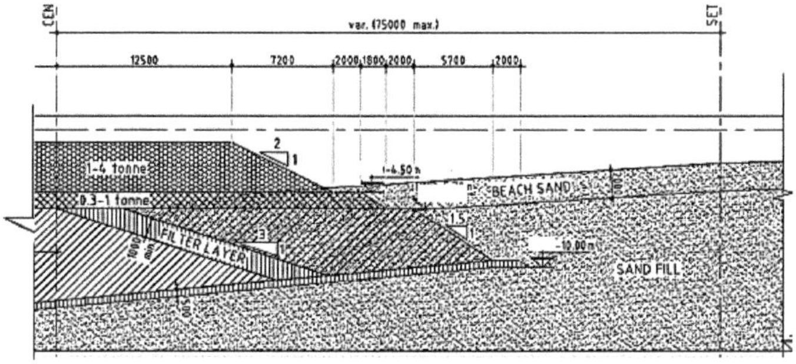

(d)

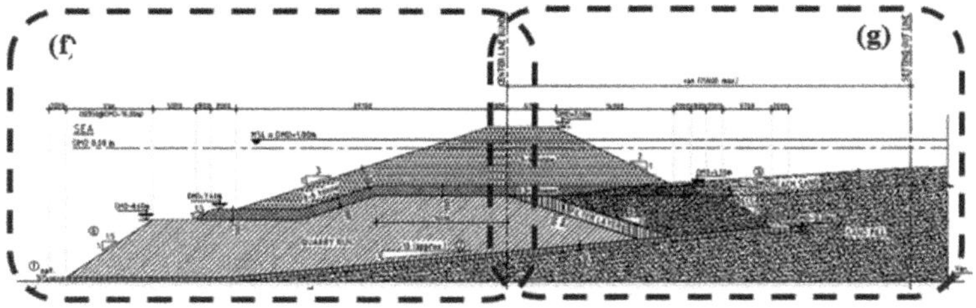

(e)

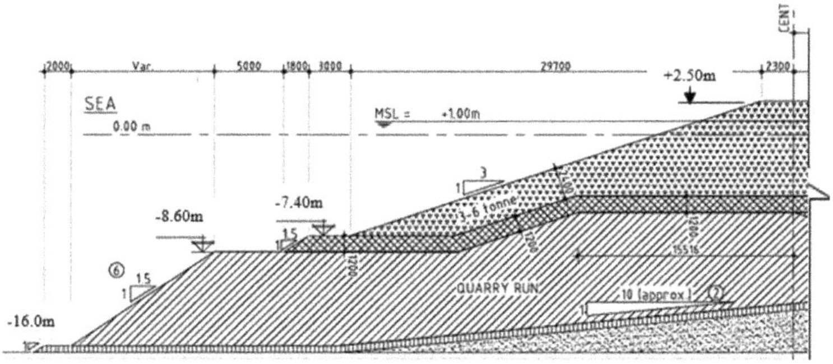

(f)

Figure 22.18. (*Continued*)

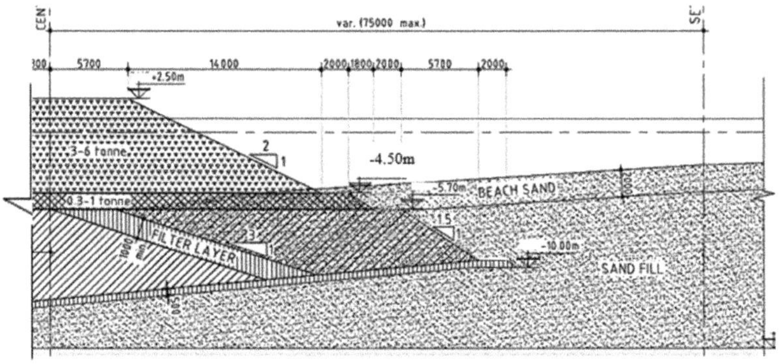

(g)

Figure 22.18. (*Continued*)

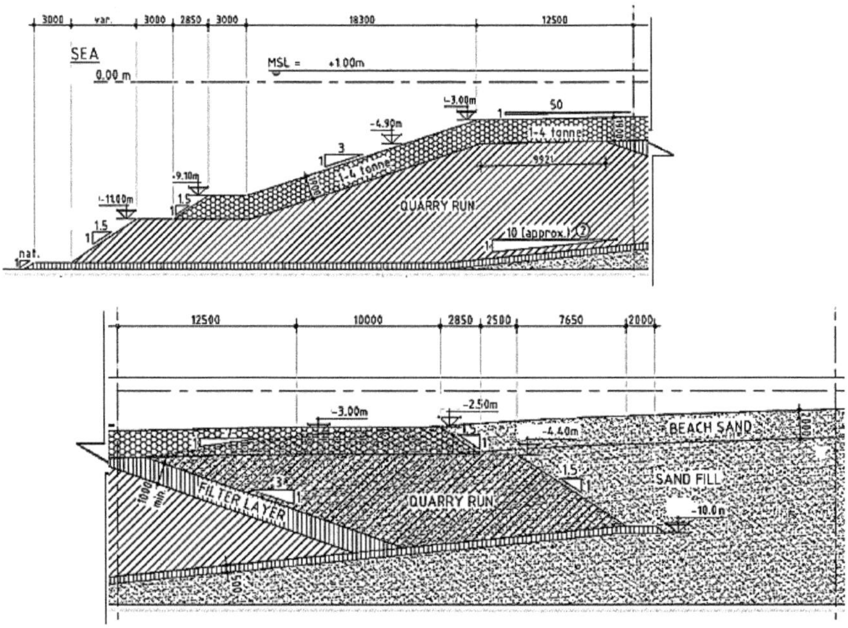

Figure 22.19. Perched beach sill with an apron.

(a) for the double pocket beaches and a cross-section for the middle part of one of the beach pockets in (b) showing the sill. Enlarged scale details for the sill are shown in (c and d), and for the side groynes, in (e). Note that the sill is just below the MLWS at –0.90 m, with a crest width of

25.0 m, and a 5.80 m apron just after the crest securing the beach toe. In the same project and at the same location, another perched beach is designed with a different concept, and that is by using an even lower sill (from –0.9 m to –3.0 m), as shown in Figure 22.19. In order to compensate for the height, an extended apron is used, adding extra 10.0 m to the total width of the crest and the apron.

Chapter 23

Dredging Near Marine Structures

23.1. Introduction

This chapter is not dedicated to dealing with the dredging requirements to accommodate a certain ship size for safe navigation and manoeuvring. However, it presents the interface and interaction between the dredged channels and basins with the port structures, in particular, rubble mound structures. This interface should be carefully evaluated in the design of various port, marina or any other marine facility to ensure the safety of the structure (breakwaters, revetments, wharves, ...etc.).

23.2. Capital, Maintenance and Upgrading Dredging Works

Capital dredging is where the natural seabed is dredged as part of a new project to create a new port basin and/or an access channel. The needed depth of water, width of channel and other possible basin dimensions to accommodate the design vessel will determine the new bed level and the extent of the dredging. Thereafter, and based on the dredging layout and depths, the layout of other structures could be finalised.

Maintenance dredging will normally be required to dredge the natural accumulation of sand and silt that are carried/moved by waves and currents. Maintenance dredging is to maintain access channels, basins and berths at their original levels and dimensions. However, this type of operation requires strict control so that it will not deepen/enlarge the channel/basin more than its original design levels and dimensions. In case

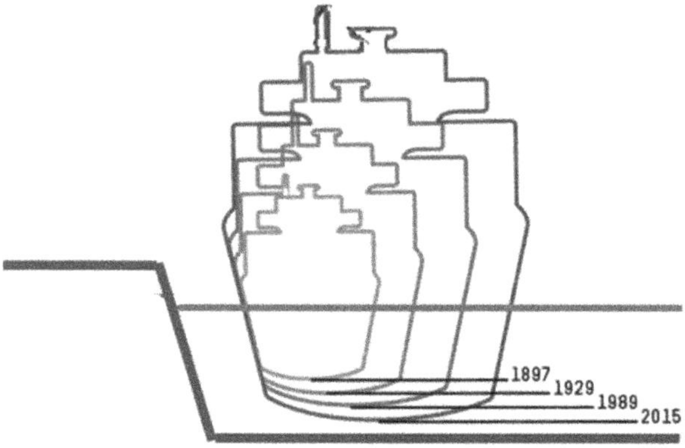

Figure 23.1. Schematic illustration to the continuous increase in vessel draft.

of over-dredging, i.e. exceeding the design levels/dimensions is to take place, then the nearby structures could get badly affected and destabilised.

Ports across the world are continually dredging not just as maintenance dredging but also often to enlarge and deepen access levels to maintain their commercial operations and to have a competitive advantage by welcoming larger vessels. This trend of the increase in the vessel draft over time and the consequent deepening of the navigational channels is schematically illustrated in Figure 23.1. Deepening the channel and the port basin will definitely affect many structures as the seabed level will be lower than what has been considered in the original design of these structures. There is no remedial action to be considered for rubble mound structures apart from respecting the dredging clearance distance from the structures as per the original design. For port quay structures, whether open piled or solid wall type, there are various strengthening/modification methods which could be adopted, but frequently with considerable cost. An elaborate work was carried out by Douairi and De Gijt (2013) and it is presented here as an example to demonstrate the complexity of the problem by different possible concepts for upgrading one type of quay walls, namely the sheet pile quays, as shown in Figure 23.2. Similar or different concepts will be required for different quay structures.

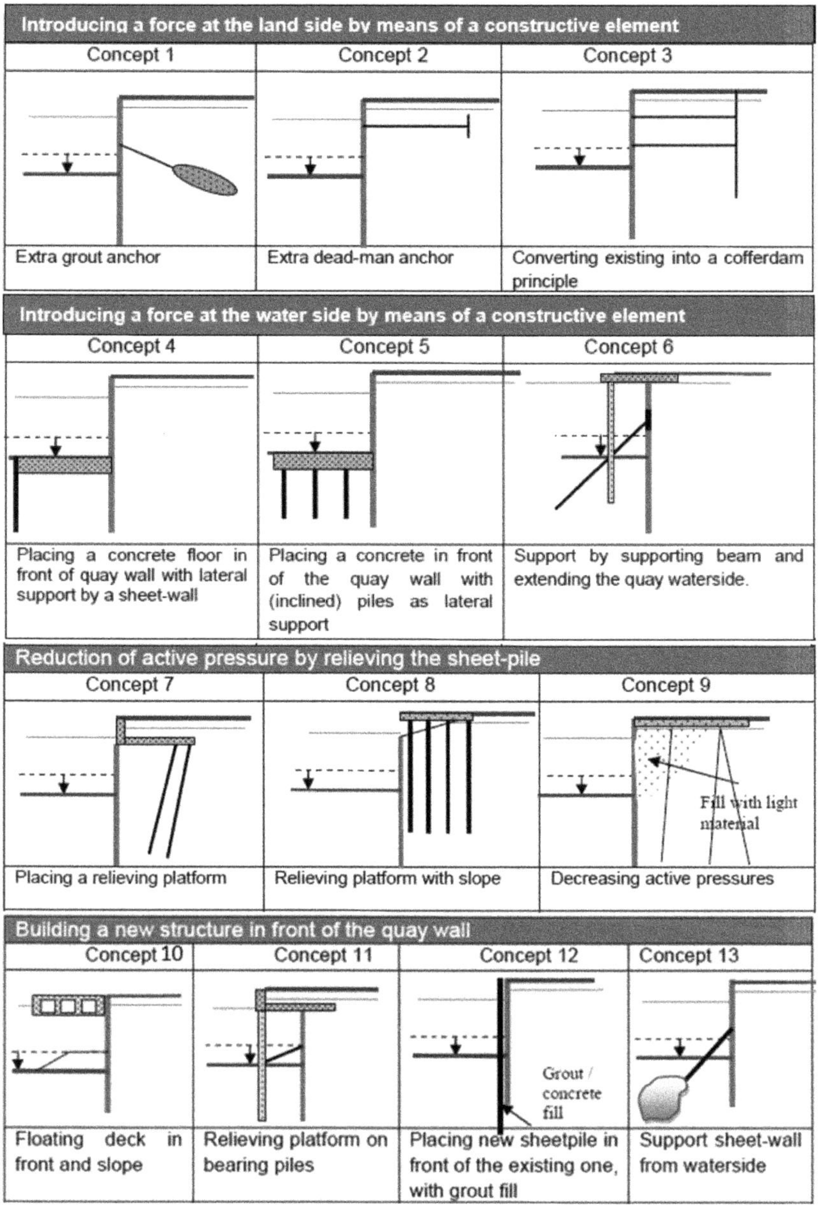

Figure 23.2. Schematic concepts for upgrading techniques for sheet pile quays (Douairi and de Gijt, 2013).

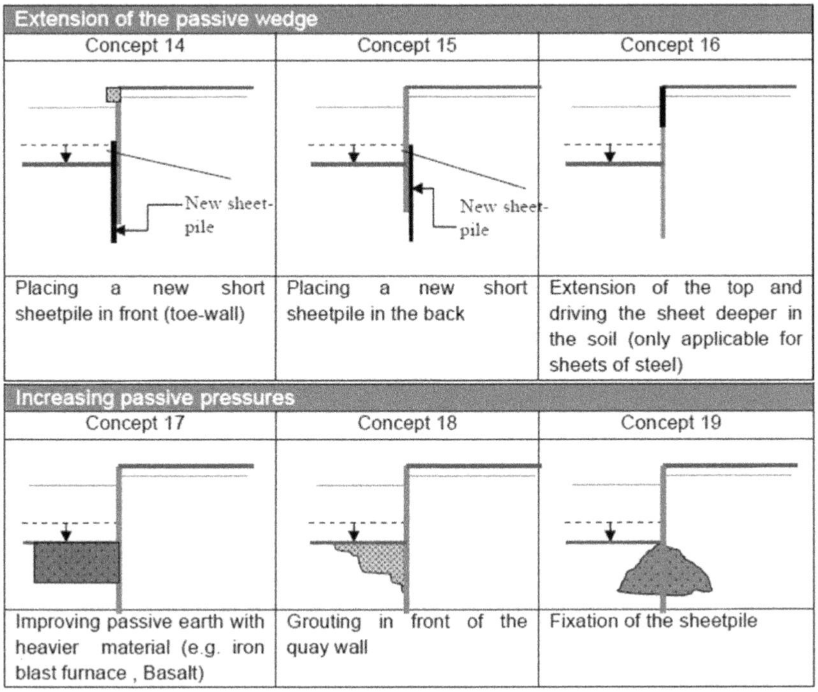

Figure 23.2. (*Continued*)

23.3. Dredging Levels

The design dredging level, width and limits for harbour approach channels shall be carefully calculated based on different factors, and this will also decide the port entrance channel dimensions. There are slightly different calculation methods, however, the commonly used are the PIANC (2014), the BS 6349-1 (2013) and the CEM-Part V (2008). The dredging depth and extent for the port basins and berthing areas are designed separately and under different factors than those for the approach channels. These are to include the vessels manoeuvring requirement, port operation guidelines and berths design criteria.

Figure 23.3 is given to illustrate the various terms associated with the channel depth, noting that all levels should be referenced to the same datum, which is normally the Lowest Astronomical Tide (LAT). What is important from this sketch, which needs to be identified in the design of the rubble mound structures, is what bed level is to be shown on the

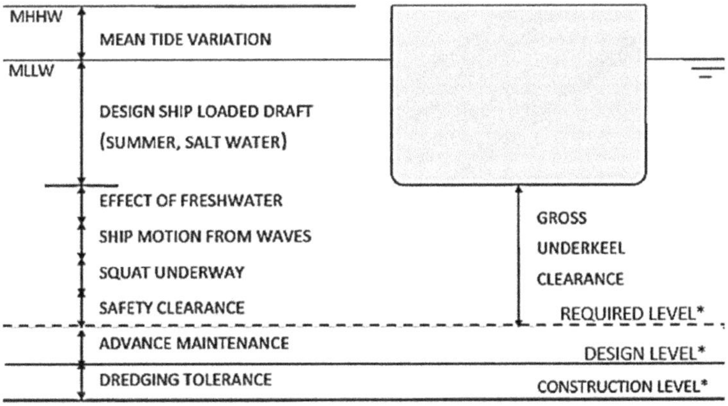

Figure 23.3. Channel depth factors (CEM – Part V, 2008).
(* added by the author)

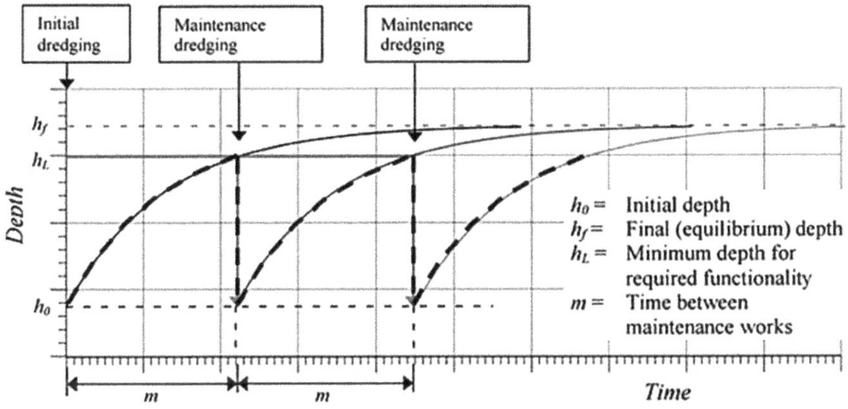

Figure 23.4. Maintenance dredging cycles (Silveira, 2017).

design drawings as the "design level", and what bed level is actually considered in evaluating the stability of the structure. The design level, as shown in the sketch, shall include all items required for safe navigation, including the advance maintenance dredging depth as explained in Figure 23.4. However, the adjacent rubble mound structures should be designed considering the "construction bed level", which is usually specified in the project technical specifications and not on the design drawings.

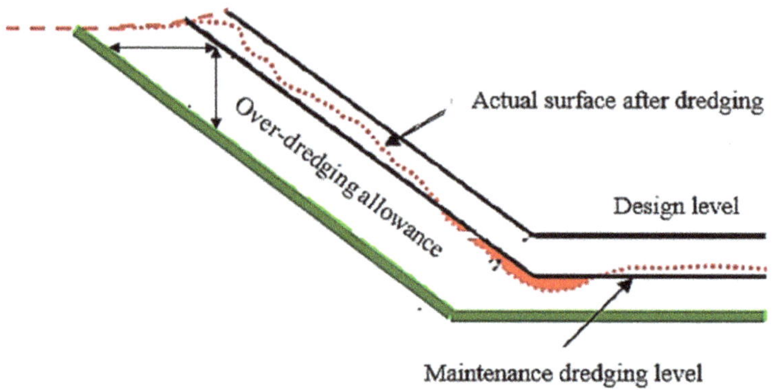

Figure 23.5. Illustrative sketch for the allowable over-dredging limits.

Dredging accuracies vary with the type of the used dredging equipment (mechanical or hydraulic), its interaction with the site-specific hydrodynamic conditions (tides, currents and waves) and the type and thickness of sediment or rock to be dredged. Because of these variables and the resulting dredging inaccuracies, dredging contractors normally slightly increase the depth and width of the dredging design limits. This is known as "allowable over-dredging", as illustrated in Figure 23.5. The stipulated allowable over-dredging limits in the project technical specifications should also be considered in the design, of course, with some additional safety margin.

23.4. Dredging Clearance from Rubble Mound Structures

There are no guidelines for the dredging clearance from the rubble mound structures as this clearance is project-specific. The required clearance depends on many factors, which could be summarized as follows: the depth of the dredging, the seabed material, the type of the rubble mound (size, side slope, type of armour units, superstructure) and the width of the toe (including the apron/scour protection). However, a safe design clearance distance could be determined as per the sketch shown in Figure 23.6, noting that possible future deepening of the seabed to accommodate larger vessels is not considered in this sketch. However, if this is to take place, then the same side slope of the dredging should be maintained while

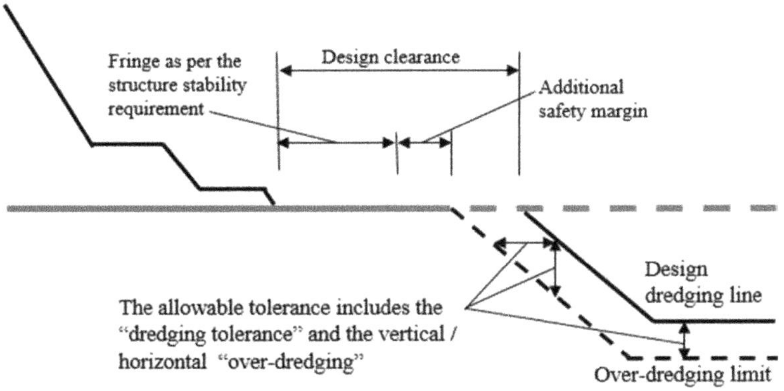

Figure 23.6. Dredging clearance from rubble mound structures.

deepening the seabed (the dredging side slope depends of course on the seabed material, it could be 1:1 for bed rock material or normally about 1:6 for loose sand material).

It is clear from Figure 23.6 that the design clearance together with the dredging side slope extends a considerable distance from the toe line, especially if the dredging is deep and in sandy material. Therefore, in locations where the designer cannot have such a wide dredging clearance, an alternative measure is normally adopted by bringing the dredging side slope inside the rubble mound structure. Such a situation is usually encountered in port entrances where providing a dredging clearance means a wider entrance opening up for more undesirable wave energy entering the port basin.

23.5. Design Examples

Different design examples are given in this section including plans and cross-sections to illustrate how to present the design requirements, including the dredging clearance, from the structures.

A simple clarification example is presented in Figure 23.7 illustrating the dredging plan inside a small temporary construction harbour that has only been constructed to serve the construction activities of a nearby large project. This example is only given to demonstrate the dredging clearance for rubble mound structures. A plan view is given in (a), which shows a wide harbour entrance and a partly dredged basin. A cross-section in the

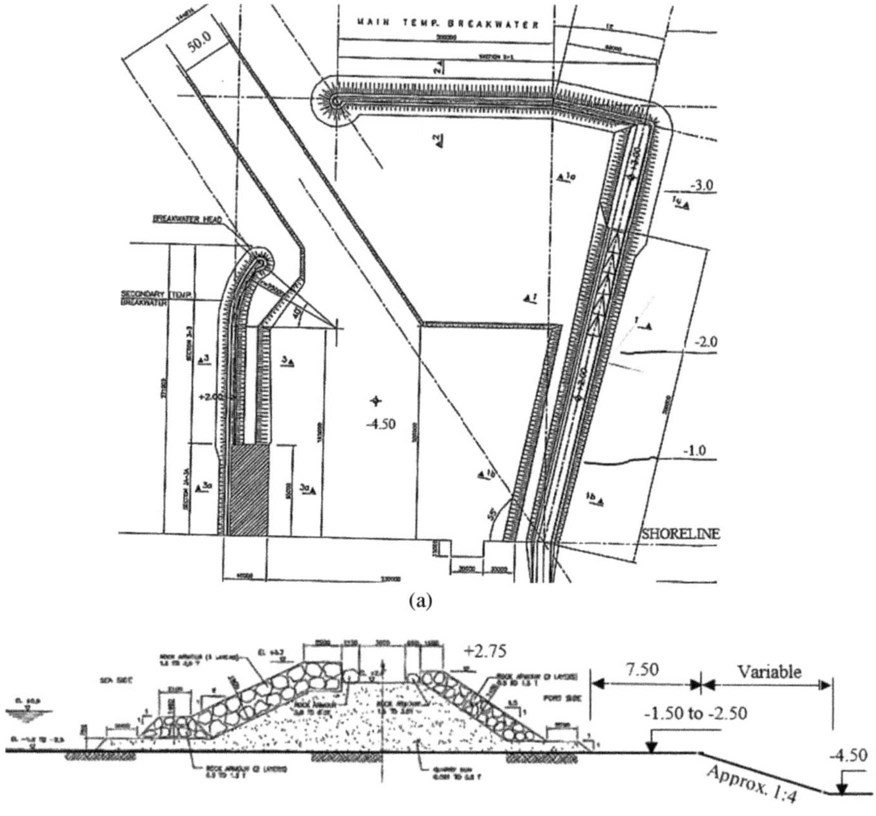

(a)

(b)

Figure 23.7. Temporary construction harbour with dredging plan: (a) Plan; (b) Cross-section 3-3.

short breakwater, C-S 3-3, is given in (b) showing the seaside and the harbour side of the breakwater.

In Figure 23.8, three different dredging patterns are given for port entrances. In (a), the roundhead is constructed on the natural seabed with the dredging carried out around the roundhead with the necessary clearance from the toe line. In (b), the required dredged level is provided at the toe line of the roundhead with the dredging side slope being inside the breakwater structure. In a narrow marina entrance as in (c), the entrance has to be fully dredged to the required design level, i.e. similar to (b). However, a wider channel leading to the entrance is a preferred practice

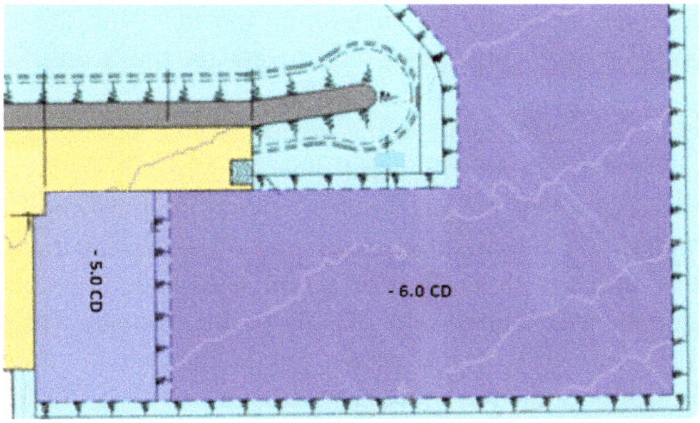

(a)

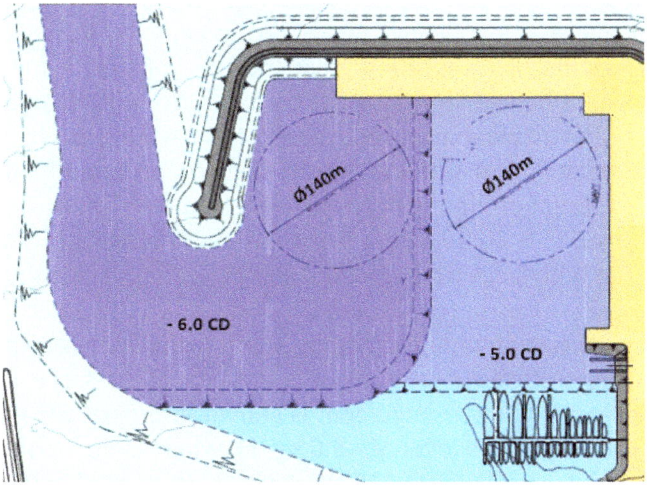

(b)

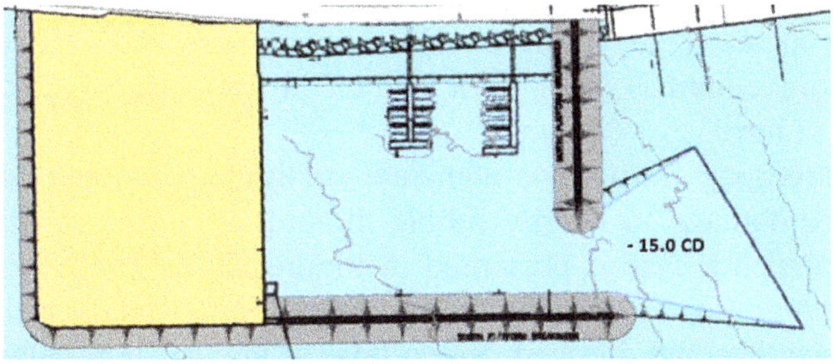

(c)

Figure 23.8. Dredging patterns at ports' entrances: (a) Roundhead constructed on natural seabed; (b) Roundhead constructed on partly dredged seabed; (c) Roundheads constructed on dredged seabed.

by some designers specially for such small marinas, and this could push the dredging slope beyond the roundhead to the trunk, as in this example.

Different dredging arrangements are illustrated in Figure 23.9 for a port project. Part of the port general plan is shown in (a) including the different dredging zones, the access channel and the location of the different cross-sections. A 10.0 m clearance is kept between the main breakwater roundhead and the channel, whereas no clearance is provided at the secondary breakwater roundhead. C-S 2-2 in the secondary breakwater is shown in (b) to illustrate the dredging clearance and the dredging side slope before reaching the required basin dredged level of –14.0 mCD. C-S 7-7 in (c) shows that the access channel dredged level at –22.0 mCD is required up to the toe line of the secondary breakwater roundhead. The dredging of the access channel needs however to be extended slightly inside the roundhead footprint so that the toe rests on a flat bed. The access channel boundary limit is set to be 5.0 m from the scour protection to provide sufficient clearance from the structure while carrying out maintenance dredging in the future. The inclined dashed line in the section indicates the design dredging slope, which could be made as one slope or a stepped slope. C-S 20-20 in (d) is for an inner breakwater showing the dredging from both sides of the breakwater extending inside the breakwater body. C-S 21-21 as in (e) is for the roundhead of the inner breakwater showing the toe resting on the dredging slope, which needs to be made as a stepped slope so that the toe rests on a flat part of the slope.

A different dredging arrangement is shown in Figure 23.10 where the port basin is dredged to a level of –8.0 mCD, whereas the berthing pocket is dredged to a deeper level of –10.0 mCD. This arrangement is normally used to save the dredging cost by allowing larger draft vessels, or when loaded, to enter and leave the port during the high water window. This type of arrangement is to be well-written in the port operation manual with strict control.

Over-dredging during maintenance dredging, in depth and width, could cause damage to nearby rubble mound structures. An illustrative example to such a case is presented in Figure 23.11. Part of the general plan is shown in (a) for an offshore breakwater and an approach channel. The approach channel dredged design level is about 2.0 m below the surrounding natural seabed level at the roundhead, with a dredging clearance of 55.0 m. During successive maintenance dredgings, the channel depth has been increased to reach a level of –15.5 m, which is 1.5 m deeper than

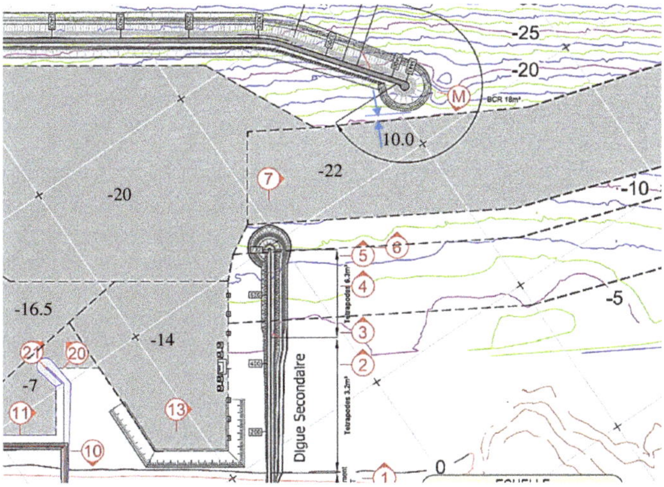

(a)

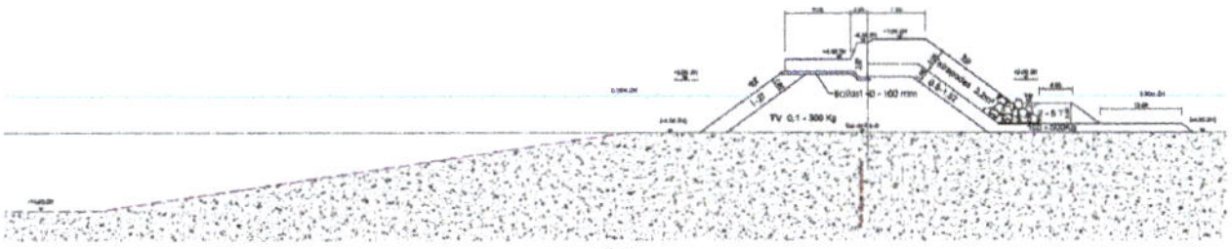

(b)

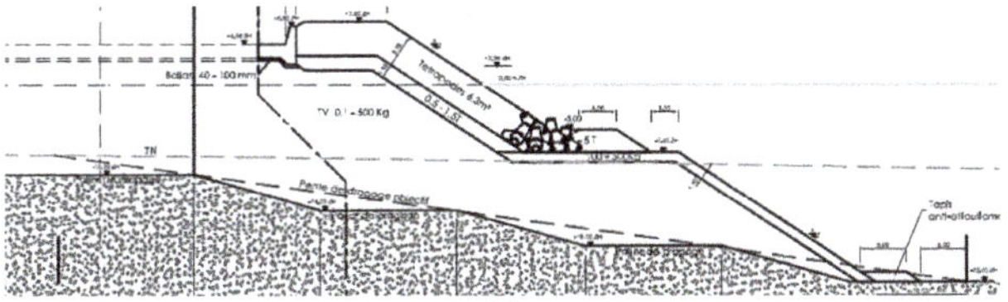

(c)

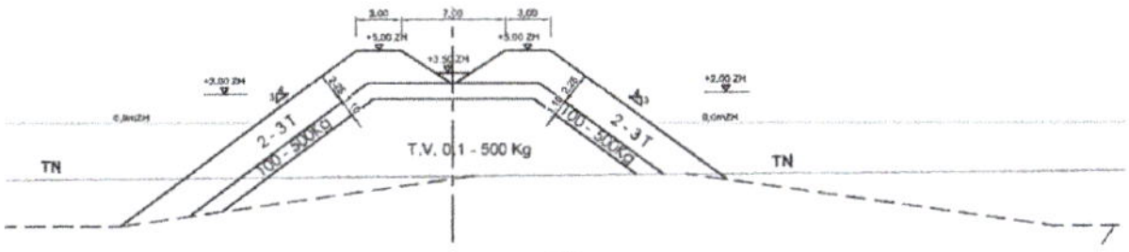

(d)

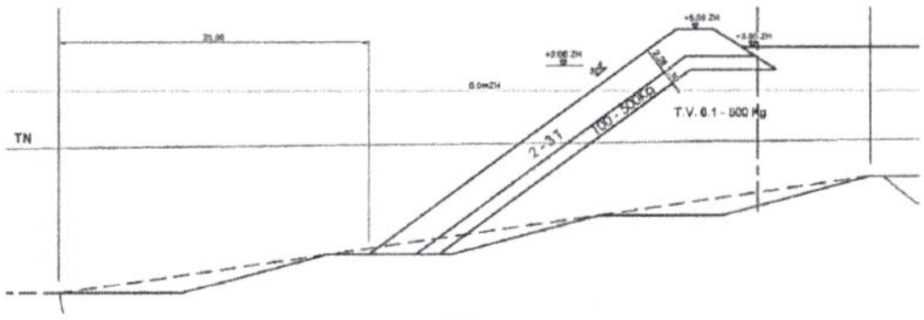

(e)

Figure 23.9. Illustrative sections for dredging near rubble mound structures: (a) Plan; (b) Cross-section 2-2; (c) Cross-section 7-7; (d) Cross-section 20-20; (e) Cross-Section 21-21.

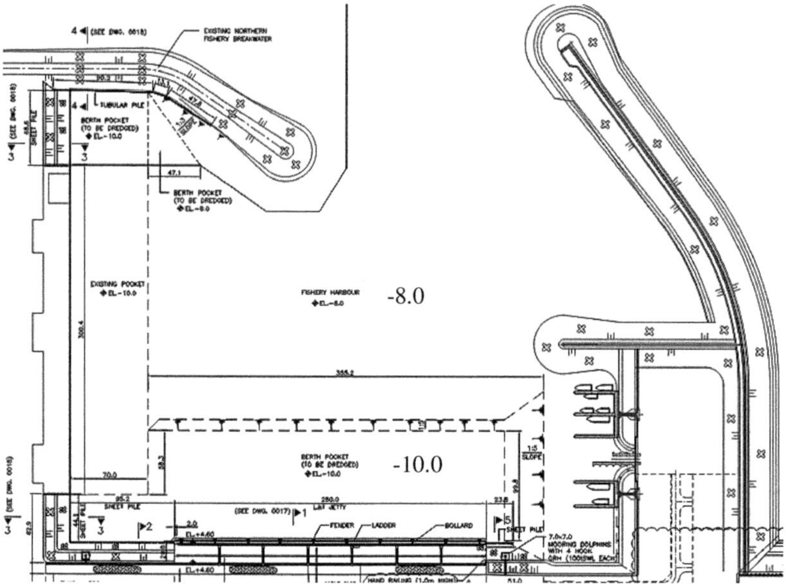

Figure 23.10. Port basin with berthing pockets.

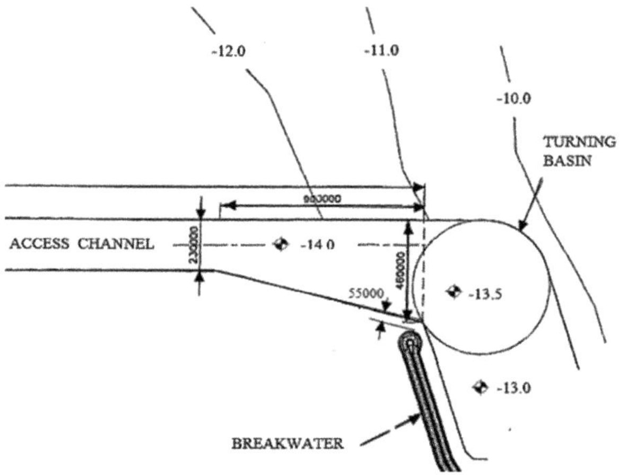

(a)

Figure 23.11. Breakwater roundhead damage due to nearby maintenance over-dredging: (a) Plan — (breakwater, access channel and turning basin); (b) Damage location at the breakwater roundhead (aerial image and multi-beam survey image); (c) Enlarged multi-beam survey image illustrating the damaged area; (d) 3D multi-beam survey image of the roundhead; (e) A photograph of the damaged area.

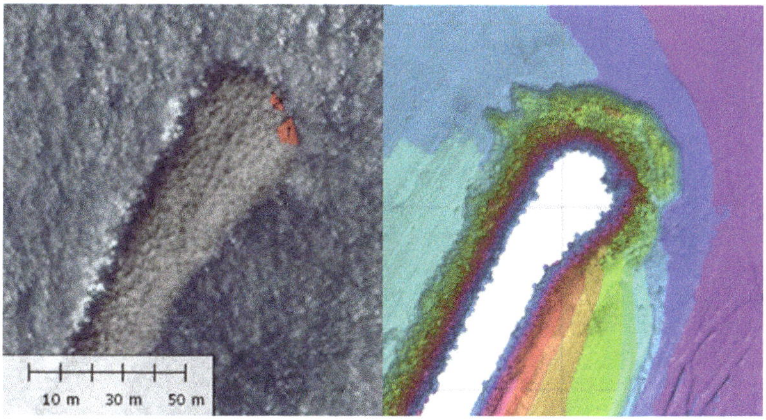

(b)

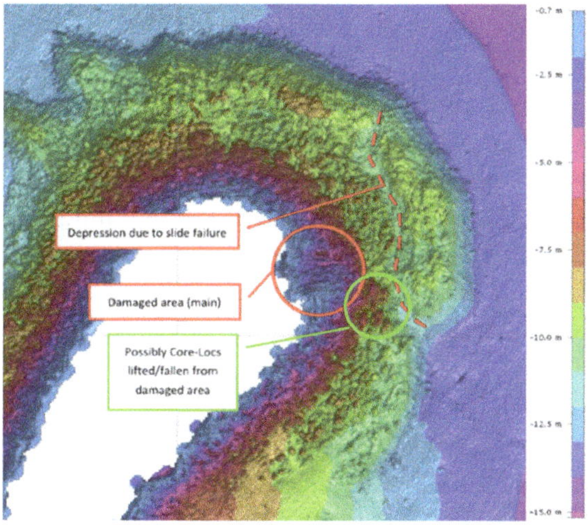

(c)

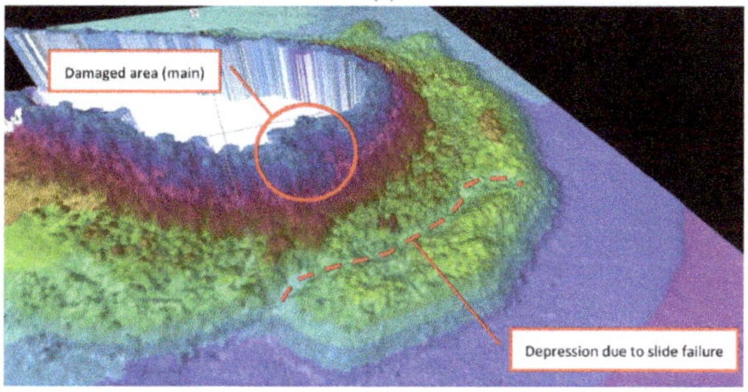

(d)

Figure 23.11. (*Continued*)

(e)

Figure 23.11. (*Continued*)

the design level, and consequently the clearance width was reduced to only 20.0 m. This has caused a geotechnical instability, localised sliding failure and general depression under the toe to the affected area. This subsidence in the toe level has caused the Core-loc units to open up exposing the underlayer (see b, c, d, and e) and reducing the interlocking between the units. Accordingly, this structure is prone to suffer from a serious damage if not repaired.

Chapter 24

Construction Progress Monitoring

24.1. Introduction

The construction of a rubble mound structure (RMS) does incorporate, apart from the design, many other different important aspects related to the activities of the contractor. This issue has been well-covered in Chapters 9 and 10 of CIRIA-CUR-CETMEF (2007), however, certain activities still need to be further illustrated and clarified.

The contractor normally provides several documents to present and record the various construction activities and to have the intermediate and final construction approvals from the site engineer. Some key documents are of special importance that are to be carefully reviewed to ensure the adherence of the contractor to the project design and technical specifications. These are related to the following list of activities:

- Construction programme
- Construction method statements
- Quality and compliance of construction material
- Surveying and datum
- Construction progress records

Most of the above documents are beyond the scope of this book, however, the last item, i.e. the construction progress records, is what shall be presented in this chapter. The reason for this is the lack of knowledge among many site engineers on how the contractor should present and document the construction progress clearly and not by ambiguous and imprecise text descriptions.

24.2. Satellite and Aerial Monitoring

Construction progress monitoring could be divided into two main categories. The first is to provide a general overview of the construction progress, and the second is to provide an exact and accurate record of the construction progress.

The overview monitoring methods are predominantly photographic and normally used for large projects. The commonly used method in this category is the satellite images, which can clearly show the whole area of the project with good resolution. Monitoring through satellite images has proved to be an accurate and reliable method, specially for large reclamation projects. Two examples are presented to illustrate this method, the World Project and Palm Deira (see Chapter 3). The World Project, shown in Figure 24.1, is composed by a series of islands surrounded by a perimeter breakwater protecting the whole project area with no edge structures for the individual islands. As the sand reclamation progresses much faster than the construction of the perimeter breakwater, it was decided to connect the outer islands into two long islands to shelter the project area against the principal northwest wave direction. These two long islands will then be reshaped into smaller ones at a later stage. With such an intricate construction methodology, a close and an accurate monitoring needs to be ensured with the help of satellite imaging.

Palm Deira, as shown in Figure 24.2, is another mega reclamation project, however, unlike the World, it will be protected by revetments/edge structures around the crescent of the Palm and along the perimeter of the large five islands at the base. The inside fronds are all with sand beaches. The progress in sand reclamation, similar to the situation in the World Project, is much faster than that of the rock revetment construction, hence a close monitoring by satellite imaging is required to ensure having the sand well inside the footprint of the project.

An example for a medium-sized project is shown in Figure 24.3, illustrating the construction progress in the edge perimeter revetment and the filling of the basins by reclamation sand. Such project satellite images are frequently used for presentation purposes; however, they could also be used to document the construction progress.

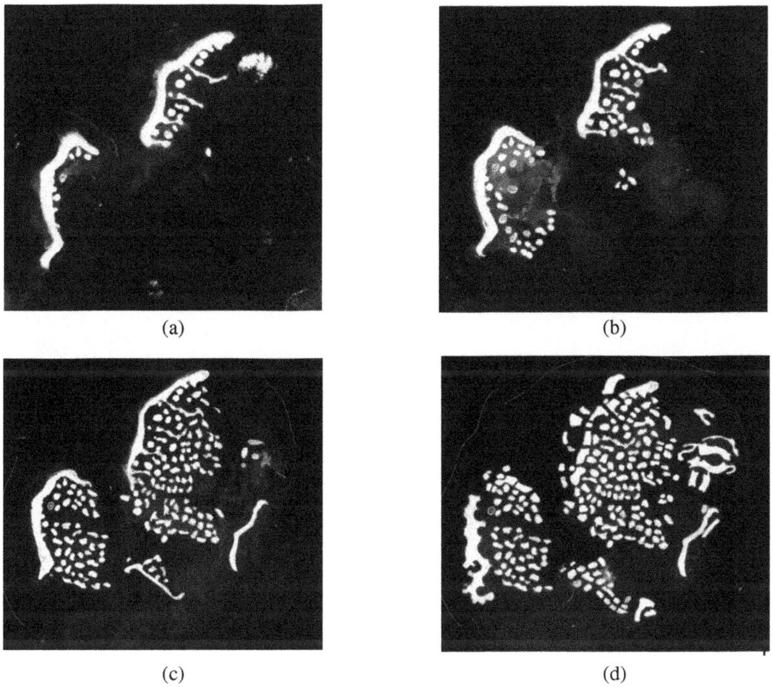

Figure 24.1. Satellite images for the construction progress of the World Project.

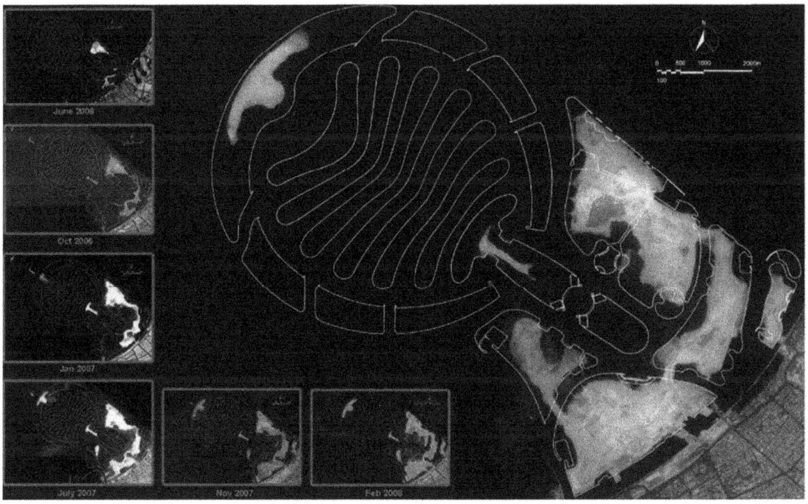

Figure 24.2. Satellite images for the construction progress of Palm Deira.

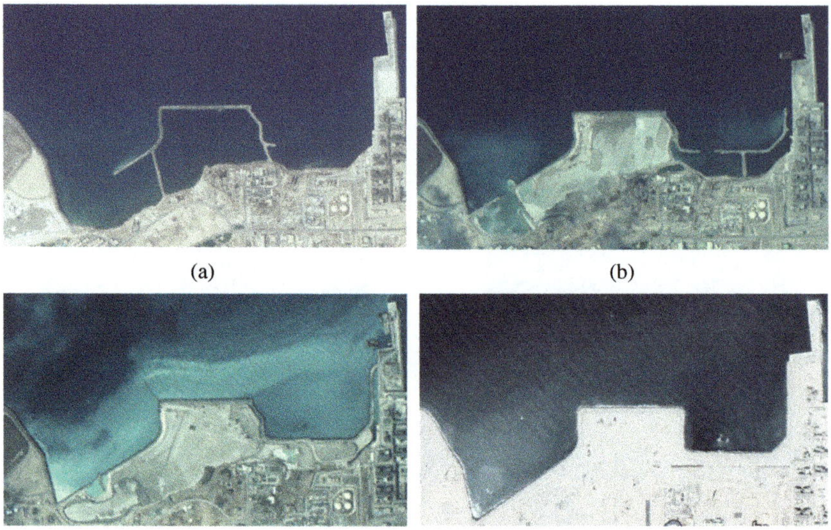

(a) (b) (c) (d)

Figure 24.3. Satellite images for the construction progress of a medium-sized project: (a) Edge structures partially completed; (b) Partial reclamation; (c) Edges completed, partially reclaimed; (d) Fully completed.

(a) (b)

Figure 24.4. Aerial photography for detailed overview: (a) Revetment construction and rock stockpile; (b) Reclamation erosion.

Aerial photography is also used either to give an overview of the progress or to give a more detailed view of special features that cannot be captured in the satellite images. Two examples are shown in Figure 24.4 for a rock revetment under construction and for the sand erosion at the unprotected sand reclamation.

Figure 24.5. Drone photography for monitoring large coastal projects.

Recently, the use of drones has become increasingly popular in supervision, inspection and progress monitoring across various construction sites, including coastal structures (Lin, 2018). This type of monitoring is extremely helpful in providing an overall view for large projects, see Figure 24.5, and also close views to the structure from the water side without the need for a boat to perform the inspection.

24.3. Longitudinal Construction Progress

Construction progress is to be reported and documented regularly by the construction contractor, usually on a weekly basis, so that other related activities could be planned accordingly (e.g. supervision) or timely executed (fx. approvals) to avoid possible construction delay. Progress reporting should be detailed in both sectional, i.e. for the different elements of the cross-section, and longitudinal views, i.e. by giving reference to the relevant chainage line (CHL). Therefore, each element/layer or construction stage (i.e. change in construction methodology) should be identified by using a simple colour coding. This colour coding is to be used on the cross-section to illustrate the construction sequence in terms of layers or stages. Then, the same colour coding is, preferably, to be followed in reporting the construction progress along the relevant chainage line (CHL).

A revetment example is shown in Figure 24.6, where the final cross-section is given first with a designated colour for each layer, i.e. type of

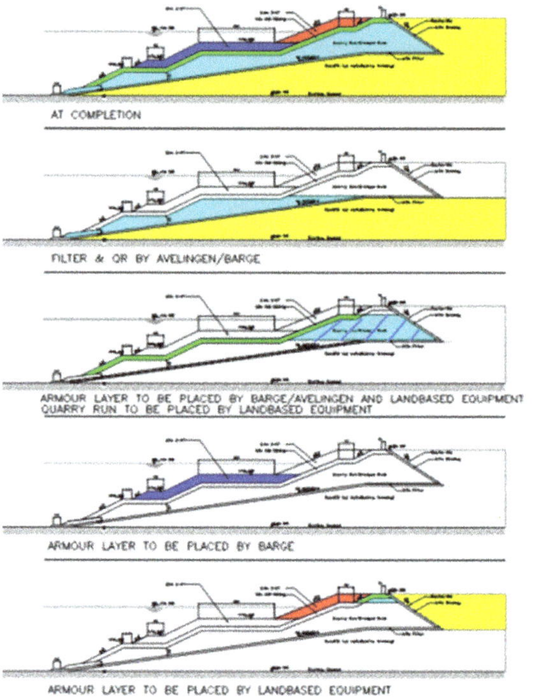

Figure 24.6. Revetment cross-section with colour coding identification.

material. Then, the different layers/construction stages are illustrated according to the construction sequence. The used colour coding is to identify a sand core, a granular filter layer, quarry run core, 1–4 t armour rock for the berm, 0.3–1.0 t underlayer rocks and the main armour of 3–6 t rocks. The example not just illustrates the different layers but also the different construction stages. The construction stages refer to the possibility of dividing the cross-section into different segments that are not based on different materials or layers. In this case, the cross-section is divided into two stages because two different construction methodologies are used, marine and land based. Other possibility to have construction stages is to delay the placement of part of the material until reaching an acceptable bed settlement rate or could be for any other reason.

The first stage in the revetment example is to be constructed by marine equipment by dumping the sand core, placing the filter layer by a side dumping vessel, and then dumping the quarry run to a level just below the WL. This will provide a platform that enables the construction team to start using the land-based equipment on a construction roadway to place the subsequent layers. It is, however, important to note that although the designated colour for the quarry run is turquoise, plain and a striped patterns are used to differentiate between the different construction stages and/or methodologies.

The same revetment cross-section example given in Figure 24.6 is used to demonstrate the reporting of the construction progress along the CHL in Figure 24.7, unfortunately with different colour coding to that of the cross-section. However, the main purpose of this progress reporting sketch is to clearly identify the construction extent of each layer along the CHL. It is worth noting that the presented sketch uses one turquoise band for the quarry run, whereas it is preferable to use two bands, plain and striped, to indicate the two layers of the quarry run. This revetment

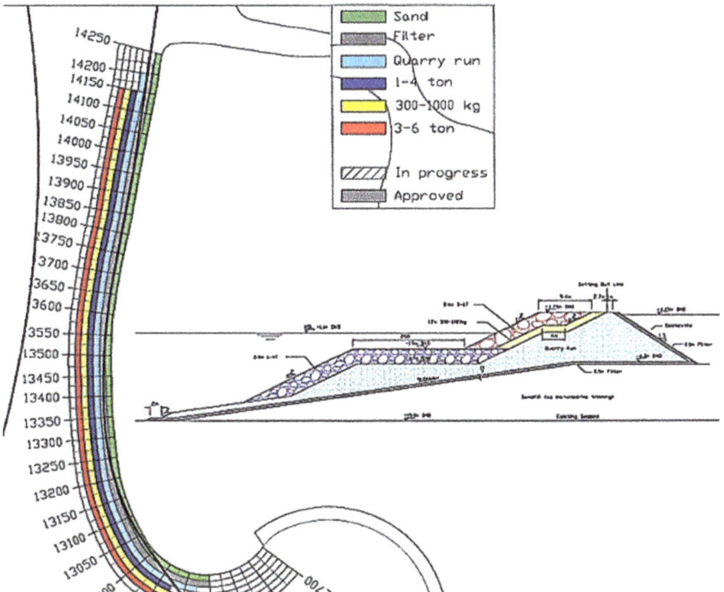

Figure 24.7. Revetment longitudinal construction progress with colour coding identification bands.

cross-section has no geotextile, but if it has, it will also be represented by a separate colour band.

A second example for the longitudinal progress bands is shown in Figure 24.8, though this is for an entire reclaimed island. The construction takes place on different fronts as different teams are working simultaneously on different tasks along the perimeter of the island. The progress on different stretches is shown on the plan, which also shows the corresponding design cross-section for the relevant stretch. One example cross-section is shown in the figure.

The importance of the longitudinal construction progress sketch is not only for the reporting but also for monitoring and the construction activities. Different examples could be given to demonstrate the importance of

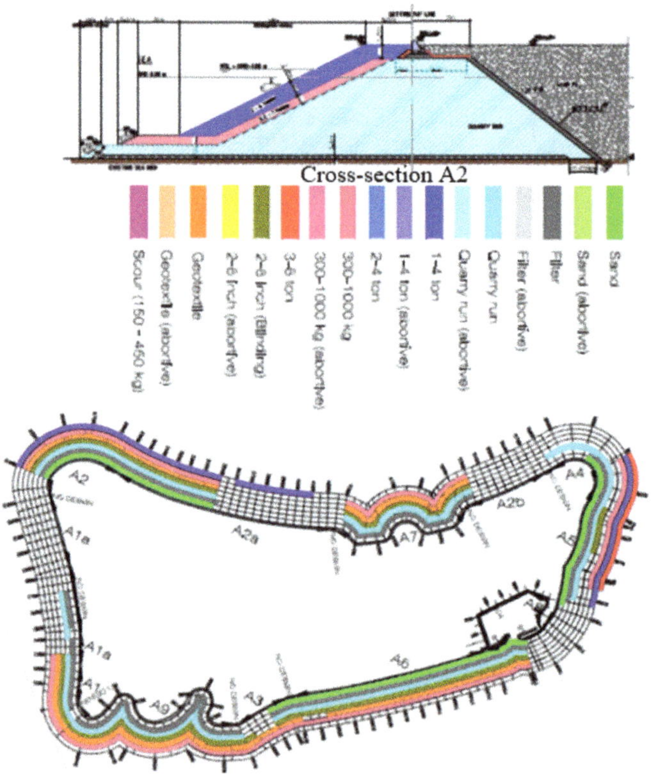

Figure 24.8. Longitudinal construction progress for the perimeter edge structures of a reclaimed island.

using this progress sketch to monitor the construction contractor activities and the adherence to the project technical specifications.

The first example is for the monitoring of the geotextile exposure time. This is a very important time-limited activity to keep the geotextile exposure time within the acceptable limits before being covered by the subsequent layer, see Section 13.3. This could easily be checked by examining the longitudinal progress sketch. Another example is for monitoring the construction advancement limitations. This is important for structures in exposed locations, where the core material or the underlayer is only permitted to advance a certain length before being covered by the subsequent protective layer, as illustrated in Figures 9.53 and 9.59 of CIRIA-CUR-CETMEF (2007). This is of course to limit possible damages to the under-construction structure. The extent and advancement of each layer can easily be checked to be within the acceptable limits from the progress sketch. One more example is for monitoring that no layer is placed before the survey profile approval of the subsequent layer, as explained in Section 24.4.

24.4. Sectional Construction Progress

The placement of the different layers is to be carefully monitored by verifying the as-built survey against the design profile, and with diver/above water inspection, as required. The construction is to be checked against the acceptable tolerances as specified in the project's technical specification. The compliance with the acceptable tolerances is to be checked for every layer, and for defined cross/longitudinal intervals. It is only after the Engineer approval that the construction contractor can proceed with the placement of the subsequent layer. This is important to ensure the smoothness of structure surfaces for a better overall integrity, with no sudden bumps or depressions beyond the relevant acceptable tolerance limits. Any major departure from the design profile at the core will be reflected on the final surface as the subsequent layers of filter/underlayer and armour normally have a defined number of layers with almost a defined thickness.

The profile survey longitudinal intervals depend mainly on the structure configuration, widely spaced for straight stretches and more closely spaced for curved stretches. The acceptable profile survey frequencies are normally in the range of 10 m to 30 m, and 5.0 m to 10.0 m for straight and curved stretches, respectively. The spacing of the measurements points for the sectional survey normally ranges between 1.5 m to 3.0 m, depending on the type of material. Closely spaced for rock armour and more widely spaced for quarry run.

An example is given in Figure 24.9 for profile surveys of the different layers of the revetment shown in Figures 24.6 and 24.7.

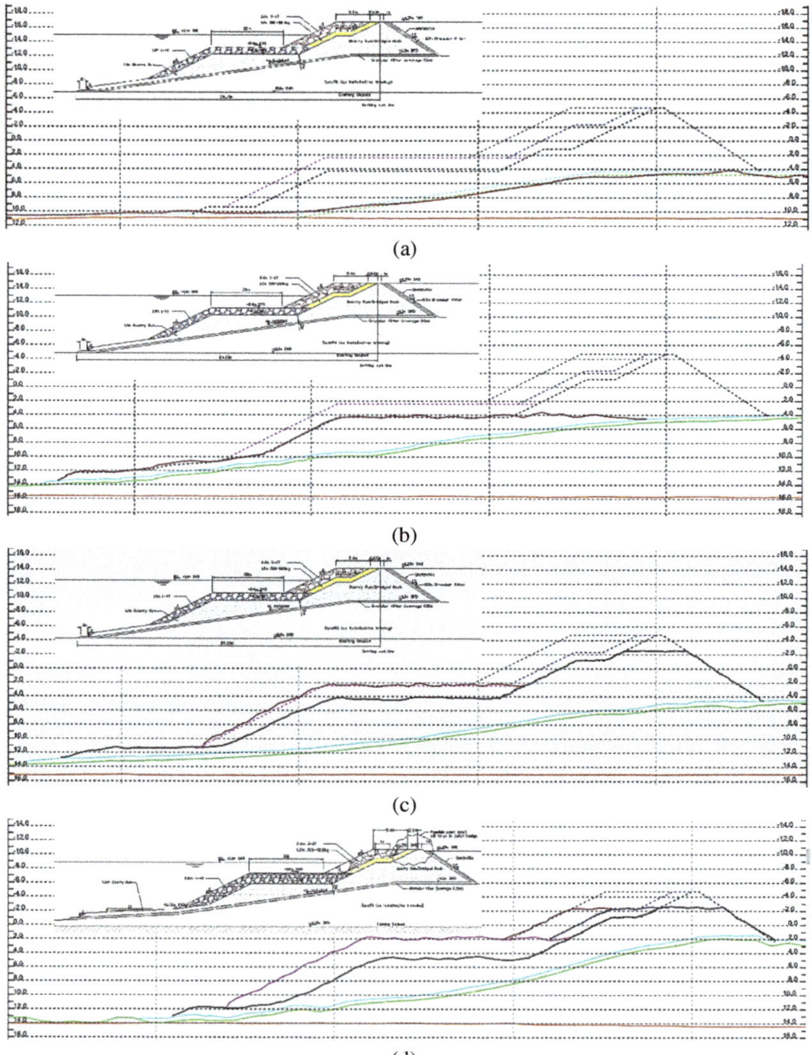

Figure 24.9. Revetment sectional profile surveys.

For deep sections, the profile survey could be a combination of a multi/single-beam echo sounder boat survey and a grab/excavator-RTK GPS survey, as shown in Figure 24.10. Note that the grab/excavator survey should be calibrated against the sphere survey measurement, as shown in Figure 24.11. For more details regarding the survey techniques and frequencies, reference could made to Section 9.9.8 of *CIRIA-CUR-CETMEF* (2007).

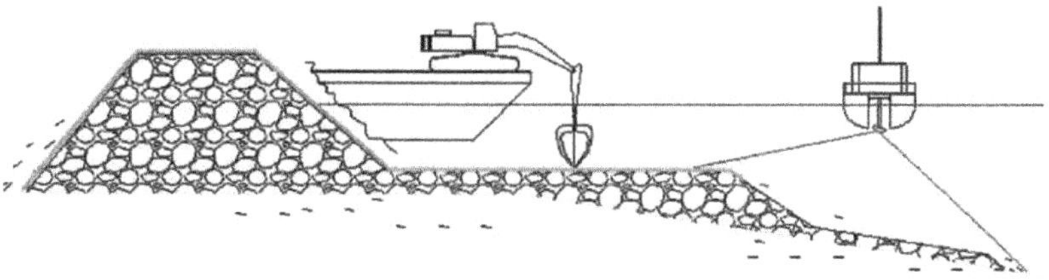

Figure 24.10. Profile survey.

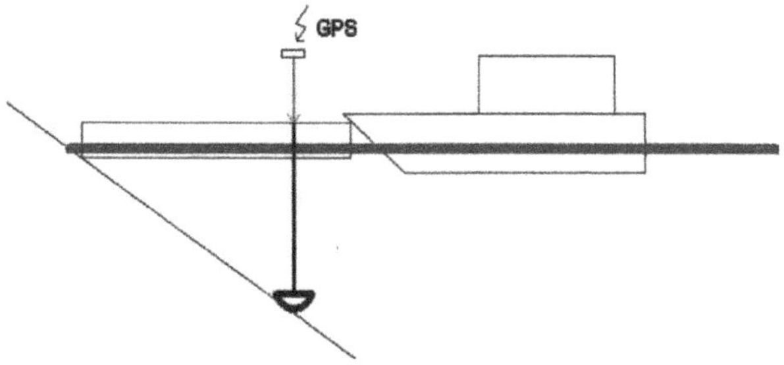

Figure 24.11. $0.5D_{n50}$ sphere profile measurement.

References

Allsop, N.W.H. (1988). "Concrete Armour Units for Rubble Mound Breakwaters and Sea Walls: Recent Progress", Report SR 100, Hydraulics Research, Wallingford, UK.

Allsop, N.W.H. and Jones, R.J. (1996). "Guidelines for Single Layer Hollow Cube Armour Systems for Breakwaters and Related Marine Structures", Report SR 482, HR Wallingford.

Allsop, N.W.H. (2006). "Use of Hollow Cube Concrete Armour", HR Wallingford.

Baart, B., Ebbens, R., Nammuni-Krohn, J. and Verhagen, H.J. (2010). "Toe Rock Stability for Rubble Mound Breakwater", 32rd International Conference Coastal Engineering (ICCE), Shanghai, China.

Baelus, L., Mercelis, P., Leyssen, G., Van Holland, G., Dufour, M., Gebelin, A.A., Caldas, B. and Sas, M. (2014). "Design Aspects of the Breakwater for the New LNG Terminal in Montevideo", PIANC World Congress, San Francisco, USA.

BAW (2010). "Principles for the Design of Bank and Bottom Protection for Inland Waterways (GBB)", Code of Practice, Bundesanstalt für Wasserbau (BAW), Germany.

Besley, P. and Denechere, M. (2009). "Single Layer Armour Systems: Toe, Crest and Roundhead Details", Report HRPP 431, HR Wallingford, UK.

Braestrup, M.W. (Ed.), Anderson, J.B., Anderson, L.W., Bryndum, M., Christensen, C.J. and Nielson, N.R. (2005). "Design and Installation of Marine Pipelines", Blackwell Science Ltd.

Bretler, M.K. and Pilarczyk, K.W. (1998). "Alternative Revetments", Ch.16 in *Dikes and Revetments*, Pilarczyk, K.W. (Ed.), A.A. Balkema Publishers.

BS 6349-7 (1991). British Standard for Maritime Structures- Part 7: Guide to The Design and Construction on Breakwaters.

BS 6349-1 (2013). British Standards for Maritime Works — Part 1-1: General — Code of Practice for Planning and Design for Operations.

Burchart, H.F. (1987). "The Lessons from Recent Breakwater Failures, Development of Breakwater Design", Technical Congress on Inshore Engineering, Vancouver, Canada.

Burcharth, H.F. (1993a). "The Design of Breakwaters", Department of Civil Engineering, Aalborg University, Denmark.

Burcharth, H.F. and Liu, Z. (1993b). "Design of Dolos Armour Units", 23rd International Conference Coastal Engineering, ASCE, Chapter 79, Venice, Italy.

Burcharth, H.F., d'Angremond, K., van der Meer, J.W. and Lin, Z. (2000). "Empirical Formula for Breakage of Dolosse and Tetrapods", *Coastal Engineering* 40, 183–206.

Burcharth, H.F. (2007). "Aspects Related to Design and Construction of Breakwaters in Deep Water", Symposium Design and Construction of Deep Water Maritime Works, Gijon, Spain.

Carver, R.D. and Davidson, D.D. (1977). "Dolos Armor Units Used on Rubble-Mound Breakwater Trunks Subjected to Nonbreaking Waves with no Overtopping", Tech. Rep. H-77-19, Hydraulics Laboratory, U.A. Army Waterways Experiment Station.

CEM (2003). "Coastal Engineering Manual" EM 1110-2-1100, Coastal Engineering Research Centre, U.S. Army Corps of Engineers. (Different parts of the CEM may have been revised with different issue dates).

CEM (2008). "Coastal Engineering Manual" EM 1110-2-1100 (Part V), Coastal Engineering Research Centre, U.S. Army Corps of Engineers.

CEM (2011). "Coastal Engineering Manual" EM 1110-2-1100 (Part VI), Coastal Engineering Research Centre, U.S. Army Corps of Engineers.

Ceuppens, T.R.C. (2007). "The Ras Laffan Port Expansion Project — Qatar", https://www.linkedin.com/in/tomceuppens/.

CIRIA-CUR-CETMEF (2007). "The Rock Manual-The Use of Rock in Hydraulic Engineering," Publication C683, CIRIA.

CIRIA (2020). "Groynes in Coastal Engineering," Publication C793.

CLAS Certification. http://www.clascertification.com/.

CLI (2012). "Core-Loc™ Design Guide Table-2012_0_0" Concrete Layer Innovations.

Close, D., Brett, D. and AL Dakhil, O. (2009). "Manifa Field Causeway and Islands, Saudi Arabia," J. Terra et Aqua, No. 114.

Concrete Layer Innovations (CLI). https://www.concretelayer.com/.

Corredor, A., Santos, M., Gomez-Martin, M.E., and Medina, J.R. (2012). "Placement of Cubipod Armor Units in San Andres Breakwater, Port of Malaga, Spain", 33rd International Conference Coastal Engineering (ICCE), Santander, Spain.

Coutos, M. and Hong, K.W. (2016). "Toe Berm Design for Rubble-Mound Breakwaters: Example of The Safi Power Plant Project", 35th International Conference Coastal Engineering (ICCE), Antalya, Turkey.

Cubipod (2016). SATO, www.cubipod.com.

CUR Report 243, "Durability of Geosynthetics", CUR Building & Infrastructure, 2012.

Danel, P. (1953). "Tetrapods", 4th International Conference Coastal Engineering, ASCE, Chapter 28, Chicago, USA.

d'Angremond, K. and van Roode, F.C. (2004). "Breakwaters and Closure Dams", Spon Press.

Derun, A.B., Altinisik, I., Almusharafi, R.B. and Gunbak, A.R. (2005). "Design and Model Experiments for Intake and Outfall Structures in Sohar Industrial Area", Arabian Coast 2005, 1st International Conference Coastal Management and Engineering, Dubai, U.A.E.

Douairi, M. and De Gijt, J. (2013). "Upgrading Techniques for Quay Walls", IABSE (International Association for Bridge and Structural Engineering) Conference, Report Vol.99, Rotterdam, The Netherland.

Dubai Municipality (2010). "Coastal Development Guidelines for Dubai Coast", Environmental Department, Coastal Zone & Waterways Management Section.

Dupray, S. and Roberts, J. (2009). "Review of the Use of Concrete in the Manufacture of Concrete Armour Units", International Conference on Coasts, Marine Structures and Breakwaters, ICE, Edinburgh, UK.

EAU (2012). "Recommendations of the Committee for Waterfront Structures, Harbours and Waterways", 9th Edition, Issued by the Committee for Waterfront Structures of the German Port Technology Association and the German Geotechnical Society.

Eldrup, M.R., Andersen, T.L. and Burchart, H.F. (2019). "Stability of Rubble Mound Breakwaters — A Study of the Notional Permeability Factor Based on Physical Model Test", *Journal of Water*, 11(5), 934, MDPI.

EN 12224 (2000). "Geotextiles and Geotextile-Related Products", *European Standards*.

EN 13383 (2013). "Armourstone -Part 1: Specificaton, -Part 2: Test Methods", *European Standards*.

EurOtop (2018). "Manual on Wave Overtopping of Sea Defences and Related Structures", 2nd Edition.

Fabião, J., Trigo-Teixeira, A. and Araujo, M.A.V.C. (2013). "Hydraulic Stability of Tetrapod Armour Layers, Physical Model Study", 6th SCACR-International Short Course/Conference on Applied Coastal Research, Lisbon, Portugal.

Feng, G., Hau-Ping, Z., Ci-Heng, Z. and Long-Zai, G. (2016). "Study on 3-D Stability and Overtopping of Breakwater under Random Wave Action", 6th

International Conference on the Application of Physical Modelling in Coastal and Port Engineering and Science (Coastlab16), Ottawa, Canada.
Fowler, J., Stephens, T.C., Santiago, M. and Bruin, P. (2002). "Amwag Islands Constructed with Geotubes, Bahrain", *Journal of World Dredging, Mining and Construction*, 38, 6–40.
Freitas, P., Trigo-Teixeira, A. and Araujo, A.V.C. (2013). "Hydraulic Stability of Antifer Block Armour Layer", 6th SCACR-International Short Course/ Conference on Applied Coastal Research, Lisbon, Portugal.
Frens, A.B. (2007). "The Impact of Placement Method on Antifer Block Stability", M.Sc. Thesis, Delft University of Technology, The Netherlands.
Frens, A.B., van Gent, M.R.A. and Olthof, J. (2008). "Placement Methods for Antifer Armour Units", 31st International Conference Coastal Engineering (ICCE), Hamburg, Germany.
Galal, E.M. (2017). "A Numerical Study for Upgrading the Container Terminal of Port-Said West Port", *Port Said Engineering Research Journal, Faculty of Engineering, Port Said University*, 21(2), 88–97.
Gierlevsen, T., Nielsen, C.Q., Lauridsen, H., Jensen, O.J. and Clausen, E. (2013). "Economically Optimised Breakwater Design at Exposed Location", International Conference on Coasts, Marine Structures and Breakwaters, ICE, Edinburgh, UK.
Gierlevsen, T. (2019). "Expansion of Port Hanstholm — New Western Breakwater", Conference NordPIANC, Hirtshals, Denmark.
Glennerster, M., Everitt, S., Pobjoy, E., Hunt, T. and Gambrill, J. (2013). "Khalifa Port Breakwaters and Protection of Coral Reefs", International Conference on Coasts, Marine Structures and Breakwaters, ICE, Edinburgh, UK.
Goda, Y. (2000). "Random Seas and Design of Maritime Structures", Advanced Series on Ocean Engineering-Volume 15, World Scientific.
Gravesen, H. (2008). "Rubble Mound Breakwaters", Draft Lecture Notes, DTU, Denmark.
Halleett, D.K., Pritchard, D.J. and Edbury, A.I. (1995). "The Design and Construction of Major Extensions in The Port of Limassol for the Development of Modern Container Terminal", International Conference Coastal and Port Engineering in Developing Countries, Brazil.
Hawkswood, M., Lafeber, F.H. and Hawkswood, G.M. (2014). "Berth Scour Protection for Modern Vessels", PIANC World Congress, San Francisco, USA.
Hedges, T.S. (1983). "The Core and Underlayers of Rubble Mound Structures", Conference Proceeding Breakwaters, Design and Construction, ICE (Institution of Civil Engineers).
Hellinga, L.B. (2016). "Stability of Single Layer Cubes on Breakwater Rear Slopes", M.Sc. Thesis, Delft University of Technology, The Netherlands.

Hoff, J.V. and van der Kolff, A.N. (Eds.) (2012). "Hydraulic Fill Manual, For Dredging and Reclamation Works", CUR and CIRIA Publication, CRC Press.

H R Wallingford (2000). "The Joint Probability of Waves and Water Levels: JOIN-SEA", Report SR 537.

Hoff, J.V. and Kolff, A.N.V. (Ed.) (2012). "Hydraulic Fill Manual — For Dredging and Reclamation Works," CRS Press.

International Breakwater Directory, T U Delft and H R Wallingford, www.kennisbank-waterbouw.nl/breakwaters/.

Jensen, O.J. (1984). "A Monograph on Rubble Mound Breakwaters", DHI (Danish Hydraulic Institute), Denmark.

Jensen, O.J. (2013). "Safety of Breakwater Armour Layers with Special Focus on Monolayer Armour Units", International Conference on Coasts, Marine Structures and Breakwaters, ICE, Edinburgh, UK.

Kasai, Y., Mori, N., Kashima, H., Kakuno, S. and Orihashi, T. (2006). "Fully Reinforced 80T Dolos and Sloping Top Caisson in Hososhima Port", 30th International Conference Coastal Engineering, Vol.5, Ed. Smith, J.M.

Kik, R., Van der Bos, J.P., Maertens, J., Verhagen, H.J. and Van der Meer, J.W. (2012). "Notional Permeability", 33rd International Conference Coastal Engineering (ICCE), Santander, Spain.

Kirby, A.M., Roca, M., Kitchen, A., Escarameia, M. and Chesterton, O.J. (2015). "Manual on Scour at Bridges and Other Hydraulic Structures", 2nd edition, CIRIA C742.

Koerner, R.M. (2005). "Designing with Geosynthetics", 5th Edition, Pearson Prentice Hall.

Lin, A. (2018). "Use of Drone Technology on Commercial Construction Projects", Digital Commons @ Cal Poly, California Polytechnic State University, California, USA.

Maciňeira, E., Peňa, E., Sande, J. and Figuero, A. (2019). "Dynamic Calculation of Breakwater Crown Walls under Wave Action: Influence of Soil Mechanics and Shape of the Loading State", *Journal of Water*, 11, 1149, MDPI.

Maddrell, R.J. (2005). "Lessons Re-Learnt from the Failure of Marine Structures", International Conference on Coastlines, Structures and Breakwaters (ICE), London, UK.

Mahalingaiaha, A.V., Tayade, B.R., Gokhale, N.V. and Kudale, M.D. (2015). "Design of Submerged Offshore Reefs for the Coastal Protection Measures", Aquatic Procedia 4, Elsevier B.V.

Mangor, K. (2007). "General Guidelines for Good Quality Artificial Beaches and Lagoons, and Case Stories", Proceeding 18th World Dredging Congress (WODCON XVIII), Florida, USA.

Mangor, K., Drønen, N.K., Kærgaard, K.H. and Kristensen, S.E. (2017). "Shoreline Management Guidelines" DHI (Danish Hydraulic Institute), Denmark.

Mansour, C., Steinberg, A. and Matasovic, N. (2004). "Analysis, Design and Construction of the Supporting Structure and Wharf Retrofit for a New Shiploader at the Port of Long Beach, California", Proceeding: Ports 2004-Port Development in the Changing World, Houston, Texas, USA.

Martin, F.L., Losada, M.A. and Medina, R. (1999). "Wave Loads on Rubble Mound Breakwater Crown Walls", *Journal of Coastal Engineering*, 37, 149–174, Elsevier.

McConnell, K. and Allsop, W. (1998). "Revetment Systems Against Wave Attack — A Design Manual", Thomas Telford.

Medina, J.R. and Gómez-Martin, M.E. (2016). "Cubipod® Manual 2016", UPV [S], Universitat Politècnica de València, Spain.

Min, E.-J. (2015). "Experimental Study of Stability Coefficient of Breakwater Armour Layer Depending on Placement Methods", M.Sc. Thesis, Seoul National University, Seoul, South Korea.

Mohammad, L.A. and Jensen, O.J. (2002). "The Development of a Breakwater Design at Caleta La Mision Port, Argentina", 3rd International Conference Maritime Engineering & Ports, WIT Press, Rodes, Greece.

Mohammad, L.A. (2005). "Low-crested Low-overtopped Breakwaters — An Overview of Selected Nakheel Projects", 1st International Conference Arabiancoast, Dubai, UAE.

Mohammad, L.A. (2008a). "Palm Deira-The Largest Man-made Island", Paper No. M-11, COPEDEC VII, Dubai, UAE.

Mohammad, L.A. (2008b). "Describing Next Generation of Waterfront Developments-The Dubai Waterfronts", Urban Waterfronts Conference, by Marcus Evans, Dubai, UAE.

Musumeci, R.E., Cavallaro, L. and Foti, E. (2012). "Performance of Perched Beach Nourishments", International 33rd Conference Coastal Engineering (ICCE), Santander, Spain.

Neelissen, R., Tanis, A. and Van Gool, V. (2010). "Dredging Rock with a Hopper Dredger: The Road to The Ripper Draghead", J. Terra et Aqua, No. 118, International Association of Dredging Companies (IADC).

Nozu, A., Ichii, K. and Sugano, T. (2004). "Design of Port Structures", *Journal of Japan Association for Earthquake Engineering*, 4(3) (Special Issue).

Nurmohamed, N., Steendam, G.J. and van der Meer, J. (2006). "Weight and Stability Assessment of Single Layers Orderly Placed or Pitched Natural Rock", 30th International Conference Coastal Engineering (ICCE), San Diego, California, USA.

Oumeraci, H., Kortenhaus, A., Allsop, N.W.H., De Groot, M.B., Crouch, R.S., Vrijling, J.K. and Voortman, H.G. (2001). "Probabilistic Design Tools for Vertical Breakwaters", MAST III/PROVERBS, A.A. Balkema Publishers.

Pardo, V., Herrera, M.P., Molines, J. and Medina, J.R. (2012). "Placement Grids, Porosity and Randomness of Armor Layers", International 33rd Conference Coastal Engineering (ICCE), Santander, Spain.

Pardo, V., Herrera, M.P., Molines, J. and Medina, J.R. (2014). "Placement Tests, Porosity and Randomness of Cube and Cubipod Armor Layer", *Journal Waterway, Port, Coastal, Ocean Engineering*, 140(5), 0401417.

PIANC, Working Group 12 (1992). "Analysis of Rubble Mound Breakwaters".

PIANC, Working Group 21 (1992a). "Guidelines for the Design and Construction of Flexible Revetments Incorporating Geotextile in Marine Environment".

PIANC, Working Group 38 (2001). "Seismic Design Guidelines for Port Structures", A.A. Balkema Publishers.

PIANC, Report no. 103 (2008). "Life Cycle Management of Port Structures — Recommended Practice for Implementation".

PIANC, Report no. 121 (2014). "Harbour Approach Channels — Design Guidelines".

PIANC, Report no. 180 (2015). "Guidelines for Protecting Berthing Structures from Scour Caused by Ships".

Pilarczyk, K.W. (1996). "Design and Application of Geotubes and Geocontainers", Road and Hydraulic Division of the Rijkswaterstaat, Delft, the Netherlands.

Pilarczyk, K.W. (Ed.) (1998). "Dikes and Revetments: Design, Maintenance and Safety Assessment", A.A. Balkema Publishers.

Pilarczyk, K.W. (2000a). "Geosynthetics and Geosystems in Hydraulic and Coastal Engineering", A.A. Balkema Publishers.

Pilarczyk, K.W. (2000b). "Critical Review of Geosystems in Hydraulic and Coastal Engineering Applications", Atti II European Geosynthetics Conference, Bologna, Italy.

Pilarczyk, K.W. (2003). "Design of Revetments", Road and Hydraulic Division of the Rijkswaterstaat, Delft, the Netherlands.

Pilarczyk, K.W. (2010). "Design of Alternative Revetments", in *Handbook of Coastal and Ocean Engineering*, pp. 479–520. Kim, Y.C. (ed.), World Scientific, Singapore.

Port of Nappier, Wharf 6 (2020). www.napierport.co.nz.

Port Works Design Manual - Part 5 (2003). "Guide to Design of Beaches", Civil Engineering Department, The Government of Hong Kong Special Administration Region.

Premalatha, P.V., Kumar, S.S. and Baskar, K. (2017). "Influence of Change in Pile Diameter at Various Locations of a Pile Group in a Berthing Structure", *Indian Journal of Geo Marine Sciences*, 46(6), 1198–1209.

REMR (1994). Technical Note CO-RA-1.2, "Concrete Armour Unit Performance Survey", US Army Corps of Engineers, pp. 479–520.

Schiereck, G.J. and Verhagen, H.J. (2012). "Introduction to Bed, Bank and Shore Protection", 2nd Ed., Publ. by VSSD.

Shinde, S., Ukalkar, A., Tambe, A. and Prakash, A. (2017). "Design of Breakwaters for the Development of Fishing Harbour – A Case Study", *International Research Journal of Engineering and Technology (IRJET)*, 4(4), 2043–2047.

Silveira, L., Gomes, G., Taiani, L., Pinheiro, V., Benedet, L., Marques, A. and Loureiro, C. (2017). "Integrated Method for Optimal Channel Dredging Design", J. Terra et Aqua, No. 146, IADC (International Association of Dredging Companies).

Singh, K.Y. (1968). "Stabits, A New Armour Block", Chapter 52, 11th International Conference Coastal Engineering, London, UK.

Smith, G.M., van der Klis, P., Ockeloen, W. and Mootaz, K. (2013). "Das Island Phase III — Final Design for Wave Overtopping", ICE Conference Breakwaters, Edinburgh, UK.

SPM (1984). "Shore Protection Manual" Coastal Engineering Research Centre, U.S. Army Corps of Engineers.

Stabits (1979). "Stabits, A New Type of Artificial Armour Block for Breakwaters and Sea Defence Works", Stabits Limited, Columbia House 69, Aldwych, London, U.K.

Stewart, T.P., Newberry, S.D., Latham, J.P. and Simm, J.D. (2003). "Packing and Voids for Rock Armour in Breakwaters", Report SR 621, HR Wallingford.

Takahashi, S. (2002). "Design of Vertical Breakwaters" PARI (Port and Airport Research Institute), Japan.

Thoresen, C.A. (2014). "Port Designer's Handbook", 3rd Edition, ICE Publishing, London, UK.

Treadwell, D.D. and Magoon, O.T. (2006). "Reinforced Concrete Armour Units for Coastal Structures", 30th International Conference Coastal Engineering (ICCE), San Diego, California, USA.

Turk, G.F. and Melby, J.A. (1997). "CORE-LOC™ Concrete Armour Units: Technical Guidelines" US Army Corps of Engineers, Miscellaneous paper CHL-97-6.

UKOPA (2015). "Impact Protection Slab", UN Onshore Pipeline Operators' Association — Industry Good Practice Guide, UKOPA/GP/006.

Van der Lem, J.C., Stive, R.J.H. and van Gent, M.R.A. (2016). "Sal Rei Breakwater with Single Layer Cubes", PIANC — COPEDEC IX, Rio de Janeiro, Brazil.

Van der Meer, J.W. (1988). "Rock Slopes and Gravel Beaches Under Wave Attack", PhD Thesis, Delft University of Technology, Delft, The Netherlands.

Van der Meer, J.W. (2014). "From Test to Practice", UNESCO-IHE Institute for Water Education, The Netherlands.

Van der Meer, J.W. and Sigurdarson, S. (2017). "Design and Construction of Berm Breakwaters", World Scientific Publication Co. Pte. Ltd.

Van Gent, M.R.A. and Luis, I. (2013). "The Application of Cubes in a Single Layer", 6th SCACR- International Short Course/Conference on Applied Coastal Research, Lisbon, Portugal.

Van Gent, M.R.A. and Van der Werf, I.M. (2014). "Toe Stability of Rubble Mound Breakwaters", 34th International Conference ICCE, Seoul, South Korea.

Van Zwicht, B., Bakker, P., Van den Bos, J. and Pol, B. (2013). "The Extension of the Point Noire Breakwater — Congo", ICE Conference Breakwaters, Edinburgh, UK.

Verhagen, H.J., Van der Hoven, M. and Thiel, B. (2003). "A New View on Falling Aprons", COPEDEC VI, PIANC Conference, Colombo, Sri Lanka.

Verhagen, H.J. and Jansen, L. (2014). "Ration between Stone Diameter and Nominal Diameter", Communications on Hydraulic and Geotechnical Engineering, TU Delft.

Vos-Rovers, I., Reedijk, B. and Ekelmans, S. (2008). "Crown-wall with Extended Slab", 31st International Conference Coastal Engineering (ICCE), Hamburg, Germany.

Wareing, A., Wright, K., Heijne, I. and Shields, W. (2013). "Beyond Early Contractor Involvement: Keeping the Sea Out — The West Sands Coast Protection Scheme", International Conference on Coasts, Marine Structures and Breakwaters, ICE, Edinburgh, UK.

Wesson, M., Provan, M., Cox, J. and Knox, P. (2018). "Physical Modelling Supporting Design and Construction of Low Crested Breakwater for the Ayia Napa Marina, Cyprus", 34th PIANC World Congress, Panama.

White, P. and Cruickshank, I. (2005). "Integrating Architecture and Engineering in Waterfront Developments International Experience in Dealing with Challenges", 1st International Conference Arabiancoast, Dubai, UAE.

WIKIPEDIA/Land Reclamation.

Xbloc (2018). "Guidelines for Xbloc Concept Designs", https://www.xbloc.com/.

Yong, M., Hayman-Joyce, J. and Kim, S.H. (2012). "Use of Single Layer Armour Units as Toe Reinforcement", 33rd International Conference Coastal Engineering (ICCE), Santander, Spain.

Zhaohua, M. and Borong, Y. (2018). "Earthquake Damage to Harbor Wharves and Shore Slopes", Engineering for Water Transport, Ed. G. Blankenship.

Index